Electricity for Refrigeration, Heating, and Air Conditioning Sixth Edition

Russell E. Smith
Athens Technical College

THOMSON

DELMAR LEARNING

Australia Canada Mexico Singapore Spain United Kingdom United States

THOMSON

DELMAR LEARNING

Electricity for Refrigeration, Heating, and Air Conditioning, 6th edition

Russell E. Smith

Executive Director:
Alar Elken

Executive Editor:
Sandy Clark

Acquisitions Editor:
James DeVoe

Development Editor:
John Fisher

Executive Marketing Manager:
Maura Theriault

Channel Manager:
Fair Huntoon

Marketing Coordinator:
Brian McGrath

Executive Production Manager:
Mary Ellen Black

Production Editor:
Ruth Fisher

Art/Design Coordinator:
Cheri Plasse

Editorial Assistant:
Mary Ellen Martino

Library of Congress Cataloging-in-Publication Data:

Smith, Russell E., 1944–
   Electricity for refrigeration, heating, and air conditioning / Russell E. Smith.—6th ed.
      p. cm.
Includes index.
   ISBN 0-7668-7337-4
   1. Electric engineering. 2. Heating. 3. Air conditioning. 4. Refrigeration and refrigerating machinery. I. Title.
   TK153 .S57 2002
   621.3'024697—dc21
                                 2002066490

NOTICE TO THE READER

Table of Contents

Preface

Electricity for Refrigeration, Heating, and Air Conditioning was initially written because there was no text that adequately covered the electrical principles and practices required of an installation or service technician in the refrigeration, heating, and air conditioning industry. This text is written with a blend of theory and practice suitable for the vocational/technical student or the industry practitioner who wishes to upgrade his or her knowledge and skills. The purpose of this text is to assemble concepts and procedures that will enable the reader to work successfully in the industry.

ORGANIZATION

It is difficult to organize an electrical text to be used in refrigeration, heating, and air conditioning programs in educational institutions because of the many different types of programs and the variety of the delivery of information. The information covered in this text is organized from the very basics to the circuitry and troubleshooting of control systems in the industry. The organization is industry-driven because of the correlation of industry standards and the many new developments that continue to be made. Electrical devices are covered in detail in a systematic order with the troubleshooting of the components following an explanation of how they work. Troubleshooting control systems should be the objective of most students and industry personnel using this text and is covered in detail.

FEATURES OF THIS EDITION

There are new features as well as existing features of this text that are advantages to students and instructors alike. Each chapter begins with **objectives** that should be mastered as the student progresses through each chapter. **Key terms** are emphasized at the beginning of each chapter in order for the student to know what information is ahead, and the key terms are highlighted in color in the body of the text. Each chapter is concluded with a **summary** that allows the student to review information that has been covered in the chapter. Many chapters use **service calls** to reinforce service procedures that are commonly used in the industry along with some procedures that the student has the opportunity to solve. Important elements of the text are highlighted in color in this edition, including circuits that are being discussed and important concepts and safety cautions. In the back of the text there is a reference chart of electrical symbols including switches, thermostats, contactors, and relays.

NEW IN THIS EDITION

Three new chapters have been added to the text. Chapter one covers the introductory safety basics that are needed with a person's entry into the industry. Sections on reading wiring diagrams have been removed from an existing chapter and expanded to make up a new Chapter six. Chapter six now covers reading schematic wiring diagrams. This new chapter covers schematic diagram basics along with step-by-step explanations of reading 12 schematic diagrams that are common to the industry. Chapter seventeen emphasizes procedures that are used for troubleshooting refrigeration, heating, and air conditioning systems. It is the desire of the author that this stimulates the users of this text to develop a troubleshooting method that suits their personality.

SUPPLEMENTS

Available to instructors and students are two supplements: an Instructor's Guide and a Lab Manual. Included in the Instructor's Guide is a short description of the material covered in each chapter, unit objectives, transparency masters, safety notes, lab notes, and answers to the questions in the text and Lab Manual. The Lab Manual has a section for each chapter in the text. Included in the Lab Manual are chapter overviews, key terms, review test, and lab exercises, including many of the elements required to complete the lab. The Lab Manual should prove to be extremely helpful to the new instructor.

ACKNOWLEDGMENTS

I would like to thank God for the skills and knowledge He has given me and the ability and desire to write this text. I thank my wife for 35 wonderful years of marriage and her encouragement, which is priceless. I thank my family for their encouragement and support through each and every edition. I would like to thank my fellow colleagues, especially Carter Stanfield, at Athens Technical College for their support and encouragement. I would like to thank past and present students for suggestions that have made each edition easier to write.

I would like to express appreciation to the following people for their input as reviewers of this edition:

Neal Broyles, Rolla Technical Institute,
Rolla, MO

William H. Burklo, Indian River Community
College, Ft. Pierce, FL

John Demree, Burlington County Institute of
Technology, Medford, NJ

Gary M. DeWitt, Monroe Community
College, Rochester, NY

Eugene C. Dickson, Indian River Community
College, Ft. Pierce, FL

Herb Haushahn, College of DuPage,
Glen Ellyn, IL

James Mendieta, Western Technical Institute,
El Paso, TX

Robert Reynard, Western Technical Institute,
El Paso, TX

Darius Spence, North Virginia Community
College, Woodbridge, VA

Richard Wirtz, Columbus State Community
College, Columbus, OH

I would like to thank each of the following manufacturers and manufacturer's representatives who have helped with the photographs and artwork.

A.W. Sperry Instrument Co., Inc.
Acme Transformer
Aerovox Inc.
American-Standard Air Conditioning
Amprobe Instruments, Inc.
BICC Industrial Cable Company
Carrier Corporation
Chromolox, Wiegand Industrial, Divison
   of Emerson Electric Co.
Copeland Corporation
Delmar Publishers
Dunkirk Radiator Corporation
Duracell, Inc.
John Fluke Mfg. Co., Inc.
Furnas Electric Company
Gould Shawmut
Honeywell, Inc.
Indeeco
Inter-City Products Corp.
Johnson Controls/Penn
Lennox Industries, Inc.
Magnetek, Inc.

Motors & Armatures, Inc.
National Fire Protection Association
Paragon Electric Co., Inc.
Peerless Pump Company
Ranco Controls
Reliance Electric Company, a Rockwell
   Automation Business
Sealed Unit Parts Co.
Seimens Electromechanical Components, Inc.,
   Potter & Brumfield Products Division
Simpson Electric Co.
Sporlan Valve Company
Square D Company
Tecumseh Products Co.
Texas Instruments, Inc.
ThermoPride Williamson Co.
Therm-O-Disc
The Trane Company, a Division
   of American Standard, Inc.
White-Rodgers Division, Emerson
   Electric Co.
York International Corporation

Electrical Safety

OBJECTIVES

After completing this chapter, you should be able to

- Explain the effect of electric current on the human body.
- Understand the injuries that are possible from an electrical shock.
- Know the basic procedures in the event of an electrical shock.
- Understand the importance of properly grounding tools and appliances.
- Safely use electrical hand tools and electrical meters.
- Follow the principles of safety when installing and servicing heating and air-conditioning equipment.

KEY TERMS

Cardiopulmonary resuscitation (CPR)	Fuse
	Ground
Circuit breaker	Ground fault circuit interrupter (GFCI)
Circuit lockout	Grounding adapter
Conductor	*Live* electrical circuit
Double insulated	*National Electrical Code®* (*NEC®*)
Electrical shock	Three-prong plug
Electromotive force	

INTRODUCTION

Electricity is very commonplace in our environment today; in fact it's hard for us to envision life without electricity. No matter what part of our lives we examine, electricity plays an important role, from our home life to our places of employment. Our homes are filled with personal electric appliances like toothbrushes and hair dryers, small electric appliances like mixers and toasters, major appliances like washers and refrigerator/ freezers, and large equipment that heats and cools our living spaces. Many people work in environments that use large electrical equipment that is powered by an extremely high-voltage source. No matter what a person does, they are likely to come near to electrical power sources that are dangerous.

The single most important element to remember when dealing with electrical circuits is to respect them. It is impossible for a service technician to adequately troubleshoot heating and air conditioning with the electrical power turned off, so it is imperative to use safe procedures when the power is on. Many troubleshooting procedures can be performed with the electric power to the equipment interrupted, such as checking the condition of electric motors, relays, contactors, transformers, and other electrical devices. However, there are other times when troubleshooting requires a connection to the power source—checking power available to the equipment, checking power available to a specific electrical device, or checking the voltage drop across a set of contacts in a relay, for example. The important thing for a HVAC/R technician to know is when it is necessary to have the power to the unit on or off.

CAUTION Always perform repairs with the power off.

One of the most important things that a service technician must learn is how to safely work around equipment when the power is being supplied to the equipment. Good service technicians cannot fear being shocked, but they must always pay attention to what they are doing and not get careless when they are working around live electrical circuits. A "live" electrical circuit is one that is being supplied with electrical energy. It is possible for an installation technician to completely install a heating and air-conditioning system without the power being turned on until it is time to check the system for proper operation. No matter what part of the heating, ventilating, and air-conditioning industry a person works in, it is imperative they respect electricity and know how to properly work around it without being injured.

1.1 ELECTRICAL INJURIES

Electrical shocks and burns are common hazards to personnel who are employed in the heating and air-conditioning industry. It is impossible to install or troubleshoot air-conditioning equipment without working close to electrical devices that are being supplied with electrical energy. It is the responsibility of the technician to develop a procedure to work around live electric circuits without coming in contact with conductors and electrical components that are being supplied with electrical power.

Electrical shock occurs when a person becomes part of an electrical circuit. When electricity passes through the human body, the results can range from death to a slight, uncomfortable stinging sensation, depending upon the amount of electricity that passes through the body, the path that the electricity takes, and the amount of time that the electricity flows. Technicians should never allow themselves to become the conductor between two wires or a hot and a ground in an electrical circuit.

The amount of electrical energy needed to cause serious injury is very small. The electrical energy supplied to an electrical circuit is called **electromotive force**, and it is measured in volts. In the heating and air-conditioning industry, the technician oftentimes is in close proximity to 24 volts, which is used for the control circuits of most residential systems, 120 volts, which is used to operate most fan motors in gas furnaces, 240 volts, which is used to operate compressors in residential condensing units, and much higher voltages used to operate compressors in commercial and industrial cooling systems. The heating and air-conditioning technician is often around voltages that can cause serious injury or even death.

Your body can become part of an electrical circuit in many ways. First, your body can become part of an electrical circuit if you come in contact with both a conductor that is being supplied with power and the neutral conductor at the same time, as shown in Figure 1.1. Another way that you can become part of an electrical circuit is to come in contact with both a conductor that is being supplied with power and with the ground, as shown in Figure 1.2. A conductor is a wire or other device that is used as a path for electrical energy to flow. You may become part of the electrical circuit if you touch two conductors that are being supplied with electrical energy, as shown in Figure 1.3.

The severity of injury from electric shock is directly related to the path that current flow takes in the body. The current flow is the amount of electrons flowing in a circuit and is measured in amperes. For example, if the thumb and index finger of the same hand come in contact with a conductor that is supplied with electrical energy and a neutral as shown in Figure 1.4, then the path would only be from the thumb to the index finger. If you touch a **conductor** being supplied with electrical energy with one hand and another conductor being supplied with electrical energy with the other hand,

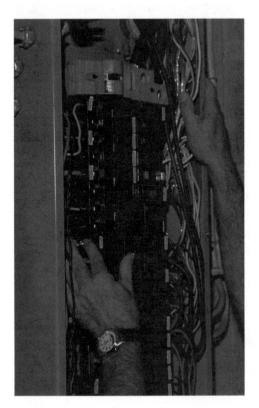

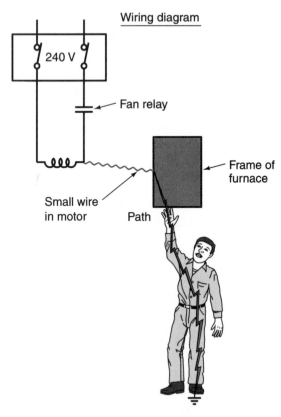

FIGURE 1.1 Technician becoming part of an electrical circuit by coming in contact with $L1$ and neutral

FIGURE 1.2 Technician coming in contact with a conductor (shorted fan motor) and ground

then the electrical path would be from one hand up the arm and across the heart to the other arm and to the hand, as shown in Figure 1.5. If the path is through an arm and a leg, then it would also cross or come near to the heart. When the path of electrical flow crosses the heart, the risk of serious injury increases. Most fatal electrical accidents happen when the electrical flow is passed near or through the heart. When the electrical path crosses near or through the heart for only a short period of time, it can cause ventricular fibrillation of the heart, in which the heart only flutters instead of beats and the blood flow to the body stops. Unless the heartbeat is returned to normal quickly, the person will usually die without immediate medical attention.

The other injury caused by electrical shock is burns to the body. This usually occurs when the technician is shocked with high voltage. Electrical burns can come from an electrical arc, such as the arc from a high-voltage transformer, the arcing of high voltage, and a short circuit to ground, where electrons are allowed to flow unrestricted. For example, if

FIGURE 1.3 Technician touching $L1$ and $L2$ in an electrical panel

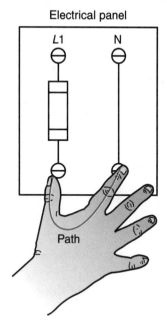

FIGURE 1.4 Electrical path from technician's thumb to index finger

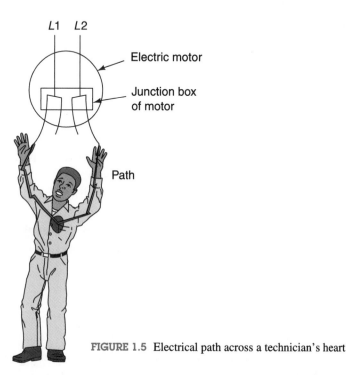

FIGURE 1.5 Electrical path across a technician's heart

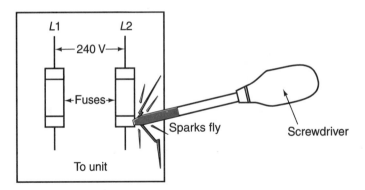

FIGURE 1.6 Screwdriver shorted between $L1$ and ground

you are working in an electrical panel with a screwdriver and allow the blade of the screwdriver to touch a ground while in contact with a conductor that is being supplied with electrical energy, the potential difference is tremendous, and sparking will usually occur, as shown in Figure 1.6. If the resistance is very small, then the current flow in the circuit will be very large. A current flow through the body of 0.015 ampere or less can prove fatal. By comparison, the current draw of a 60-watt light bulb is only .50 ampere.

Another danger of electrical shock is a person's reaction when shocked. For example, if you are working on a ladder and get shocked, you could fall off the ladder. If you are using an electrical-powered hand tool and a short occurs then you might drop the tool, causing personal injury to yourself or others. Technicians should keep in mind that their reactions when getting shocked could endanger others, so they must be cautious and attentive when working near live electrical circuits.

Technicians should be aware of the danger of electrical shock when using ladders that conduct electricity, such as aluminum ladders. If at all possible, the technician should use nonconductive ladders on all jobs. The two primary types of nonconductive ladders used today are wood and fiberglass. Nonconductive ladders work as well as the aluminum ladders, except that they lack the same ease of handling because of their added weight. Whenever you are using a ladder, you should make sure that you do not position the ladder under electrical conductors that you might accidentally come in contact with when climbing the ladder.

1.2 DEALING WITH SHOCK VICTIMS

The first concern when assisting an electrical shock victim who is still in contact with an electrical source is personal safety. If an electrical accident occurs, personnel trying to assist a shock victim should not touch a person who is in contact with an electrical source. The rescuing party should think fast, proceed with caution, and request medical assistance.

Often when someone receives an electrical shock, they cannot let go of the conductor that is the source of the electrical energy. The person who is trying to help should never come in direct contact with the victim. If you try to remove a shock victim from an electrical source that is holding the victim, you become part of the circuit, and there will be two victims instead of one. Rescuers should think before they act. If the switch to disconnect the power source is close by, then turn the switch off. If the switch to disconnect the electrical power source is not close by or cannot be located, then use some nonconductive material to push the victim away from the electrical source. The material used to remove the victim from the electrical source should be dry to reduce the hazard of shock to the person attempting the rescue. If there are wires lying close to the victim and the rescuer is unsure if they are still connected to a power source, then they should be moved with a nonconductive material. When moving conductors or a victim who is still connected to a power source, you should never get too close to the conductors or the person.

As soon as the shock victim is safely away from the electrical source, the rescuer should start first aid procedures. The rescuer should see if the victim is breathing and has a heartbeat. If these vital signs are absent, then **cardiopulmonary resuscitation (CPR)** should be started as soon as possible, or permanent damage may occur. At least one person on each service or installation truck should be trained to perform CPR in case of an accident requiring it. You should be trained before administering CPR.

1.3 *NATIONAL ELECTRICAL CODE*®

The *National Electrical Code*® and *NEC*® are registered trademarks of the National Fire Protection Association, Inc., Quincy, MA 02269. The *National Electrical Code*® specifies the minimum standards that must be met for the safe installation of electrical systems. The *NEC*® is revised every four years. Technicians should make sure when using the *NEC*® that the latest edition is being used. The information in the *NEC*® and local codes must be followed and adhered to when making any type of electrical connection in a structure. The *NEC*® is made up of nine chapters, with each of the first eight chapters divided into articles. Chapter 9 is made up of miscellaneous tables used in the design of electrical systems. The following is a list of the main topics of the eight chapters.

Chapter 1	General
Chapter 2	Wiring and Protection
Chapter 3	Wiring Methods and Materials
Chapter 4	Equipment for General Use
Chapter 5	Special Occupancies
Chapter 6	Special Equipment
Chapter 7	Special Conditions
Chapter 8	Communications Systems
Chapter 9	Tables

Chapters 1 through 4 are directly related to the electrical standards of the refrigeration, heating, and air-conditioning industry. Articles is Chapter 4 that apply directly to the industry include:

Article 400	Portable Cords and Cables
Article 422	Appliances
Article 424	Fixed Electric Space-Heating Equipment
Article 430	Motors, Motor Controls, and Controllers
Article 440	Air Conditioning and Refrigeration Equipment

1.4 ELECTRICAL GROUNDING

The **ground** wire is used in an electrical circuit to allow current to flow back through the ground instead of through a person and causing electrical shock. For example, if a live electrical conductor touched the frame or case of an air-conditioning unit and was not grounded, then whoever touched that air-conditioning unit would become part of the electrical circuit if they provided a ground. In other words, they would receive an electrical shock, which could cause bodily harm or even death. This condition is shown in Figure 1.7. The ground wire forces the path of electrical current flow to pass through the electrical device that is used to protect the circuit, such as a fuse or circuit breaker. The ground wire is identified by the color green in almost all cases.

If an electrically powered tool requires a ground, it is equipped with a **three-prong plug** as shown in Figure 1.8. On this type of plug, the semicircular prong is the grounding section of the plug and should never be cut off or removed. The same goes for extension cords; the grounding prong should never be removed for convenience. It is important when using a power tool that requires a ground that the technician make certain that the receptacle is grounded. Electrical tools or cords with a ground prong that is altered should be taken out of service until replaced or repaired. A **grounding**

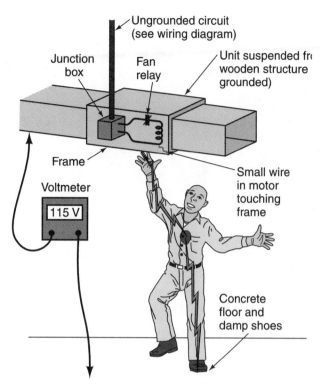

FIGURE 1.7 Technician receives electrical shock from grounded fan motor

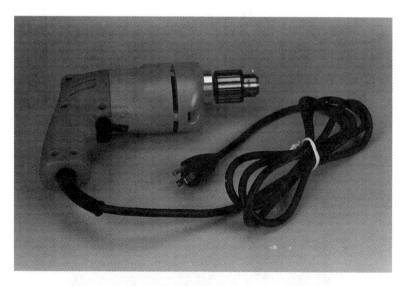

FIGURE 1.8 Electrical drill with three-prong grounded plug

FIGURE 1.9 Double insulated drill with two-prong plug

adapter is a device that permits the connection of a three-prong plug to a two-prong receptacle. A grounding adapter should not be used on a power tool with a three-prong plug unless there is a sure ground that the grounding wire can be attached to. The technician should use caution when using grounding adapters, because in many older structures grounding is not provided at the receptacle box. Most late-model power tools are **double insulated** and do not require a ground. This type of tool will have a plug with only two prongs, as shown in Figure 1.9.

A **ground fault circuit interrupter (GFCI)** is an electrical device that will open the circuit, preventing current flow to the receptacle when a small electrical leak to ground is detected. Figure 1.10 shows a ground

FIGURE 1.10 Ground fault circuit interrupter receptacle *(Photo by Bill Johnson)*

FIGURE 1.11 Ground fault circuit interrupter breaker *(Courtesy of Square D Company)*

fault receptacle with an extension cord plugged into it. This type of receptacle is recommended for use with portable electric power tools. Ground fault circuit interrupters are also available in the form of circuit breakers, as shown in Figure 1.11. Portable ground fault interrupters are available for use where permanent units are not available, such as on job sites. They are designed to help protect the operator from being shocked. Use ground fault circuit interrupters when required by the *National Electrical Code®*.

1.5 CIRCUIT PROTECTION

Electrical circuits in structures are designed to operate at or below a specific current (ampere) rating. Each electrical circuit should be protected, according to the *NEC®*. The wire or conductor of each circuit should be protected to prevent a higher current than it is designed to carry. The electrical components in the circuit are also a consideration when protection is concerned. The standard wire used for receptacles in most residences is #12 TW. The maximum current protection for this type of wire according to the *NEC®* is 20 amperes. However, if there is an electrical component in the circuit that requires protection at 10 amperes, the circuit protection should be at 10 amperes. If the current in the circuit becomes greater than the rating of the protective device, the device opens, disrupting the power source from the circuit.

The most common methods of circuit protection in structures are **fuses**, as shown in Figure 1.12, and **circuit breakers**, as shown in Figure 1.13. These devices protect the circuit by interrupting the flow of electrical energy to the circuit if the current in the circuit exceeds the rating of the fuse or circuit

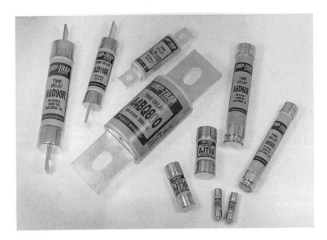

FIGURE 1.12 Fuses *(Courtesy of Gould Shawnot, Newburyport, MA)*

breaker. There are many types of fuses available today with special designs for particular purposes, but the primary purpose of any fuse is protection. Fuses are made with a short strip of metal alloy called an element that has a low melting point, depending on the rating of the fuse. If a larger current flow passes through the fuse than is designed to pass through the element, the element will melt and open the circuit. Circuit breakers look a lot like ordinary light switches placed in an electrical panel. If the current in the circuit that a circuit breaker is protecting exceeds the breaker's rating, then the switch of the circuit breaker will trip and interrupt the electrical energy going to the circuit. Fuses and circuit breakers should be sized for the par-

FIGURE 1.13 Circuit breakers

ticular application according to the *National Electrical Code®*. Technicians should never arbitrarily adjust the size of the fuse or circuit breaker without following the standards in the *NEC®* and local codes. Use only electrical conductors that are the proper size for the load of the circuit according to the *NEC®* to avoid overheating and possible fire.

1.6 CIRCUIT LOCKOUT PROCEDURES

Circuit lockout is a procedure that is used to interrupt the power supply to an electrical circuit or equipment. When a technician is performing work on a circuit where there is a possibility that someone might accidentally restore electrical power to that circuit, the technician should place a padlock and/or a warning label on the applicable switch or circuit breaker. When you are working in a residence, the chance of the homeowner closing switches that might affect your safety is remote but still possible, so use some type of warning tag or verbally inform the homeowner. When working in a structure where there are many people who could open and close switches, you should make absolutely certain that the electrical energy is disconnected from the circuit. Once the circuit is opened, mark the circuit so that others will not turn the circuit on while the repair is under way. In a commercial and industrial setting, this can be accomplished by using safety warning tags, padlocks, or locking devices made for that purpose.

1.7 ELECTRICAL SAFETY GUIDELINES

1. Follow the *National Electrical Code®* as a standard when making electrical connections and calculating wire sizes and circuit protection.
2. Make sure the electrical power supply is shut off at the distribution or entrance panel and locked out or marked in an approved manner.
3. Always make sure that the electrical power supply is off on the unit that is being serviced unless electrical energy is required for the service procedure.
4. Always keep your body out of contact with damp or wet surfaces when working on live electrical circuits. If you must work in damp or wet areas, make certain that some method is used to isolate your body from these areas.
5. Be cautious when working around live electrical circuits. Do not allow yourself to become part of the electrical circuit.
6. Use only properly grounded power tools connected to properly grounded circuits.

7. Do not wear rings, watches, or other jewelry when working in close proximity to live electric circuits.
8. Wear shoes with an insulating sole and heel.
9. Do not use metal ladders when working near live electrical circuits.
10. Examine all extension cords and power tools for damage before using.
11. Replace or close all covers on receptacles that house electrical wiring and controls.
12. Make sure that the meter and the test leads being used are in good condition.
13. Discharge all capacitors with a 15,000 ohm 2 watt resistor before touching the terminals.
14. When attempting to help someone who is being electrocuted, do not become part of the circuit. Always turn the electrical power off or use a nonconductive material to push the person away from the source.
15. Keep tools in good condition, and frequently check the insulated handles on tools that are used near electrical circuits.

SUMMARY

Electricity cannot be seen but it certainly can be felt. It only takes a small amount of electricity to cause injury or even death. It is imperative that heating and air-conditioning technicians respect and be cautious around electrical circuits. It only takes a slip or careless move to find oneself in danger of electrocution or injury. The technician must be careful and cautious around live electrical circuits.

It would be ideal if you never had to work in close proximity with live electrical circuits, but that is not possible, especially when you are called on to troubleshoot heating and air-conditioning systems and equipment. You will be responsible for your own safety, and you should learn to respect and work carefully around live electrical circuits.

REVIEW QUESTIONS

1. True or False: A heating and air-conditioning service technician can usually troubleshoot heating and air-conditioning systems without the voltage being supplied to the equipment.

2. What is a "live" electrical circuit?

3. Which of the following voltages will a refrigeration, heating, and air-conditioning technician come in contact with in the industry?
   a. 24 volts
   b. 120 volts
   c. 240 volts
   d. all of the above

4. Electrical shock occurs when a person _____.
   a. touches an insulated wire
   b. touches an electric motor
   c. becomes part of an electric circuit
   d. touches a conductor that has power applied to it, but is making contact with a ground

5. What are the important elements of electrical safety when working around live circuits?

6. Which of the following conditions is the most dangerous and likely to cause serious injury?
   a. The technician touches a ground with his thumb and a live wire with his index finger.
   b. The technician touches a live wire with his hand but is standing on an insulated platform.
   c. The technician touches a live wire with his right hand and accidentally touches his right elbow on the metal part of the same unit.
   d. The technician touches a live conductor with his right hand and touches a ground with his left hand.

7. Which of the following is the standard by which electrical installations are measured in the United States?
   a. National Electrical Code®
   b. United Electrical Code®
   c. Basic Electrical Code®
   d. none of the above

8. True or False: A current flow of 0.1 amperes or less could be fatal.

9. What type of ladder should the technician use on the job?
   a. aluminum
   b. fiberglass
   c. wood
   d. both a and b
   e. both b and c

10. What precautions should be taken when you see a coworker receiving an electrical shock?

11. True or False: It is recommended that at least one person on a truck know CPR.

12. True or False: The correct fuse size for an electrical circuit is one that is sized twice as large as needed for circuit protection.

13. What is the difference between a two-prong plug and a three-prong plug?

14. Which prong on a three-prong plug is the ground?
    a. the left flat prong
    b. the right flat prong
    c. the center semicircular prong
    d. none of the above

15. True or False: A grounding adapter does no good if it is not connected to an electrical ground.

16. An electrical device that will open an electrical circuit, preventing current flow to the circuit if a small leak to ground is detected, is called a _____.
    a. GFCI
    b. common circuit breaker
    c. fuse
    d. receptacle

17. True or False: Receptacles used on the job site should be protected with a GFCI.

18. What precautions should you use when working in an area with a large number of people and you must disconnect the power from an appliance you are working on?

19. What is the difference between a fuse and a circuit breaker?

20. List at least five electrical safety rules that should be followed by refrigeration, heating, and air-conditioning technicians.

LAB MANUAL REFERENCE

For experiments and activities dealing with material covered in this chapter, refer to Chapter 1 in the Lab Manual.

2

Basic Electricity

OBJECTIVES

After completing this chapter, you should be able to

- Briefly explain the atomic theory and its relationship to physical objects and electron flow.
- Explain the flow of electrons and how it is accomplished.
- Explain electrical potential, current flow, and resistance and how they are measured.
- Explain electrical power and how it is measured.
- Explain Ohm's law.
- Calculate the potential, current, and resistance of an electrical circuit using Ohm's law.
- Calculate the electrical power of a circuit and the Btu/hour rating of an electrical resistance heater.

KEY TERMS

Alternating current
Ampere
Atom
Compound
Conductor
Current
Direct current
Electric energy
Electric power
Electric pressure
Electricity
Electrode

Electrolyte
Electron
Element
Field of force
Free electron
Insulator
Kilowatthour
Law of electric charges
Matter
Molecule
Neutron
Nucleus

Ohm
Ohm's law
Power factor
Proton
Resistance
Seasonal energy efficiency ratio
   (SEER)

Static electricity
Volt
Voltage/Potential
   difference/Electromotive force
Watt

INTRODUCTION

Most control systems used in the heating, cooling, and refrigeration industry use electrical energy to maintain the desired temperature. Electrical components in systems that require rotation, such as compressors and fan motors, use electric motors to accomplish this rotation. Many other devices such as electric heaters, solenoid valves, and signal lights that are incorporated into equipment also require electrical energy for operation. The use of electricity can be seen in all aspects of the industry.

Along with all the electric devices used in systems today come problems that are, in most cases, electrical and that must be corrected by field service technicians. Thus, it is essential for all industry technicians to understand the basic principles of electricity so that they can perform their jobs in the industry.

We begin our study of electricity with a discussion of atomic structure.

2.1 ATOMIC THEORY

Matter is the substance of which a physical object is composed, whether it be a piece of iron, wood, or cloth, or whether it is a gas, liquid, or solid. Matter is composed of fundamental substances called **elements**. There are 105 elements that have been found in the universe. Elements, in turn, are composed of atoms. An **atom** is the smallest particle of an element that can exist alone or in combination. All matter is made up of atoms or a combination of atoms, and all atoms are electrical in structure.

Suppose a piece of chalk is broken in half and one piece discarded. Then the remaining piece is broken in half and one piece discarded. If this procedure is continued, eventually the piece of chalk will be broken into such a small piece that by breaking it once more there will no longer be a piece of chalk but only a molecule of chalk. A **molecule** is the smallest particle of a substance that has the properties of that substance. If a molecule of chalk is broken down into smaller segments, only individual atoms will exist, and they will no longer have the properties of chalk. The atom is the basic building block of all matter. The atom is the smallest particle that can combine with other atoms to form molecules.

Although the atom is a very small particle, it is also composed of several parts. The central part is called the **nucleus**. Other parts, called **electrons**, orbit around the nucleus. Each electron is a relatively small, negatively charged particle. The electrons orbit the nucleus in much the same way that the planets orbit the sun.

The nucleus, the center section of an atom, is composed of protons and neutrons. The **proton** is a heavy, positively charged particle. The proton has an electric charge that is opposite but equal to that of the electron. All atoms contain a like number of protons and electrons. The **neutron** is a neutral particle, which means that it is neither positively nor negatively charged. The neutrons tend to hold the protons together in the nucleus.

The simplest atom that exists is the hydrogen atom, which consists of one proton that is orbited by one electron, as shown in Figure 2.1(a). Not all atoms are as simple as the hydrogen atom. Other atoms have more particles. The difference in each different atom is the number of electrons, neutrons, and protons that the atom contains. The hydrogen atom has one proton and one electron. The oxygen atom has 8 protons, 8 neutrons, and 8 electrons, as shown in Figure 2.1(b). The silver atom contains 47 protons, 61 neutrons, and 47 electrons. The more particles an atom has, the heavier the atom is. Since there are 110 elements, but millions of different types of substances, there must be some way of combining atoms and elements to form these substances.

When elements (and atoms) are combined, they form a chemical union that results in a new substance, called a **compound**. For example, when two hydrogen atoms combine with one oxygen atom, the compound *water* is formed. The atomic structure of one molecule of water is shown in Figure 2.1(c).

The chemical symbol for a compound denotes the atoms that make up that compound. Refrigerant 22 (R-22) is a substance commonly used in refrigeration systems. A refrigerant is a fluid that absorbs heat inside the conditioned area and releases heat outside the conditioned area. The chemical symbol for one molecule of R-22 is $CHCl_2F_2$. One molecule of the refrigerant contains one atom of carbon, one atom of hydrogen, two atoms of fluorine, and one atom of chlorine. The chemical name for R-22 is dichlorodifluoromethane. All materials can be identified according to their chemical makeup, that is, the atoms that form their molecules.

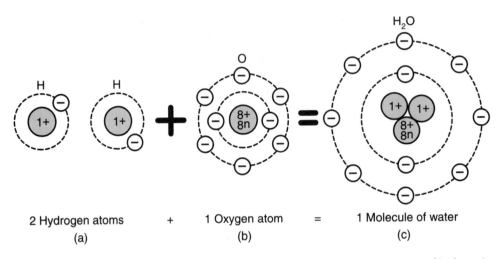

FIGURE 2.1 Atomic structure of a water molecule (one atom of oxygen and two atoms of hydrogen)

2.2 POSITIVE AND NEGATIVE CHARGES

An atom usually has an equal number of protons and electrons. When this condition exists, the atom is electrically neutral because the positively charged protons exactly balance the negatively charged electrons. However, under certain conditions an atom can become unbalanced by losing or gaining an electron. When an atom loses or gains an electron, it is no longer neutral. It is either negatively or positively charged, depending on whether the electron is gained or lost. Thus, in an atom a charge exists when the number of protons and electrons is not equal.

Under certain conditions some atoms can lose a few electrons for short periods. Electrons that are in the outer orbits of some materials, especially metals, can be easily knocked out of their orbits. Such electrons are referred to as **free electrons**, and materials with free electrons are called **conductors**. When electrons are removed from the atom, the atom becomes positively charged, because the negatively charged electrons have been removed, creating an unbalanced condition in the atom.

An atom can just as easily acquire additional electrons. When this occurs, the atom becomes negatively charged.

Charges are thus created when there is an excess of electrons or protons in an atom. When one atom is charged and there is an unlike charge in another atom, electrons can flow between the two. This electron flow is called **electricity**.

An atom that has lost or gained an electron is considered unstable. A surplus of electrons in an atom creates a negative charge. A shortage of elec-

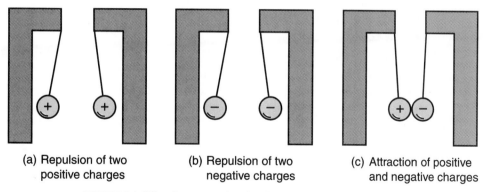

(a) Repulsion of two
    positive charges

(b) Repulsion of two
    negative charges

(c) Attraction of positive
    and negative charges

FIGURE 2.2 Like charges repel and unlike charges attract each other.

trons creates a positive charge. Electric charges react to each other in different ways. Two negatively charged particles repel each other. Positively charged particles also repel each other. Two opposite charges attract each other. The **law of electric charges** states that like charges repel and unlike charges attract. Figure 2.2 shows an illustration of the law of electric charges.

All atoms tend to remain neutral because the outer orbits of electrons repel other electrons. However, many materials can be made to acquire a positive or negative charge by some mechanical means, such as friction. The familiar crackling when a hard rubber comb is run through hair on a dry winter day is an example of an electric charge generated by friction.

2.3 FLOW OF ELECTRONS

The flow of electrons can be accomplished by several different means: friction, which produces static electricity; chemical, which produces electricity in a battery; and magnetic (induction), which produces electricity in a generator. Other methods are also used, but the three mentioned here are the most common.

Static Electricity

The oldest method of moving electrons is by **static electricity**. Static electricity produces a flow of electrons by permanently displacing an electron from an atom. The main characteristic of static electricity is that a prolonged or steady flow of current is not possible. As soon as the charges between the two substances are equalized (balanced), electron flow stops.

Friction is usually the cause of static electricity. Sliding on a plastic seat cover in cold weather and rubbing silk cloth on a glass rod are two examples of static electricity produced by friction. Static electricity, no matter what the cause, is merely the permanent displacement or transfer of electrons. To obtain useful work from electricity, a constant and steady flow of electrons must be produced.

Electricity Through Chemical Means

Electricity can also be produced by the movement of electrons due to chemical means. A battery produces an electron flow by a chemical reaction that causes a transfer of electrons between two **electrodes**. An electrode is a solid conductor through which an electric current can pass. One electrode collects electrons and one gives away electrons. The dry cell battery uses two electrodes made of two dissimilar metals inserted in a paste-like **electrolyte**. Electricity is produced when a chemical reaction occurs in the electrolyte between the electrodes, causing an electron flow. The construction of a dry cell battery is shown in Figure 2.3. A dry cell battery is shown in Figure 2.4.

The container of a dry cell battery, which is made of zinc, is the negative electrode (gives away electrons). The carbon rod in the center of the dry cell is the positive electrode (collects electrons). The space between the electrodes is filled with an electrolyte, usually manganese dioxide paste.

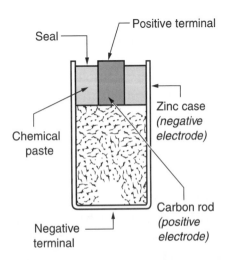

FIGURE 2.3 Construction of a dry cell battery

FIGURE 2.4 Dry cell battery *(Courtesy of Duracell)*

The acid paste causes a chemical reaction between the carbon electrode and the zinc case. This reaction displaces the electrons, causing an electron flow. The top of the dry cell is sealed to prevent the electrolyte from drying and to allow the cell to be used in any position. The dry cell battery will eventually lose all its power, because energy is being used and not being replaced.

The storage battery is different from a dry cell battery because it can be recharged. Thus, it lasts somewhat longer than a dry cell battery. But it, too, will eventually lose all its energy.

The storage battery consists of a liquid electrolyte and negative and positive electrodes. The electrolyte is diluted sulfuric acid. The positive electrode is coated with lead dioxide and the negative electrode is sponge lead. The chemical reaction between the two electrodes and the electrolyte displaces electrons and creates voltage between the plates. The storage battery is recharged by reversing the current flow into the battery. The storage battery shown in Figure 2.5 is commonly used in automobile electric systems.

Electricity Through Magnetism

The magnetic or induction method of producing electron flow uses a conductor to cut through a magnetic field, which causes a displacement of electrons. The alternator, generator, and transformer are the best examples

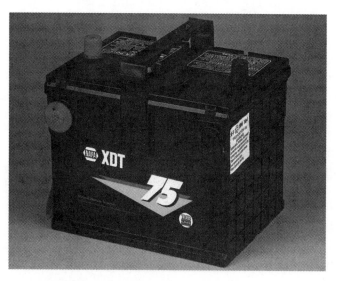

FIGURE 2.5 Common storage battery used in automobile electrical system

of the magnetic method. The magnetic method is used to supply electricity to consumers.

The flow of electrons in a circuit produces magnetism, which is used to cause movement, or thermal energy, which in turn is used to cause heat. A magnetic field is created around a conductor—an apparatus for electrons to flow through—when there is a flow of electrons in the conductor. The flow of electrons through a conductor with a resistance will cause heat, such as in an electric heater.

The heating, cooling, and refrigeration industry uses magnetism to close relays and valves and to operate motors by using coils of wire to increase the strength of the magnetic field.

2.4 CONDUCTORS AND INSULATORS

The structure of an atom of an element is what makes it different from the atom of another element. The number of protons, neutrons, and electrons and the arrangement of the electrons in their orbits vary from element to element. In some elements, the outer electrons rotating around the nucleus are easily removed from their orbits. As stated earlier, elements that have atoms with this characteristic are called *conductors*. A conductor can transmit electricity or electrons.

Most metals are conductors, but not all metals conduct electricity equally well. The most common conductors are silver, copper, and aluminum. The high cost of silver prevents it from being used widely. Its use is largely limited to contacts in certain electrical switching devices such as contactors and relays. Copper, almost as good a conductor as silver, is usually used because it is less expensive.

Materials that do not easily give up or take on electrons are called **insulators**. An insulator retards the flow of electrons. Glass, rubber, and asbestos are examples of insulators. Thermoplastic is one of the best insulators used to cover wire today. How well an insulator prevents electron flow depends on the strength of the potential applied. If the potential is strong enough, the insulator will break down, causing electrons to flow through it.

There is no perfect insulator. All insulators will break down under certain conditions if the potential is high enough. Increasing the thickness of the insulation helps overcome this problem.

Conductors and insulators are important parts of electric circuits and electric systems. They are widely used in all electric components in the industry.

2.5 ELECTRIC POTENTIAL

In a water system, water can flow as long as pressure is applied to one end of a pipe and the other end of the pipe is open. The greater the pressure in a water system, the greater the quantity of water that will flow. Similarly, in an electrical system electrons will flow as long as electric pressure is applied to the system. **Voltage, potential difference,** and **electromotive force** are all terms used to describe electric pressure.

Recall that the law of electric charges states that unlike charges attract. Consequently there is a pull, or *force,* of attraction between two dissimilarly charged objects. We call this pull of attraction a **field of force.**

Another way of looking at this is to picture excess electrons (the negative charge) as straining to reach the point where there are not enough electrons (the positive charge). If the two charges are connected by a conductor, the excess electrons will flow to the point where there are not enough electrons. But if the two charges are separated by an insulator, which prevents the flow of electrons, the excess electrons cannot move. Hence, an excess of electrons will pile up at one end of the insulator, with a corresponding lack, or deficiency, of electrons at the other end.

As long as the electrons cannot flow, the field of force between the two dissimilarly charged ends of the insulator increases. The resulting strain between the two ends is called the **electric pressure.** This pressure can become quite great. After a certain limit is reached, the insulator can no longer hold back the excess electrons, as discussed in the previous section. Hence, the electrons will rush across the insulator to the other end.

Electric pressure that causes electrons to flow is called *voltage.* Voltage is the difference in electric potential (or electric charge) between two points. The **volt** (V) is the amount of pressure required to force one **ampere** (A, the unit of measurement for current flow) through a resistance of one **ohm** (Ω, the unit of measurement for resistance; Ω is the Greek letter omega). In the industry, voltage is almost always measured in the range of the common volt. In other areas, the voltage may be measured on a smaller scale of a millivolt (mV), or one-thousandth of a volt. For larger measurements of the volt, the kilovolt (kV), equal to 1000 volts, is used.

$$1 \text{ millivolt} = 0.001 \text{ volt}$$

$$1 \text{ kilovolt} = 1000 \text{ volts}$$

To maintain electric pressure we must have some way to move electrons in the same manner that water pressure moves water. In an electric circuit,

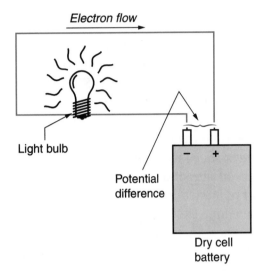

FIGURE 2.6 A dry cell battery supplying electric potential (voltage) to an electric circuit

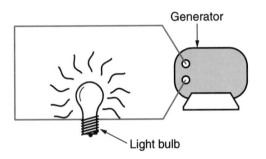

FIGURE 2.7 A generator supplying electric potential (voltage) to an electric circuit

this can be maintained by a battery, as shown in Figure 2.6, or by a generator or alternator, as shown in Figure 2.7. The battery forces electrons to flow to the positive electrode and causes electric pressure. A generator causes electric pressure by transferring electrons from one place to another.

Electromotive force can be produced in several ways. The easiest method to understand is the simple dry cell battery discussed in Section 2.3. The most popular method of producing an electromotive force is by using an alternating current generator. The alternating current generator is supplied with power from another source. Then a wire loop is rotated through the magnetic field created by the voltage being applied, and an electromotive force is produced through the wire loop. We will discuss these ideas in more detail in succeeding sections.

2.6 CURRENT FLOW

Electrons flowing in an electric circuit are called **current**. Current flow can be obtained in an electric circuit by a bolt of lightning, by static electricity, or by electron flow from a generator. Figure 2.8 shows an electric system with electric pressure; the quantity of electrons flowing is also given.

There are two types of electric current: **direct current** and **alternating current**. Direct current flows in one direction. It is the type of current produced by dry cell batteries. Direct current is rarely used in the industry as a main power source but is used in some modern control circuits.

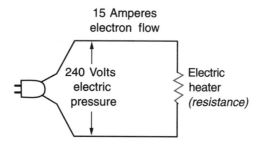

FIGURE 2.8 An electric system; electric potential forces electrons through a wire (a conductor) to a load (electric heater)

Alternating current flows back and forth. It is the type of current supplied to most homes by electric utility companies. It is the most commonly used source of electric potential in the heating, cooling, and refrigeration industry. Alternating current will be discussed in more detail in Chapter 7.

The current in an electric circuit is measured in amperes (A). An ampere is the amount of current required to flow through a resistance of one ohm with a pressure of one volt. An ampere is measured with an ammeter. In the industry, the ampere is used almost exclusively. If a smaller unit of ampere measurement is required, the milliampere (mA), which is one-thousandth of an ampere, can be used. For larger measurements of amperes, the kilo-ampere (kA) can be used. One kiloampere equals 1000 amperes.

$$1 \text{ milliampere} = 0.001 \text{ ampere}$$

$$1 \text{ kiloampere} = 1000 \text{ amperes}$$

The current that an electric device draws can be used as a guide to the correct operation of the equipment by installation and service technicians. The electric motor is the largest current-drawing device in most heating, cooling, and refrigeration systems. The larger the electric device (load), the larger the current flow. Any electric device that uses electricity requires a certain current when operating properly.

2.7 RESISTANCE

Resistance is opposition to the flow of electrons in an electric circuit and is measured on ohms. The electrical device or load in an electrical circuit that will produce some useful work is known as or represents the resistance of that circuit. Figure 2.9 shows two electric systems with different resistances. One ohm is the amount of resistance that will allow one ampere to flow with a pressure of one volt. The industry uses simple ohms

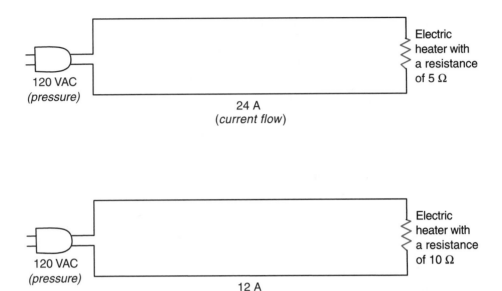

FIGURE 2.9 Two electric systems with different resistances

for resistance in most cases because the scale is broad enough for most applications. In some special cases, the microhm (μΩ), which is one-millionth of an ohm, is used for extremely small resistance readings. Larger resistance readings are read in megohms (MΩ); one megohm is equal to one million ohms.

$$1 \text{ microhm} = 0.000001 \text{ ohm}$$

$$1 \text{ megohm} = 1,000,000 \text{ ohms}$$

All electric devices will have a certain resistance. That resistance depends on the size and purpose of the device. As service technicians, you will have to become familiar with this value in components. If the resistance deviates far from the specified value or the estimated value, the device can be considered faulty.

2.8 ELECTRIC POWER AND ENERGY

When electrons move from the negative to the positive end of a conductor, work is done. **Electric power** is the rate at which the electrons do work. That is, electric power is the rate at which electricity is being used. The power of an electric circuit is measured in **watts** (W). A watt of electricity is one ampere flowing with a pressure of one volt. The electric power of a circuit is the voltage times the amperage.

$$\text{electric power} = \text{voltage} \times \text{amperage}$$

In an alternating current circuit, the voltage and current are not in phase. To obtain the correct power consumed by the circuit, the product of the voltage times the amperage must also be multiplied by a power factor. The power factor is the true power (measured with a wattmeter) divided by the calculated power and is expressed as a percentage. In a direct current circuit, the product of the voltage times the amperage gives the power of the circuit; the power factor is not needed.

The industry uses the units of watts for devices that consume a small amount of power. Examples of such devices are small electric motors and small resistance heaters. Other units used are the horsepower (hp) and the British thermal unit (Btu). One horsepower is equal to 746 watts. One watt is equal to 3.41 Btu per hour.

$$1 \text{ horsepower} = 746 \text{ watts}$$

$$1 \text{ watt} = 3.41 \text{ Btu/hour}$$

These conversion figures are often used in the industry to calculate the Btu rating of an electric heater if the watt rating is known. The horsepower conversion is used to calculate the horsepower of a motor only if the watts are known.

The rate at which electric power is being used at a specific time is called electric energy. Electric energy is measured in watthours (Wh). For example, a one-horsepower motor uses 746 watts in one hour. It therefore uses 746 watthours.

The wattage rating or consumption of any electric device only denotes the amount of power the device is using. However, time must be considered when calculating electric energy, that is, the power being consumed over a definite period. Watthours give the number of watts used for a specific period of time.

The units of kilowatts (kW) are usually used to determine the amount of electricity consumed. Thus, an electric utility calculates the power bill of its customers using kilowatthours (kWh), because the watthour readings would be extremely large. The kilowatthour reading is relatively small. One thousand watts used for one hour equals one kilowatthour. All electric meters used to measure the consumption of electricity record consumption in kilowatthours.

Heating and air-conditioning technicians are often required to make a calculation for the output of an electric heater in Btu's rather than in watts. The industry rates electric heating equipment in watts or kilowatts, which does not give consumers figures that they can understand. Most consumers are familiar with the term Btu and know basically what it means in terms of heat output. Therefore, the customer often will request the Btu output rather than the wattage of a heating appliance. The Btu output can be easily calculated by multiplying the number of watts by the conversion factor of 3.41 Btu/h. Figure 2.10 shows an electric data plate for an electric heater. The wattage on the data plate is 2.50 kilowatts. Therefore the Btu output is

$$2500 \times 3.41 \text{ Btu/h} = 8525 \text{ Btu/h}$$

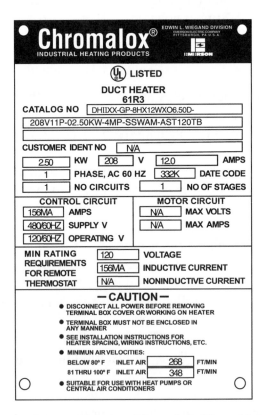

FIGURE 2.10 Data plate of an electric heater (*Courtesy of Chromalox*®, *Wiegand Industrial Division of Emerson Electric Co.*)

Another term that is used because of the high cost of energy is the **seasonal energy efficiency ratio (SEER)** of an air-conditioning unit or heat pump, measured in Btu's per watt. The SEER is the Btu output of the equipment divided by the power input with a seasonal adjustment. For example, if an air-conditioning unit has a SEER of 10, it will produce 10 Btu of cooling per watt of power consumed by the equipment. All air-conditioning manufacturers use SEER ratings for their equipment. The standards for SEER are set by the Air Conditioning and Refrigeration Institute.

2.9 OHM'S LAW

The relationship among the current, electromotive force, and resistance in an electric circuit is known as **Ohm's law.** In the nineteenth century, George Ohm developed the mathematical comparisons of the major factors in an electric circuit. Stated in simple terms, Ohm's law says it will take 1 volt of electrical pressure to push 1 amp of electrical current flow through 1 ohm of electrical resistance; in other words, the greater the voltage the greater the current, and the greater the resistance the lesser the current. Ohm's law is represented mathematically as "current is equal to the electromotive force divided by the resistance." The following equation expresses Ohm's law:

$$I = \frac{E}{R}$$

In the equation, I represents the current in amperes, E represents the electromotive force in volts, and R represents the resistance in ohms. Ohm's law can also be expressed by the following two formulas:

$$E = IR \qquad R = \frac{E}{I}$$

In any of the three formulas, when two elements of an electric circuit are known, the unknown factor can be calculated. Ohm's law calculations do not apply to alternating current circuits because coils of wire produce different effects on alternating current. Alternating current will be discussed in a later chapter. However, the general ideas or principles of Ohm's law apply to alternating current circuits.

The following examples show the relationships of the voltage, current, and resistance in an electric current.

Example 1   What is the current in the circuit shown in Figure 2.11?

Step 1: $I = \dfrac{E}{R}$

Step 2: $I = \dfrac{120}{10}$

Step 3: $I = 12$ A

FIGURE 2.11 Simple electric circuit for Example 1

Example 2   What is the resistance of a 100-watt light bulb if the voltage is 120 volts and the current is .83 amperes?

Step 1: $R = \dfrac{E}{I}$

Step 2: $R = \dfrac{120}{.83}$

Step 3: $R = 145\ \Omega$

Example 3   What is the voltage supplied to the circuit in Figure 2.12?

Step 1: $E = IR$

Step 2: $E = 5 \times 48$

Step 3: $E = 240$ V

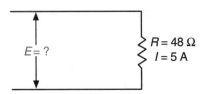

FIGURE 2.12 Simple electric circuit for Example 3

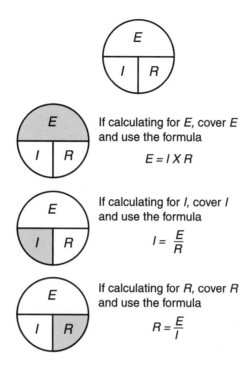

FIGURE 2.13 Using Ohm's law (not applicable on AC inductive circuits)

Ohm's law allows the calculation of the missing factor if the other two factors are known or can be measured. Figure 2.13 shows a simple method for remembering Ohm's law. If one of the factors in the circle is covered, the letters remaining in the circle give the correct formula for calculating the covered factor.

2.10 CALCULATING ELECTRIC POWER

Electric power can be calculated by using the formula $P = IE$. Two other formulas can be used to calculate the electric power of an electric circuit by substituting in the following equations:

$$P = \frac{E^2}{R} \qquad P = I^2 R$$

The letter designations in these formulas are the same as in Ohm's law, with P representing power in watts.

The following three examples show the electric power calculations of three electric circuits.

EG **Example 4**

What is the power consumption of an electric circuit using 15 amperes and 120 volts?

Step 1: $P = IE$

Step 2: $P = 15 \times 120$

Step 3: $P = 1800 \text{ W}$

EG **Example 5**

What is the current of an electric heater rated at 5000 watts on 240 volts?

Step 1: $I = \dfrac{P}{E}$

Step 2: $I = \dfrac{5000}{240}$

Step 3: $I = 21.7 \text{ A}$

EG

Example 6

What is the power of an electric circuit with 5 amperes current and 10 ohms resistance?

Step 1: $P = I^2 R$

Step 2: $P = 5^2 \times 110$

Step 3: $P = 25 \times 10$

Step 4: $P = 250 \text{ W}$

SUMMARY

Everything—solids, liquids, and gases—is composed of matter. Matter can be broken down into molecules (the smallest particles of physical objects) and atoms (the smallest particles of an element that can exist alone or in combination). An atom is composed of a nucleus (the central part) and electrons (negatively charged) that orbit around the nucleus, much like the planets orbit the sun. The nucleus is composed of protons (positively charged) and neutrons (no charge). The number of protons is usually equal to the number of electrons, making the atom electrically neutral. When an atom loses electrons, it becomes positively charged. When it gains electrons, it becomes negatively charged. The law of charges states that like charges repel and unlike charges attract. Materials can be made to acquire positive or negative charges.

Electrons can be made to flow by the use of friction, chemicals, and magnetism. A conductor is a material capable of transmitting electrons or electricity. Most metals are conductors. An insulator is a material that resists or prevents electron flow.

There are four important factors in any electric circuit: electromotive force, current, resistance, and power. The electromotive force of an electric circuit is the actual pressure in the circuit, much like water pressure in a water system. The electromotive force in an electric circuit is measured in volts. The voltage (pressure) must be sufficient to overcome the resistance of the circuit. Alternating current is used almost exclusively in the industry to supply electric power to equipment.

The number of electrons flowing in an electric circuit is called the current flow. The current flow of an electric circuit is measured in amperes. An ampere is the amount of current that will flow through a resistance of one ohm with a pressure of one volt.

The resistance of an electric circuit is measured in ohms. All electric loads have some resistance. Electric power is the rate at which electric energy is being used in an electric circuit. Electric power is measured in watts and kilowatts. The electric utilities use kilowatthours in most cases to charge their customers for the electric energy they have consumed. The kilowatthour is a measure of electric energy and takes into consideration the amount of time and the power consumption. One thousand watts used for a period of one hour equals one kilowatthour. Voltage, amperage, resistance, and wattage often use the prefixes kilo- or milli- to represent larger or smaller quantities of these factors and to avoid the use of extremely large or small numbers.

Ohm's law gives the relationship among the current, electromotive force, and resistance in an electric circuit. Ohm's law states the relationship mathematically. When any two factors in an electric circuit are known or can be measured, the formulas for Ohm's law can be used to find the third factor. Electric power can be calculated by using the formula $P = IE$.

REVIEW QUESTIONS

1. All physical objects are composed of

   _____

   a. substances
   b. matter
   c. neutrons
   d. ketons

2. What is an atom?

3. Which of the following is a part of the atom?

   a. electron
   b. proton
   c. neutron
   d. all of the above

4. What is static electricity?

5. Name three ways electricity can be produced.

6. What part do protons and electrons play in the production of electricity?

7. Which of the following is the simplest atom that exists?
   a. carbon
   b. hydrogen
   c. oxygen
   d. sulphur

8. What are the four most important factors in an electric circuit?

9. What is electromotive force?

10. Electromotive force is commonly measured in _____
    a. amperes
    b. ohms
    c. volts
    d. watts

11. What is current?

12. How is current measured?
    a. amperes
    b. ohms
    c. volts
    d. watts

13. What is resistance?

14. How is resistance commonly measured?
    a. amperes
    b. ohms
    c. volts
    d. watts

15. What is electrical power?

16. How is electrical power commonly measured?
    a. amperes
    b. ohms
    c. volts
    d. watts

17. Where do electrons exist in an atom, and what is their charge?

18. True or False: All atoms tend to lose electrons.

19. State the law of electric charges.

20. What is a proton? Where does it normally exist in an atom, and what is its charge?

21. Describe briefly the method a dry cell battery uses to produce voltage.

22. What is a conductor?

23. Which of the following is the best conductor?
    a. wood
    b. thermoplastic
    c. copper
    d. cast iron

24. What is an insulator?

25. Which of the following is the best insulator?
    a. wood
    b. thermoplastic
    c. copper
    d. cast iron

26. Why do metals make the best conductors?

27. How do electrical utilities charge customers for electricity?

28. What is the meaning of SEER when used in conjunction with an air-conditioning unit?

29. State Ohm's law.

30. True or false: Ohm's law applies to all types of electrical circuits.

31. What is the ampere draw of a 5000-watt electric heater used on 120 volts?

    a. 40 amps
    b. 45 amps
    c. 50 amps
    d. 55 amps

32. What is the resistance of the heating element of an electric iron if the ampere draw is 8 amperes when 115 volts are applied?

    a. 10 ohms
    b. 12 ohms
    c. 14 ohms
    d. 16 ohms

33. What is the voltage of a small electric heater if the heater is drawing 12 amperes and has a resistance of 10 ohms?

    a. 120 volts
    b. 240 volts
    c. 60 volts
    d. none of the above

34. What is the Btu/hour output of an electric heater rated at 15 kW?

    a. 51.15 Btu
    b. 51,150 Btu
    c. 5115 Btu
    d. none of the above

35. What is the kilowatt output of an electric heater that has an ampere draw of 50 A and a voltage source of (a) 208 volts? (b) 240 volts?

LAB MANUAL REFERENCE

For experiments and activities dealing with material covered in this chapter, refer to Chapter 2 in the Lab Manual.

3

Electric Circuits

OBJECTIVES

After completing this chapter, you should be able to

- Explain the concepts of a basic electric circuit.
- Explain the characteristics of a series circuit.
- Explain the characteristics of a parallel circuit.
- Describe how series circuits are utilized as control circuits in the air-conditioning industry.
- Describe how parallel circuits are utilized as power circuits in the air-conditioning industry.
- Explain the relationship and characteristics of the current, resistance, and electromotive force in a series circuit.
- Explain the relationship and characteristics of the current, resistance, and electromotive force in a parallel circuit.
- Calculate the current, resistance, and electromotive force in a series circuit.
- Calculate the current, resistance, and electromotive force in a parallel circuit.
- Explain the characteristics of the series-parallel circuit.
- Describe how series-parallel circuits are utilized in the air-conditioning industry.

KEY TERMS

Closed
Control circuit
Electric circuit
Open
Parallel circuit

Power circuit
Series circuit
Series-parallel circuit
Voltage drop

INTRODUCTION

The electrical circuitry in a modern heating, cooling, and refrigeration system is important to the technicians who install or work on the electrical systems. Any electrical system used in the industry is composed of various types of circuits. Each type is designed to do a specific task within the system. We will look at some commonly used types of circuits in this chapter.

The two most important kinds of circuits are parallel circuits and series circuits. A **parallel circuit** is an electric circuit that has more than one path through which electricity may flow. A parallel circuit is designed to supply more than one load in the system.

A **series circuit** is an electric circuit that has only one path through which electricity may flow. It is usually used for devices that are connected in the circuit for safety or control.

A **series-parallel circuit** is an electrical circuit that has a combination of series and parallel circuits. Most electrical systems in equipment or control systems are made up of combination circuits. Electrical loads that require electrical power for operation are usually connected in parallel, which allows them to receive full supply voltage for operation. Switches used in electrical circuits to control these loads are connected in series, breaking the circuit if the switch opens.

You must understand the circuitry in air-conditioning, heating, and refrigeration control and power systems to do an effective job of installing and servicing the equipment.

We begin our study with a discussion of the basics of electric circuits.

3.1 BASIC CONCEPTS OF ELECTRIC CIRCUITS

An **electric circuit** is the complete path of an electric current, along with any necessary elements, such as a power source and a load. When the circuit is complete so that the current can flow, it is termed **closed** or *made* (Figure 3.1). When the path of current flow is interrupted, the circuit is termed **open** or *broken* (Figure 3.2). The opening and closing of electrical switches connected in series with electrical loads control the operation of loads in the circuit.

All electric circuits must have a complete path for electrons to flow through, a source of electrons, and some electric device (load) that requires electric energy for its operation. Figure 3.3 shows a complete circuit with the basic components labeled. Electric circuits supply power and control loads through the use of switches.

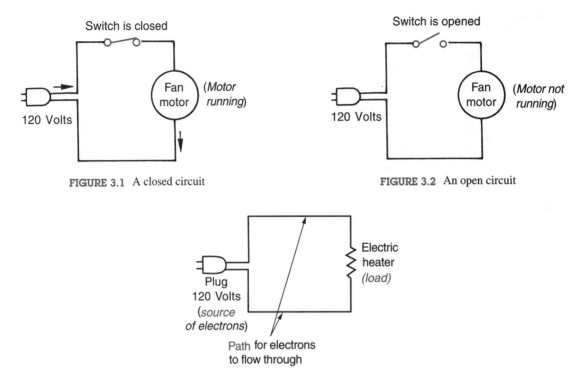

FIGURE 3.1 A closed circuit

FIGURE 3.2 An open circuit

FIGURE 3.3 Basic electric circuit with circuit components labeled

An alternating current power supply is the most common source of the electric energy needed in the electric circuits of a heating, cooling, or refrigeration system. A direct current power supply, such as the dry cell battery, is often the source of electron flow in electric meters. Other than for meters, though, direct current is rarely used as a source of electric energy. Two special applications of direct current used today are electronic air cleaners and solid-state modules used for some types of special control, such as a defrost control, ignition modules and overcurrent protection.

The purpose of most electric circuits is to supply energy to a machine that does work. The most common device supplied with energy is an electric motor. Motors are used to rotate fans, compressors, pumps, and other mechanical devices that require a rotating motion. Automatic switches also require a power source to open and close them so they can start other electric devices. Electric circuits also supply energy to transformers, lights, and timers.

There are several possible path arrangements that electrons may follow. The path arrangement is determined by the use or purpose of the circuit. The three types of circuits are series, parallel, and series-parallel. The series circuit allows only one path of electron flow. The parallel circuit has

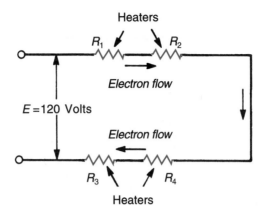

FIGURE 3.4 Series circuit with four resistance heaters

more than one path. The series-parallel circuit is a combination of the series and parallel circuits. In the following sections, we will look at each of these arrangements.

3.2 SERIES CIRCUITS

Switches and controls are commonly wired in series with each other to control one or more loads. The simplest and easiest electric circuit to understand is the series circuit. The series circuit allows only one path of current flow through the circuit. In other words, the path of a series circuit must pass through each device in the entire circuit. All devices are connected end-to-end within a series circuit. Figure 3.4 shows a series circuit with four resistance heaters.

Applications

Series circuits are incorporated into most control circuits used in heating, cooling, and refrigeration equipment. A **control circuit** is an electric circuit that controls some major load in the system. If all control components are connected in the circuit in series, the opening of any switch or component will open the circuit and stop the electric load, as shown in the circuit in Figure 3.5. (The symbols L1 and L2 in Figure 3.5 and in other figures represent the source of the voltage, the power supply. We will discuss circuit notation in more detail in Chapter 5.)

Series circuits are used in the electric circuitry of heating, cooling, and refrigeration equipment to operate the equipment and maintain a desired temperature. Any electric switch or control that is placed in series with a load will operate that particular load. Figure 3.6 shows one circuit of a

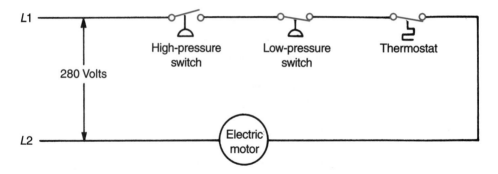

FIGURE 3.5 Series circuit with three switches controlling an electric motor

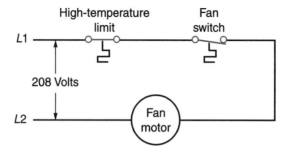

FIGURE 3.6 Control (series) circuit of a fan motor in a gas furnace. Both temperature controls must be closed for the fan to operate.

control system. The controls are connected in series with the device that is being controlled, in this case an electric motor.

The series circuit also contains any safety devices that are needed to maintain safe operation of the equipment components. Figure 3.7 shows a series circuit that is basically made up of safety devices designed to stop the compressor if an unsafe operating condition occurs. If any of the safety controls open, the circuit will open and stop the compressor. Safety devices should be connected in series to ensure that unsafe conditions will cut off the load being protected.

Characteristics of a Series Circuit and Calculations for Current, Resistance, and Voltage

The current draw in a series circuit is the same throughout the entire circuit because there is only one path for the current to follow. The current in a series circuit is shown by the following equation:

$$I_t = I_1 = I_2 = I_3 = I_4 = \cdots$$

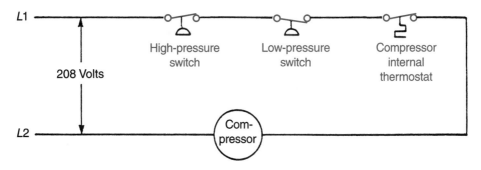

FIGURE 3.7 Control (series) circuit showing safety devices used on modern air-conditioning systems

(The centered dots $\cdots$ indicate that the equation continues in the same manner until all the elements of that particular circuit have been accounted for.)

The total resistance R_t in a series circuit is the sum of all the resistances in the circuit. The resistance of a series circuit is shown by the following equation:

$$R_t = R_1 + R_2 + R_3 + R_4 + \cdots$$

The voltage in a series circuit is completely used by all the loads in the circuits. The loads of the series circuit must share the voltage that is being delivered to the circuit. Thus, the voltage being delivered to the circuit will be split by the loads in the circuit.

The voltage of a series circuit changes through each load. This change is called the **voltage drop.** The voltage drop is the amount of voltage (electrical pressure) used or lost through any load or conductor in the process of moving the current (electron flow) through that part of the circuit. The voltage drop of any part of a series circuit is proportional to the resistance in that part of the circuit. The sum of the voltage drops of a series circuit is equal to the voltage being applied to the circuit. This is shown by the following equation:

$$E_t = E_1 + E_2 + E_3 + E_4 + \cdots$$

Ohm's law can be used for the calculations on any part of a series circuit or on the total circuit. Figure 3.8 shows a series circuit with four resistance heaters of different ohm ratings. The calculations for the total resistance, the amperage, and the voltage drop across each heater will be calculated using the circuit shown in Figure 3.8.

The total resistance can be calculated by adding the ohm rating of each heater.

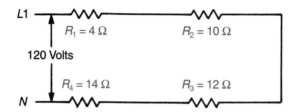

FIGURE 3.8 Series circuit containing four resistance heaters with different resistance values

Step 1: Use the formula $R_t = R_1 + R_2 + R_3 + R_4$.

Step 2: Substitute the values given in the figure into the formula:

$$R_t = 4 \text{ ohms} + 10 \text{ ohms} + 12 \text{ ohms} + 14 \text{ ohms}$$

Step 3: Solve the formula: $R_t = 40$ ohms.

We use Ohm's law to calculate the amperage draw (current) of the circuit.

Step 1: Use the formula

$$I = \frac{E}{R}$$

Step 2: Substitute the given values into the formula:

$$I = \frac{120}{40}$$

Step 3: Solve the formula: $I = 3$ amperes.

Now we use Ohm's law to calculate the voltage drop across each heater.

Step 1: Use the formula $E = IR$ for each resistance.

Step 2: Substitute the given values into the formula: $E_{d1} = 3 \times 4$. (The symbol E_{d1} means the voltage drop across resistance 1.)

Step 3: Solve the formula: $E_{d1} = 12$ volts.

Step 4: Solve for each resistance using the same procedures that were used in Steps 2 and 3:

$$E_{d2} = 3 \times 10 = 30 \text{ volts}$$
$$E_{d3} = 3 \times 12 = 36 \text{ volts}$$
$$E_{d4} = 3 \times 14 = 42 \text{ volts}$$

Note that the total voltage E_t supplied to the circuit is equal to the sum of the voltage drops.

3.3 PARALLEL CIRCUITS

The parallel circuit has more than one path for the electron flow. That is, in a parallel circuit the electrons can follow two or more paths at the same time. Electric devices (loads) are arranged in the circuit so that each is connected to both supply voltage conductors.

Parallel circuits are common in the industry because most loads used operate from line voltage. Line voltage is the voltage supplied to the equipment from the main power source of a structure and typically has a value of 115 volts or 240 volts. The parallel circuit allows the same voltage to be applied to all the electric loads connected in parallel, as indicated in Figure 3.9. Note that each load in the circuit is supplied by the line voltage of 115 volts.

Applications

Parallel circuits are used in the industry to supply the correct line voltage to several different circuits in a control system and are called **power circuits.** Figure 3.10 shows a control system with several circuits in parallel being fed line voltage. Many different paths for electron flow are in this parallel hookup. Each circuit that is connected from $L1$ to $L2$ on the wiring diagram is in parallel with all the others and is being supplied with the line voltage.

Parallel circuits are used in all power wiring that supplies the loads of heating, cooling, or refrigeration systems. The electric loads of a system must be connected to the power supply separately or in a parallel circuit to supply the load with the full line voltage.

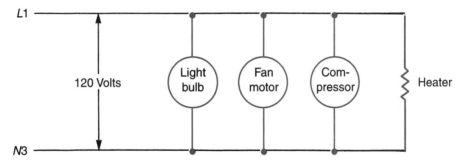

FIGURE 3.9 Parallel circuit with four components; each component is supplied with 120 volts.

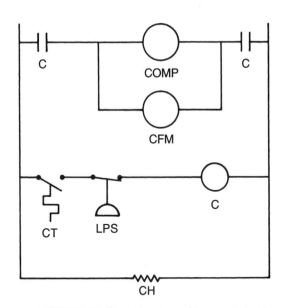

COMP:	Compressor
CFM:	Condenser fan motor
CH:	Crankcase heater
CT:	Cooling thermostat
LPS:	Low-pressure switch
C:	Contactor

FIGURE 3.10 Control system with several circuits (parallel); each circuit is supplied with line voltage.

Characteristics of a Parallel Circuit and Calculations for Current, Resistance, and Voltage

There will be few occasions when field technicians are required to make calculations for a parallel circuit. This is usually done by the designer of the equipment. However, field technicians should be familiar with the basic concepts and rules of parallel circuits.

The current draw in a parallel circuit is determined for each part of the circuit, depending on the resistance of that portion of the circuit. The total current draw of the entire parallel circuit is the sum of the currents in the individual sections of the parallel circuit. The current in each individual circuit can be calculated by using Ohm's law when the resistance and voltage are known. The total ampere draw of a parallel circuit is given by the following equation:

$$I_t = I_1 + I_2 + I_3 + I_4 + \cdots$$

The resistance of a parallel circuit gets smaller as more resistances are added to the circuit. The total resistance of a parallel circuit cannot be obtained by taking the sum of all the resistances. It is calculated by the following formula if two resistances are used:

$$R_t = \frac{R_1 \times R_2}{R_1 + R_2}$$

If three or more resistances are located in the circuit, the reciprocal of the total resistance is the sum of the reciprocals of all the resistances (the reciprocal of a number is one divided by that number). The following formula is used to calculate the resistance of a parallel circuit with more than two resistances:

$$\frac{1}{R_t} = \frac{1}{R_1} + \frac{1}{R_2} + \frac{1}{R_3} + \frac{1}{R_4} + \cdots$$

The voltage drop in a parallel circuit is the line voltage being supplied to the load. In other words, in a parallel circuit each load uses the total voltage being supplied to the load. For example, if 115 volts are supplied to a load, it will use the total 115 volts. The voltage being applied to each of the four components in Figure 3.9 is the same and is given by the following equation:

$$E_t = E_1 = E_2 = E_3 = E_4$$

Ohm's law can be used to calculate voltage, amperage, or resistance if the other two values are known. You can use Ohm's law to determine almost any condition in a parallel circuit, but pay careful attention to the individual sections of each complete circuit.

Example 1

What is the total current draw of the parallel circuit shown in Figure 3.11?

Step 1: First calculate the current draw for each individual circuit by using Ohm's law in the form

$$I = \frac{E}{R}$$

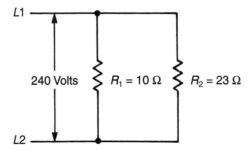

FIGURE 3.11 Parallel circuit for Example 1

Step 2: For I_1 substitute the given values for E and R_1 in the formula and solve:

$$I_1 = \frac{240}{10} = 23 \text{ amperes}$$

Step 3: For I_2 substitute the given values of that circuit in the formula and solve:

$$I_2 = \frac{240}{23} = 10 \text{ amperes}$$

Step 4: Use the formula $I_t = I_1 + I_2$.

Step 5: Substitute in the formula for I_t and solve:

$$I_t = I_1 + I_2$$
$$= 23 \text{ amperes} + 10 \text{ amperes} = 33 \text{ amperes}$$

Example 2　Find the total resistance of the parallel circuit in Figure 3.11.

Step 1: Use the formula

$$R_t = \frac{R_1 \times R_2}{R_1 + R_2}$$

Step 2: Substitute the known values in the formula:

$$R_t = \frac{10 \times 23}{10 + 23} = 6.97 \text{ ohms}$$

Example 3　What is the resistance of a parallel circuit with resistances of 3 ohms, 6 ohms, and 12 ohms?

Step 1: Use the formula

$$\frac{1}{R_t} = \frac{1}{R_1} + \frac{1}{R_2} + \frac{1}{R_3}$$

Step 2: Substitute the known values in the formula:

$$\frac{1}{R_t} = \frac{1}{3} + \frac{1}{6} + \frac{1}{12}$$

Step 3: Mathematical computation of this formula is sometimes difficult. If you have problems, consult your instructor.

$$\frac{1}{R_t} = \frac{4}{12} + \frac{2}{12} + \frac{1}{12} = \frac{7}{12}$$

$$R_t = 1.71 \text{ ohms}$$

3.4 SERIES-PARALLEL CIRCUITS

A *series-parallel circuit* is a combination circuit made up of series and parallel circuits and is used only sparingly in the industry. It is more often seen on the full wiring layout of an air-conditioning, heating, or refrigeration unit. This type of electric circuit is a combination of the series and parallel circuits, as shown in Figure 3.12. The series-parallel circuit is sometimes easier to understand when it contains only a few components. It becomes harder to understand when there are a large number of components.

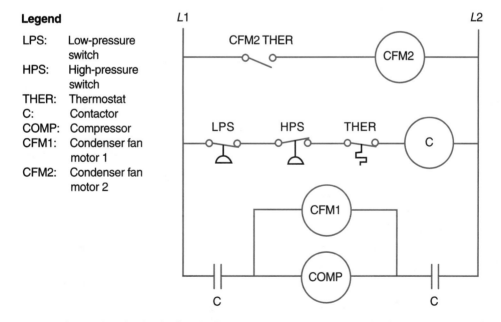

FIGURE 3.12 Series-parallel circuit with four loads and controlling switches (switches in series with loads in parallel)

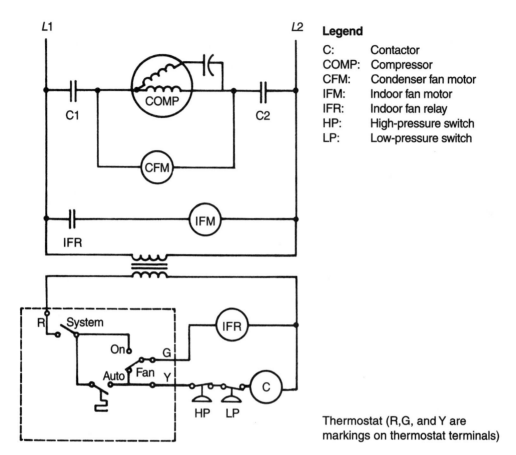

Legend

C: Contactor
COMP: Compressor
CFM: Condenser fan motor
IFM: Indoor fan motor
IFR: Indoor fan relay
HP: High-pressure switch
LP: Low-pressure switch

Thermostat (R,G, and Y are markings on thermostat terminals)

FIGURE 3.13 Wiring diagram of a packaged air-conditioning unit including series-parallel circuit arrangement where switches are in series to loads and loads are in parallel

The series-parallel circuit is often used to combine control circuits with circuits that supply power to loads, as shown in Figure 3.13. The circuit arrangement of most series-parallel circuits is designed so that all loads receive the correct voltage to operate switches and contacts in series with these loads, thus controlling the operation of the system.

Any calculation of the values in a series-parallel circuit must be performed carefully, because each portion of the circuit must be identified as series or parallel. Once the type of circuit has been determined, the calculations are made accordingly.

Applications

Series-parallel circuits are used in most types of heating, cooling, and refrigeration equipment in the industry. This type of circuit allows the required voltage to be supplied to electrical loads in the system. In the parallel circuits, the switches used to control the loads are connected in series. Refer again to Figure 3.13, that shows this type of circuit.

In a series-parallel circuit, both the series and parallel circuits are used to supply electrical power to the loads in the electrical systems. In Figure 3.13 the compressor, condenser fan motor, and indoor fan motor are connected in parallel to the supply voltage $L1$ and $L2$. The indoor fan relay coil and the contactor coil are in parallel with the transformer supplying voltage to the loads. The use of parallel circuits enables the correct voltage to be supplied to the loads in the electrical circuitry of the equipment. Series circuits are used in the electrical circuitry of air-conditioning equipment for control and safety switches that stop and start loads to maintain the correct temperature and safe operating conditions. Any switch connected in series with the loads will stop or start that load, depending on the switch position.

In Figure 3.13 the high- and low-pressure switches and thermostat are connected in series and control the contactor coil. If any one of these switches opens, the power supply to the contactor coil will be interrupted, stopping the compressor.

SUMMARY

The three types of electric circuits used in the industry are series, parallel, and series-parallel. The series circuit has only one path for electron flow. The most common type of electric circuit used in the industry is the parallel, which has more than one path for electron flow. This type of electric circuit allows line voltage to reach all electric loads. The series-parallel circuit is a combination of the series and parallel circuits.

The series circuit is used for most of the control circuits used in the industry, because if any switch in the circuit opens, the load in the circuit will stop or start, depending on the position of the switches. For example, if any switch that is connected in series with a load is open, the load will be de-energized; if all the switches in the circuit are closed, the load will operate. If a switch in the circuit used for temperature control is open, the load will be off. If the switch is closed or calling, the load will operate. Series circuits are used to control loads in the electrical system.

Parallel circuits are used in the industry and in equipment to ensure that proper voltage is supplied to the load. Parallel circuits are designed with more than one path for electron flow. In the electrical circuitry for equipment, the parallel circuit is used to supply the correct voltage to the electrical loads in the system. Parallel circuits are used in the wiring of houses to ensure that all receptacles are supplied with 115 volts.

The series-parallel circuit is a combination of series and parallel circuits. In the proper combination of these circuits, control systems supply voltage and control these circuits. Most electric control systems used to operate equipment have used the series-parallel circuit.

The series and parallel circuits have different relationships of voltage, amperage, and resistance. In a series circuit, the voltage is split among the electrical loads, which is the reason this circuit is not used to supply power to loads in equipment. The current in the series is equal in all parts of the circuit. The sum of all resistances in series is the total circuit resistance. In a parallel circuit, the voltage in all circuit parts is equal; this is the reason they are used. The sum of each current flow in the circuit is the total current. The reciprocal of the total resistance is the sum of the reciprocals of all resistances.

REVIEW QUESTIONS

1. What is a series circuit?

2. What is a control circuit?

3. Safety devices in an electrical circuit are connected in _____ with the load.

4. How are series circuits used in air-conditioning equipment?

5. Why are series circuits used for most control circuits?

6. The current draw in a series circuit is _____.

   a. the sum of each load in the circuit
   b. the same throughout the entire circuit
   c. divided by the number of loads in the circuit
   d. none of the above

7. The voltage drop of a series circuit is _____.

   a. is the amount of voltage lost through any load

   b. is proportional to the resistance in that part of the circuit
   c. is equal to the voltage being applied to the circuit
   d. all of the above

8. How would switches used as safety devices be connected with respect to the loads they are protecting?

9. Draw a series circuit with a thermostat, a low-pressure switch, and a high-pressure switch used as safety controls to protect a motor.

10. Two 115-volt loads connected in series with 240 volts would _____.

    a. burn dimly
    b. burn out immediately
    c. burn correctly
    d. none of the above

11. Draw a circuit with four electric heaters (‑ᴧᴧᴧ‑) in series.

12. What is a parallel circuit?

13. Parallel circuits are used in the air-conditioning industry to _____.
    a. supply the correct line voltage to several circuits
    b. act as a safety circuit
    c. divide the voltage between two major loads
    d. all of the above

14. Why are parallel circuits used in the air-conditioning industry?

15. True or False: The current draw of a parallel circuit is the sum of the current of each branch circuit.

16. If three circuits are connected in parallel with a power supply of 30 volts, what would be the voltage supplied to each circuit?
    a. 10 volts
    b. 90 volts
    c. 30 volts
    d. none of the above

17. If two 115-volt loads were connected in parallel with 240 volts, they would _____.
    a. burn dimly
    b. immediately burn out
    c. burn correctly
    d. none of the above

18. What is a series-parallel circuit?

19. Why are series-parallel control circuits important in the circuitry used in air-conditioning equipment?

20. Draw a series-parallel circuit with one switch (—•—⁄•—) controlling two electric heaters, each being supplied with line voltage.

21. What is the resistance of a parallel circuit with resistances of 2 ohms, 4 ohms, 6 ohms, and 10 ohms?

22. What is the resistance of a parallel circuit with resistances of 10 ohms and 20 ohms?

23. What is the total amp draw of a parallel circuit with ampere readings of 2 A, 7 A, and 12 A?

24. What is the voltage of a series circuit with four voltage drops of 30 volts?

25. What is the resistance of a series circuit with resistances of 4 ohms, 8 ohms, 12 ohms, and 22 ohms?

LAB MANUAL REFERENCE

For experiments and activities dealing with material covered in this chapter, refer to Chapter 3 in the Lab Manual.

4

Electric Meters

OBJECTIVES

After completing this chapter, you should be able to

- Describe the use of the volt-ohm meter and clamp-on ammeter in the heating, cooling, and refrigeration industry.
- Explain the operation of the basic electric analog meter.
- Explain how analog electric meters transfer a known value in an electrical circuit to the meter movement.
- Describe the operation of an analog voltmeter.
- Describe the operation of an analog and digital clamp-on ammeter.
- Describe the operation of an analog ohmmeter.
- Explain the operation of a digital volt-ohm meter.
- Give the advantages and disadvantages of the analog and digital meters.
- Describe the conditions of resistance that can exist in an electrical circuit in reference to continuity.
- Describe the source of energy for the operation of the analog voltmeter, ammeter, and ohmmeter.

KEY TERMS

Ammeter	Measurable resistance
Analog meter	Ohmmeter
Clamp-on ammeter	Open
Continuity	Short
Digital meter	Voltmeter
Magnetic field	

INTRODUCTION

Electricity is used in the control and operation of all refrigeration, heating, and air-conditioning systems. Approximately 80% of all service calls made by service technicians are diagnosed as electrical problems. Electrical meters, in most cases, are required for the proper diagnosis of these electrical problems. The large number of electrical problems encountered in the industry requires that industry technicians be able to correctly read and use electric meters. In the industry there are many types of electric meters. Some meters are for specific purposes, while others are for the general day-to-day tasks that industry technicians perform.

The installation mechanic must be able to read electric meters to properly complete the initial start-up and testing of equipment. The installation mechanic must be able to read and use electric meters to check the electrical characteristics of a newly installed refrigeration, heating, and air-conditioning system. No installation is complete without electrical checks of the system.

The service technician diagnoses and repairs problems found in refrigeration, heating, and air-conditioning systems. The service technician must be able to read and use all types of electric meters to quickly and efficiently locate electrical problems. Without electric meters and the knowledge to correctly use them, the service technician would face an almost impossible task in troubleshooting electrical problems.

An electric meter is a device used to measure the electrical characteristics of an electrical circuit. Electric meters are available in many different types and designs and must be selected for their use in the industry. Most meters are used by installation and service technicians. Meters must be durable and maintain adequate accuracy.

The electrical characteristics in a circuit that are most important to industry technicians are volts, amperes, and ohms. Although there are other important characteristics, these three are the most important. Most popular electrical meters are built around these three electrical characteristics. Most electrical meters used in the industry are capable of reading more than one electrical characteristic. The most common electrical meters used are the **volt-ohm-milliammeter** and the **clamp-on ammeter** with the ability to read volts and ohms.

The installation and service technician should be careful in selecting an electrical meter. Electrical meters should be selected for their everyday use in the industry. Technicians should select meters with the best ranges for the electrical characteristics that will be measured the most.

We will look at the basic principles used to measure electrical characteristics along with the various types and designs of electrical meters.

4.1 ELECTRIC METERS

An electric meter is a device used to measure some electrical characteristic of a circuit. The most common types of electric meters are the voltmeter, the ammeter, and the ohmmeter.

 CAUTION Use the proper electrical test equipment for the job being performed.

Basic Principles

Most electric measuring instruments make use of the magnetic effect of electric current. When electrons flow through a conductor in an electric circuit, a **magnetic field** is created around that conductor, as shown in Figure 4.1. This magnetic field is used to move the needle of a meter a certain distance, which represents the amount of the characteristic (volts, ohms, or amperes) being measured. The stronger the magnetic field, the larger the movement of the needle. The weaker the magnetic field, the smaller the movement of the needle.

If a compass is suspended next to a conductor that is not carrying an electron flow, the compass reacts only with the magnetic field of the earth and there is no other movement, as shown in Figure 4.2. However, when

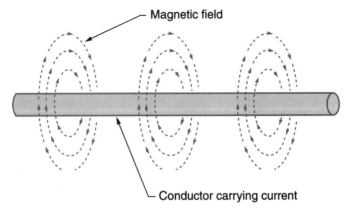

Magnetic field

Conductor carrying current

FIGURE 4.1 A magnetic field is produced around a conductor when current is flowing through the conductor.

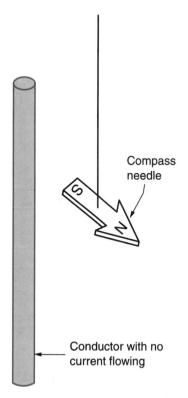

FIGURE 4.2 When there is no current flow, the compass reacts only to the magnetic field of the earth.

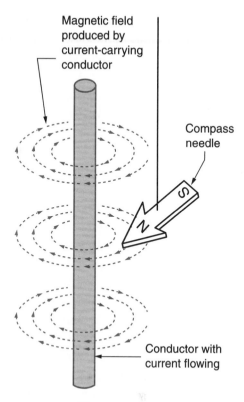

FIGURE 4.3 When a current flows through a conductor, the compass needle swings in line with the conductor's magnetic field.

electrons flow through that same conductor, the compass needle swings in line with the conductor's magnetic field, as shown in Figure 4.3. The mechanical movement of the needle is caused by the magnetic field produced by the electron flow through the conductor. The larger the current flow, the stronger the magnetic field produced and the greater the needle movement on the scale. This simple principle is the basis of the meter movement in most electric meters.

Differences Among Meters

The differences among the various electric meters are not in the meter movements, except for digital meters, but in the internal circuits of the meters and in how the magnetic fields are created. For electrons to flow in an electric circuit, an electric load must be present. This electron flow is somewhat different for the clamp-on ammeter, voltmeter, and ohmmeter. The clamp-on meter picks up the magnetic field through a set of laminated jaws on the meter. The **voltmeter** uses a resistor as a load to produce a magnetic field when voltage is applied to the circuit in the meter. The **ohmmeter** has its own power supply and uses the device being checked as the load to produce a magnetic field. All three meters use the same meter movement, unless the movement is digital. Their methods of loading and their power supplies are varied to attain the needle movement for reading the magnetic field.

Meters may be made in a combination and mounted in one case, or they may be completely separate. Figure 4.4(a) shows an analog clamp-on ammeter, and Figure 4.4(b) shows an analog volt-ohm meter. Figure 4.4(c) shows a digital clamp-on ammeter and Figure 4.4(d) shows a digital volt-ohm meter. The volt-ohm meter, as its name suggests, is used to measure either voltage or resistance, depending on the scale selected.

The meter movement created by the magnetic field around a conductor is reflected by the movement of the needle on a scale of an electric meter. This scale is usually broken down into several basic scales that have different ranges for voltage, amperage, and resistance. Some electric meters have a selector switch: the meter movement is shown on a certain point of the scale, but the scale to be used must be determined by the person using the meter.

Digital meters are becoming more popular in the industry due to the reduction in cost over the past several years and the ease in reading. Another reason for the increasing popularity of the digital meter is the increased use of electronic control, which oftentimes requires an electrical meter capable of accurately reading small voltages because of the lower signal voltages. An **analog meter** volt-ohm meter is shown in Figure 4.5 with an analog meter scale and needle shown in Figure 4.6. The mechanic must estimate the values between the lines on the meter scale. Digital meters can be read more accurately because there is no estimation, as shown from the digital reading in Figure 4.7. The ease of reading a digital meter is certainly an advantage, and many technicians are using digital meters in the industry today.

Most digital meters utilize a 3½- or 4½-digit display. Figure 4.8 shows a 3½-digit display, and Figure 4.9 shows a 4½-digit display. The ½ digit is blank (0) or one that is the left digit of the meter reading. The 3 or 4 digits represent the next digits of the meter reading. For example, a 3½-digit reading will display a 4-digit reading with the left digit blank (0) or 1, such as 1999. This resolution determines the basic accuracy of the meter. Typical accuracy for a digital volt-ohm-milliammeter is plus or minus 1% of actual reading compared with the typical accuracy of an analog meter of plus or minus 2% of full scale.

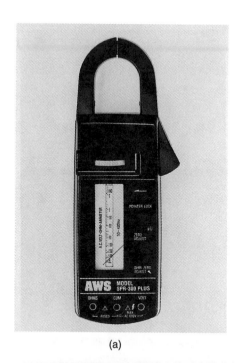

(a)

(b)

(c)

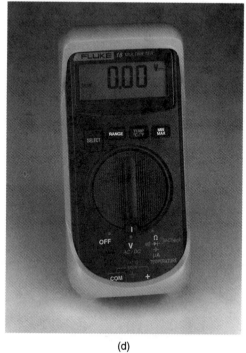

(d)

FIGURE 4.4 (a) Analog clamp-on ammeter with volt-ohm function (b) Analog volt-ohm meter *(Courtesy of A. W. Sperry Instrument Co., Inc., Hauppauge, NY)* (c) Digital clamp-on ammeter with volt-ohm function (d) Digital volt-ohm meter

FIGURE 4.5 Analog volt-ohm meter *(Courtesy of Amprobe Instruments, Inc., Lynbrook, NY)*

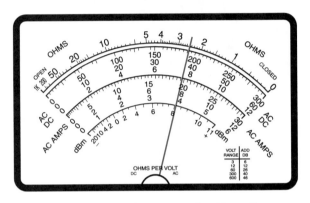

FIGURE 4.6 Analog meter scale with needle

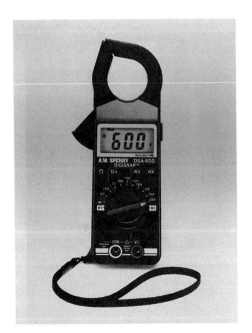

FIGURE 4.7 Digital reading *(Courtesy of A.W. Sperry Instrument Co., Hauppauge, NY)*

FIGURE 4.8 3½-digit digital display

FIGURE 4.9 4½-digit digital display

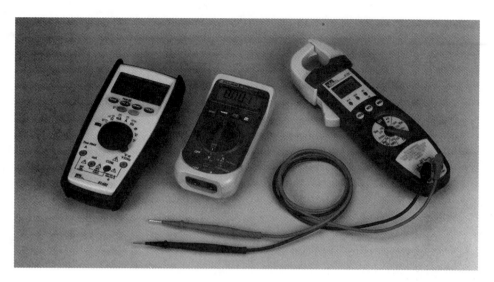

FIGURE 4.10 Digital meters

While analog meters depend on a magnetic field to move the needle, the digital meter makes use of Ohm's law to measure and display the electrical characteristics of the circuit.

Digital meters use one of two methods to protect the meter circuitry. Some digital meters are protected by internal circuits that detect an overload condition and then return the meter to normal operation. This function generally protects the ohm function from voltage overloads. Other digital meters are protected by fuses that have to be changed for the meter to again function. Several digital meters are shown in Figure 4.10.

 CAUTION Make sure electrical test equipment is in good condition.

4.2 AMMETERS

The ammeter uses the basic meter movement discussed in Section 4.1. The strength of the magnetic field determines the distance that the needle of the meter moves. Figure 4.11(a) shows the magnetic field and the meter movement when there is a high current flow through the conductor. Figure 4.11(b) shows the magnetic field and the meter movement when there is a low current flow in the conductor. The larger the current flow, the stronger the magnetic field grows and the greater the needle movement on the scale.

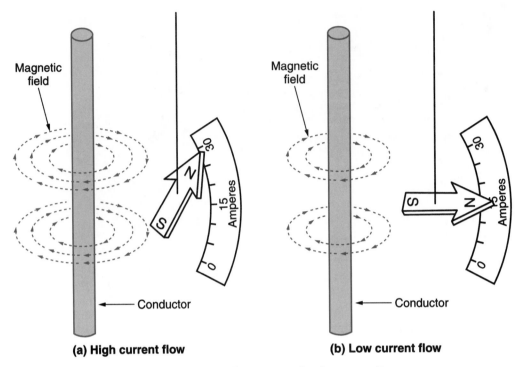

FIGURE 4.11 Scale of an ammeter showing current flow

The **ammeter** measures current flow in an electric circuit. There are basically two types of ammeters used in the industry today: the clamp-on ammeter and the in-line ammeter. The clamp-on ammeter is the most popular because it is the easiest to use. You simply clamp the jaws of the meter around one conductor feeding power to the load that is producing the current draw. The analog clamp-on ammeter is shown in Figure 4.12. Figure 4.13 shows the digital-readout clamp-on ammeter.

The clamp-on ammeter is easy to use. Just follow a few simple rules. First, select the scale that is appropriate for reading the current draw of the electrical device being checked. If the approximate current is unknown, use the highest scale until the correct scale can be determined. Most clamp-on ammeters have more than one scale and as many as five. The jaws of the clamp-on ammeter are clamped around one conductor that is supplying a load or circuit. The magnetic field created by the current flowing through the wire is picked up by the jaws of the ammeter and funneled into the internal connection of the meter. Figure 4.14 shows a clamp-on ammeter being used to read the current of an operating compressor.

Never clamp the jaws of the meter around two wires to obtain an ampere reading. If the current flows in the wires are opposite, as they often are, the

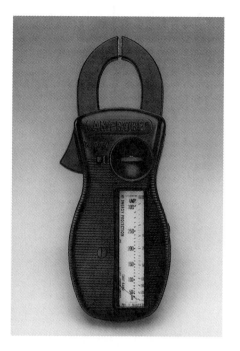

FIGURE 4.12 Clamp-on ammeter with volt-ohm functions *(Courtesy of Amprobe Instruments, Inc., Lynbrook, NY; A.W. Sperry Instrument Co., Inc., Hauppauge, NY)*

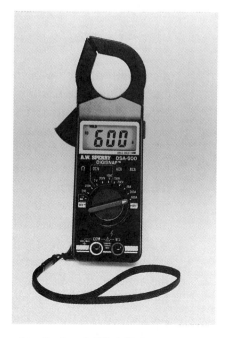

FIGURE 4.13 Digital ammeters *(Courtesy of Amprobe Instruments, Inc., Lynbrook, NY; A.W. Sperry Instrument Co., Inc., Hauppauge, NY)*

FIGURE 4.14 Technician using clamp-on ammeter to read the current of a compressor

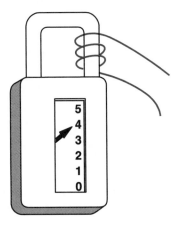

FIGURE 4.15 Taking a low-ampere reading with an ammeter by looping the wire through the jaws

meter will read zero because the current flows cancel each other out. If the current flows are not opposite, the meter will read the current draw in both conductors. In either case, you will obtain an incorrect reading, which could cause you to incorrectly diagnose the problem.

When the ampere draw is small, you may have difficulty obtaining a true reading because of the small needle movement. The problem of small needle movement can be remedied by coiling the wire around the jaws of the meter. This allows the meter to pick up a larger current flow than is actually there. The meter will be more accurate because the current reading will fall in the midrange of the scale and can be easily read.

To obtain the correct ampere reading when this method is used, divide the ampere draw read by the number of loops going through the jaws of the meter. Figure 4.15 shows an ammeter with the conductor looped through the jaws three times. The correct reading can be obtained by dividing the meter reading, which is 4, by the number of loops through the jaws, which is 3. Thus, the ampere draw of the load is 1.33 amperes. Remember: The meter reading should always be divided by the number of loops carried through the jaws of the meter.

The clamp-on ammeter is one of the most valuable tools that industry technicians can carry. When you are installing a system, knowing the ampere draw of the equipment tells you if the unit is operating properly. You can also detect many electric circuit problems with a clamp-on ammeter, and it is the easiest, quickest way to tell if a load or circuit is energized. The ammeter can be purchased with other meters built into it. For example, a single clamp-on ammeter can be purchased that will read amperage, voltage, and resistance.

4.3 VOLTMETERS

The voltmeter is used to measure the amount of electromotive force available to a circuit or load. This is an important factor to heating, cooling, and refrigeration technicians because a wide range of voltages are used in this country.

Voltmeters range from simple to complex instruments containing many scales. The simplest voltmeter available is a small, inexpensive one capable of distinguishing only between 120, 240, 480, and 600 volts (Figure 4.16). Several manufacturers build simple voltmeters that can read only voltages, but these are becoming increasingly difficult to obtain. More common is

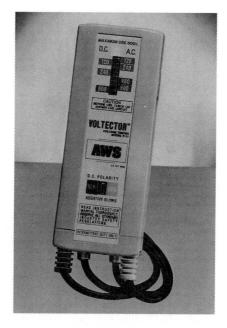

FIGURE 4.16 Simple voltmeter *(Courtesy of A.W. Sperry Instrument Co., Inc., Hauppauge, NY)*

FIGURE 4.17 Three analog volt-ohm meters used in the industry *(Courtesy of A.W. Sperry Co., Inc., Hauppauge, NY; Amprobe Instruments, Inc., Lynbrook, NY)*

the volt-ohm meter, which reads both voltage and resistance. These are available in many forms, and service technicians should follow the instructions for the particular model being used. The common volt-ohm meter has three voltage scales and several voltage ranges. Some meters also have a high-voltage jack. Figure 4.17 shows three common types of volt-ohm meters. Figure 4.18 shows a digital volt-ohm meter.

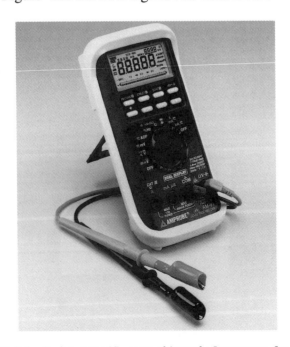

FIGURE 4.18 Digital volt-ohm meter *(Courtesy of Amprobe Instruments, Inc., Lynbrook, NY)*

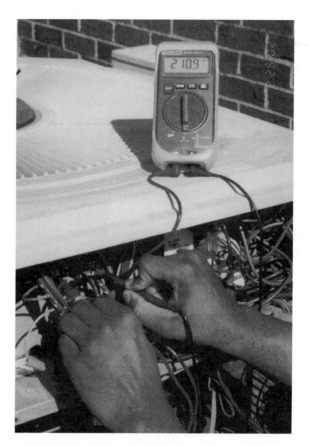

FIGURE 4.19 Technician checking the voltage at a compressor.

The voltmeter was designed much like the ammeter, but a resistor is added to the circuit to prevent a direct short and allow electrons to flow in the meter. The voltmeter uses two leads that are connected to jacks that lead to the internal wiring. To obtain a reading, the two leads must touch or be connected to the conductors supplying the load or to the circuit that transfers the electromotive force to the meter. Figure 4.19 shows a technician reading the voltage being applied to a compressor. The electrons flow through the leads into the meter through a resistor with a known ohm rating. The greater the voltage carried into the meter, the greater the magnetic field and the greater the needle movement or digital reading.

When you do not know the voltage available to the equipment being worked on, you should start with the highest scale on the meter. Then change the meter setting until the needle falls in the midrange of the voltage scale. Never abuse a voltmeter by attempting to read a voltage that exceeds the range of the meter.

The voltmeter is necessary for field technicians who have anything to do with the electrical section of equipment or with the installation or servicing of equipment. No heating, cooling, or refrigeration equipment should operate at an unsafe voltage, that is, voltage that is either too low or too high. All equipment is designed to operate at a voltage of 10% above or below the rating of the equipment. But in some cases the voltage may actually be more than the allowable figure. So field technicians should always check the supply voltage. Installation mechanics have not completed their installation unless they have checked the voltage available to the equipment. Technicians are often required to check the voltages to equipment as part of their troubleshooting job. In addition, the voltmeter can be used by technicians as a tool for diagnosing problems in the system. Thus, the voltmeter is a must for proper installation and service.

4.4 OHMMETERS

The ohmmeter is used to determine the operating condition of a component or a circuit. The ohmmeter can be used to find an open circuit, an open component, or a direct short in a circuit or component. It can also be used to measure the actual resistance of a circuit or component.

The word **continuity** is used many times when referring to the use of ohmmeters. Continuity means that a particular circuit or component has a complete path for current to follow. An open component or circuit means that there is no infinite resistance in the circuit. The term **measurable resistance** means the actual resistance that is measured with the ohmmeter. Figure 4.20 shows the three conditions as they might appear on the scale of an ohmmeter.

The ohmmeter is a valuable tool for diagnosing and correcting problems in electric circuits. In this industry, many electric devices and circuits must be checked. The ohmmeter provides an easy method for checking circuits for **opens** (i.e., open circuits) and **shorts** (i.e., short circuits) and for measuring resistance.

An open circuit causes no noticeable needle movement in an ohmmeter because there is not a complete circuit. For example, an open (circuit) could occur in a blown fuse, a motor winding (the internal portion of the motor), and in any condition where the electric circuit does not have a complete path for electrons to follow.

A direct short in an electric device or circuit causes problems because it means that two legs of the electric power wiring are touching, which causes an overload. In many cases a direct short means the wiring of the component is connected in some fashion. A closed switch is considered to

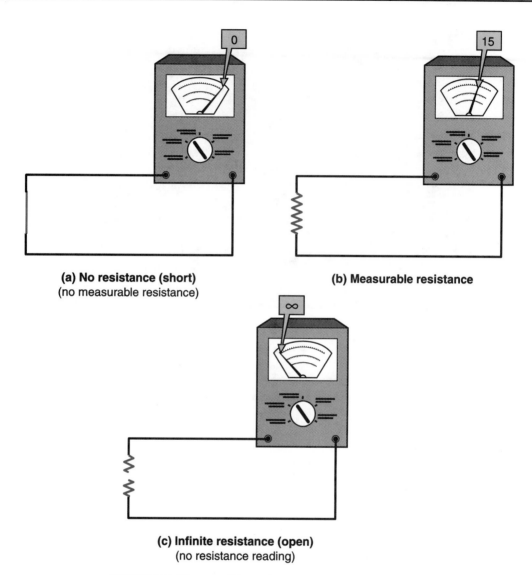

(a) No resistance (short)
(no measurable resistance)

(b) Measurable resistance

(c) Infinite resistance (open)
(no resistance reading)

FIGURE 4.20 Three conditions of a circuit as read on an ohmmeter

be a short, but without this type of short, no heating, cooling, or refrigeration system would operate properly.

In many cases, you will have to measure the resistance of a component to ensure that the component is in good operating condition. Most manufacturers make available to service technicians the exact ohmic value of motor windings and other components in the system. Figure 4.21 shows a technician reading the resistance of an electric motor.

The meter movement of an ohmmeter is designed and built for a very low current that is available from its own power source, usually a battery.

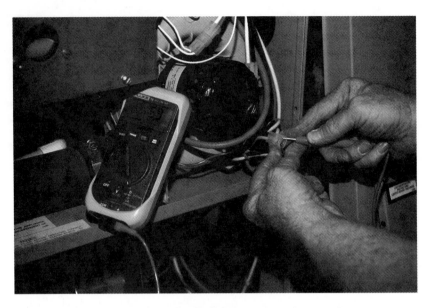

FIGURE 4.21 Technician reading the resistance of an electric motor

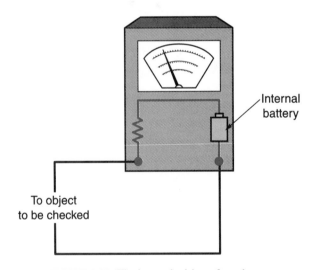

FIGURE 4.22 The internal wiring of an ohmmeter

Figure 4.22 shows the internal wiring of an ohmmeter. The ohmmeter works much like the ammeter and the voltmeter except for the small current that is supplied from the internal power source. The ohmmeter also uses a magnetic field to move the needle, but the magnetic field is created by a self-contained power source in the meter. The two leads of the ohmmeter are connected to the internal circuit of the meter, which contains a resistance and the power source. The amount of current the small battery can push through the device being tested indicates the resistance of that device and determines the needle movement.

Due to the low current that an ohmmeter is built to carry, it should never be connected to a circuit or device that is being operated. The function of the ohmmeter is merely to read the resistance of a device or circuit. Fortunately, most ohmmeters are equipped with some type of overload protection to protect the internal circuits of the meter if they are subject to line voltage.

Many types and designs of ohmmeters are available. In many cases, the ohmmeter and voltmeter are combined in a dual-purpose meter. Some manufacturers build and market combinations of voltmeters, ohmmeters, and ammeters that are inexpensive. On this three-purpose meter, the ohmmeter is a low-range ohmmeter and cannot be used for many of the jobs that a technician must do. The more expensive volt-ohm meters are more accurate and cover all ohm ranges. These meters will usually have at least three ohm scales (usually $R \times 1$, $R \times 100$, and $R \times 10,000$). Some have more ranges. Those additional ranges are useful for some troubleshooting operations. Refer again to Figure 4.17, which shows three different volt-ohm meters used today.

SUMMARY

Electricity plays an important part in the industry. Most equipment has some type of electric control system, even if it is powered by some other means. Thus, it is important for industry technicians to be familiar with the basic types of electric meters. About 80% of the problems with equipment or systems are electrical, which shows that electric meters are important.

The ammeter is used to measure the current flow in an electric circuit. Two types of ammeters are used in the industry today. The clamp-on ammeter is the most frequently used. With this type of ammeter, it is only necessary to clamp the jaws of the meter around the conductor feeding the circuit or load and to read the amperage. The in-line ammeter must be placed in series with the load or circuit to read the amperage. Because of the time required to do this, the in-line ammeter is seldom used.

The voltmeter is used to measure the voltage of an electric circuit. The voltmeter will have to be connected in parallel to the circuit to determine the voltage supplying the circuit. Voltage measurements can determine the source voltage, voltage drop, and a voltage imbalance. The first step in troubleshooting an air-conditioning system is to determine if voltage is available to the equipment or system.

The ohmmeter is used to measure the resistance of a circuit or device in ohms. The ohmmeter must be used with the circuit power off to prevent damage to the meter. Most ohmmeters have a power source built into the meter. The ohmmeter is used to determine the condition of electrical devices used in air conditioning such as motors, heaters, contactor coils, solenoid valves, and other components. A continuity check is another use of the ohmmeter, which determines if a complete path is available for current to flow.

The decrease in the cost of digital meters has made them more popular in the industry. Digital meters can be used without any interpolation of the reading, making them an advantage over analog meters where interpolation must be made with reference to the needle on the scale. Mechanics should choose the meter that best suits their needs. With the proper care, meters will give mechanics many years of good service.

REVIEW QUESTIONS

1. What are the three most common electric meters used in the industry?

2. Service technicians should be able to use electric meters because _____.
   a. 80% of field problems are electrical
   b. they are required to troubleshoot electrical control systems
   c. they must be able to determine the electrical characteristics of an electrical circuit
   d. all of the above

3. What do most analog meters use to facilitate the needle movement?
   a. Ohm's law
   b. a magnetic field that flowing electrons produce
   c. inductive resistance
   d. all of the above

4. How does an ammeter work?

5. What are the two types of ammeters? Which type is more commonly used in the industry?

6. What is the result on the meter reading of clamping the jaws of a clamp-on ammeter around two wires?

7. How can a very small ampere draw be measured with a clamp-on ammeter?

8. If a conductor was wrapped around the jaws of a clamp-on ammeter four times, the proper reading would be _____ if the clamp-on ammeter was reading 16 amps.
   a. 16 amps
   b. 8 amps
   c. 4 amps
   d. 2 amps

9. Explain the operation of an analog voltmeter.

10. Air-conditioning or refrigeration equipment can operate properly at _____% above or below its rated voltage.

11. If a technician has no idea of the voltage available to a unit, what procedure should be followed on reading the voltage?

12. What precaution should be taken when using an ohmmeter?

13. What does the term *continuity* mean?

14. What is a short circuit?

15. What is an open circuit?

16. What factors should be considered when purchasing an electric meter?

17. Match the following terms.

    ___ short circuit   a. zero ohms
    ___ open circuit   b. 22 ohms
    ___ measurable   c. infinite ohms
            resistance

18. How does the internal circuitry differ between the ohmmeter, voltmeter, and ammeter?

19. A voltmeter is connected to a circuit in (series or parallel).

20. An in-line ammeter must be connected to a circuit in (series or parallel).

21. What is the difference between an analog meter movement and a digital meter movement?

22. Give three advantages of a digital electric meter.

23. What basic concept is used in a digital meter to calculate the circuit characteristics being measured?
    a. Watt's law
    b. Volt's law
    c. Amp's law
    d. Ohm's law

24. What is the basic accuracy of most analog and digital meters?

25. How many digits will the 3½- and 4½-digit displays have?

26. How are digital meters protected?

27. The ohmmeter differs from other types of meters in that it has its own _____ _____.

28. Compare the digital and analog types of electric meters, giving advantages and disadvantages.

LAB MANUAL REFERENCE

For experiments and activities dealing with material covered in this chapter, refer to Chapter 4 in the Lab Manual.

5

Components, Symbols, and Circuitry of Air-Conditioning Wiring Diagrams

OBJECTIVES

After completing this chapter, you should be able to

- Explain what electrical loads are and their general purpose in heating, cooling, and refrigeration systems.
- Give examples of common loads used in heating, cooling, and refrigeration systems.
- Identify the symbols of common loads used in heating, cooling, and refrigeration systems.
- Explain the purpose of relays and contactors in the heating, cooling, and refrigeration systems.
- Identify the symbols of relays and contactors in heating, cooling, and refrigeration systems.
- Explain the purpose of switches and the types used in the heating, cooling, and refrigeration systems.
- Identify the symbols of switches in heating, cooling, and refrigeration systems.
- Identify the symbols and purpose of other miscellaneous controls in the heating, cooling, and refrigeration systems.
- Identify the different types of wiring diagrams used in the industry and the purpose of each.

KEY TERMS

Contactor
De-energized
Disconnect switch
Energized
Factual diagram
Fuse
Heater
Installation diagram
Load
Magnetic overload
Magnetic starter
Motor
Normally
Normally closed
Normally open

Pictorial diagram
Pilot duty device
Pole
Pressure switch
Push-button switch
Relay
Schematic diagram
Signal light
Solenoid
Switch
Thermal overload
Thermostat
Throw
Transformer

INTRODUCTION

Because of the complexity of today's air-conditioning, heating, and refrigeration systems, industry technicians should be able to read and interpret all kinds of wiring diagrams. Electric wiring diagrams contain a wealth of information about the electrical installation and operation of the equipment. The installation mechanic depends on the wiring diagram for the correct installation of the wiring to the unit. The technician uses the electrical diagrams as a guide in troubleshooting the electric system of a unit. It would be impossible for wiring diagrams to be composed of photographs of various components of the equipment. They would be too large and in many cases too complex due to the number of wires that are carried to certain devices. Thus, symbols are used in wiring diagrams to represent such system components as compressors, indoor fan motors, thermostats, pressure switches, and heaters. Industry technicians must be able to identify most symbols and know where to look up the remainder. Most manufacturers use similar symbols for each type of electric component, although there are some minor differences in symbols between some major manufacturers. Thus, a knowledge of the basic symbols is essential if you are to be successful in the industry.

We begin our study with a discussion of the various types of electric loads found in the industry and the basic symbol used for each device.

5.1 LOADS

Loads are electric devices that consume electricity to do useful work. Loads are devices such as motors (Figure 5.1), solenoids (Figure 5.2), resistance heaters (Figure 5.3), and other current-consuming devices. The sizes of loads vary from devices with a small current draw, such as a light bulb, a small fan motor, and solenoids, to large motors that could use upwards of 100 amperes.

FIGURE 5.1 Electric motor *(Courtesy of Magnetek, Inc., St Louis, MO)*

FIGURE 5.2 A solenoid used to operate a contactor

FIGURE 5.3 A resistance heater *(Courtesy of Indeeco, St Louis, MO)*

Loads are the most important part of a heating, cooling, or refrigeration system because they do all the work in the system. Loads operate compressors, which compress and transfer refrigerant in a system. They operate fans, which move air. They operate the solenoid part of a relay, which starts and stops loads. Also, loads operate with other devices that perform useful work. Industry technicians should be able to recognize the common symbols for loads and know where to look up the symbols for little-used loads, because each electric wiring diagram is composed of symbols and their interconnecting wires.

In the following paragraphs, we will take a close look at several different kinds of loads used in the industry.

Motors

A **motor** is an electric device that consumes electric energy to rotate a device in an electric system. Motors are used in the industry to rotate devices such as compressors (Figure 5.4), condenser fan motors (Figure 5.5), pumps (Figure 5.6), and other units that require rotating movement. Motors are the largest and most important loads in heating, cooling, and refrigeration systems.

FIGURE 5.4 Compressor *(Courtesy of Copeland, Sidney, OH)*

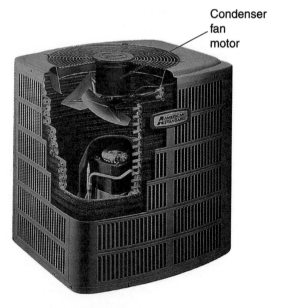

FIGURE 5.5 A condenser fan motor on a residential air-conditioning condensing unit *(Courtesy of American-Standard Air Conditioning)*

FIGURE 5.6 A centrifugal pump *(Courtesy of Peerless Pump Co., Indianapolis, IN)*

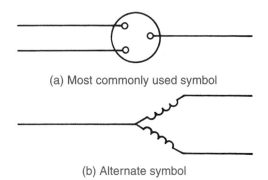

(a) Most commonly used symbol

(b) Alternate symbol

FIGURE 5.7 Symbols for an electric motor

The symbols shown in Figure 5.7 are the most common symbols used to represent motors.

A letter designation tells you what purpose the motor serves in the system. Figure 5.8 shows several symbolic representations of different uses of motors. Careful attention should be given to symbols representing motors because in some cases, a motor has an internal overload, as shown in Figure 5.8(e).

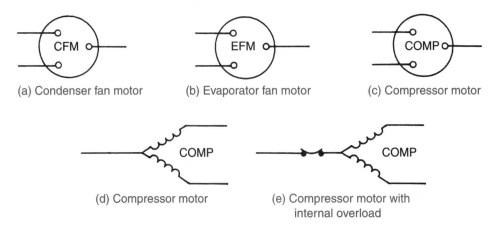

(a) Condenser fan motor (b) Evaporator fan motor (c) Compressor motor

(d) Compressor motor (e) Compressor motor with internal overload

FIGURE 5.8 Symbols representing some common uses of motors

Solenoids

The **solenoid** is a device that creates a magnetic field when energized and causes some action to an electric component such as a relay or valve. A common solenoid used to operate a relay is shown in Figure 5.9. The solenoid is considered a load because it consumes electricity to do useful work.

Solenoids are devices that open and close to control some element in a system. Solenoid valves are valves that open and close, stopping or starting a flow. Solenoid coils used in relays and contactors will be discussed later in this chapter. Some common solenoid valves are hot-gas solenoids, reversing-valve solenoids, and liquid-line solenoids. Figure 5.10(a) shows a solenoid valve and a solenoid coil, and Figure 5.10(b) shows its symbol.

Heaters

Heaters are loads that are found in many systems and wiring diagrams. A heater takes electric energy and converts it to heat. In some cases, electric resistance heaters are used to heat homes. Heaters might also be used to

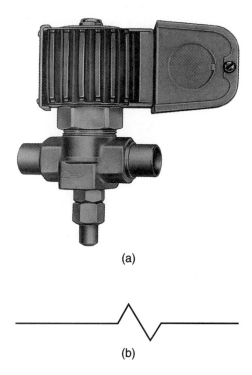

(a)

(b)

FIGURE 5.10 (a) Solenoid valve with coil *(Courtesy of Sporlan)* (b) Symbol

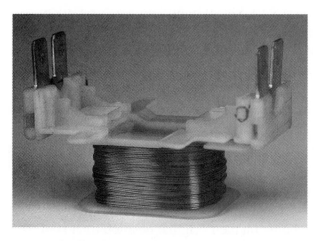

FIGURE 5.9 Solenoid coil used to operate relay

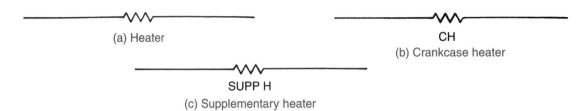

(a) Heater

CH
(b) Crankcase heater

SUPP H
(c) Supplementary heater

FIGURE 5.11 Symbols for commonly used electric heaters

(a) Red

(b) Green

(c) Blue

FIGURE 5.12 Symbols for signal lights showing the color of the light

heat a small object or area. The symbol for all heaters is the same. Only a letter designation tells you specifically why the heater is used. Figure 5.11 shows the symbols used for heaters, along with some common letter designations.

Signal Lights

A **signal light** is a light that is illuminated to denote a certain condition in a system. The letter inside the signal light symbol denotes the color of the signal light, as shown in Figure 5.12. Signal lights come in a variety of colors and are not limited to the colors shown in Figure 5.12. A signal light is used to show that a piece of equipment is operating or that it is operating in an unsafe condition. Signal lights are usually energized when a piece of equipment or component is started.

5.2 CONTACTORS AND RELAYS

Contactors and **relays** are devices that open and close a set or sets of electric contacts by the action of a solenoid coil. The contactor or relay is composed of a solenoid and the contacts. A relay is shown in Figure 5.13. A contactor is shown in Figure 5.14. When the solenoid is energized, the contacts will open or close, depending on their original position (that is, if they were open, they will close, and vice versa).

Mechanical linkage

Normally closed contacts

Solenoid coil

Normally open contacts

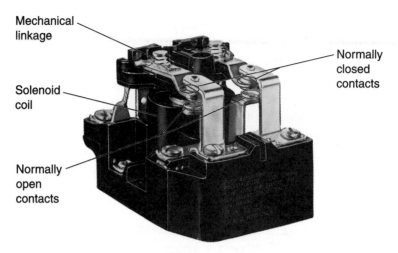

FIGURE 5.13 Relay *(Courtesy of Siemens Electromechanical Components, Inc.)*

In an air-conditioning control system, we must have some method of controlling loads. In most cases, a relay or contactor is used. Relays and contactors are widely used in control systems. Thus, it is essential that industry technicians be able to identify the symbols for relays and contactors.

The main difference between a relay and contactor is the size of the device. A contactor is simply a large relay. Usually, the devices are distinguished by their rated current flow. A contactor can carry 20 amperes or

FIGURE 5.14 Contactor

more. A relay is designed to carry less than 20 amperes. Contactors are commonly used where the ampere draw of a device is more than 20 amperes. A relay would rarely be used to carry over 20 amperes.

Contactors and relays play an important part in the control system of any air conditioner, refrigerator, or heater. For example, contactors and relays are used to stop and start different loads in a refrigeration system. Compressors, in most air-conditioning systems, are controlled by a contactor or magnetic starter. Relays can be used for pilot duty, that is, for controlling another relay or contactor. The most important fact to remember is that most control systems have many relays and at least one contactor. These relays or contactors always control some load.

Relays and contactors are composed of two parts: the contact and the coil, or solenoid. The contact makes the electrical connections. Figure 5.15 shows the symbol for a **pole**, or contact, of a relay or contactor. The term "pole" refers to one set of contacts. However, in some cases the relay or contact might have two or three poles, which means two or three sets of contacts. The coil or solenoid, the second part of the relay, is **energized** (voltage is supplied) and, through a magnetic field, closes the contact or contacts. Either symbol shown in Figure 5.17 can be used to represent a relay or contactor coil. The symbol for the relay or contactor is the same if each has the same number of poles and if their purpose is basically the same, with the exception of the ampere rating of the device.

All symbols are usually shown in the **de-energized** position. This means that there is no electric potential to the coil of the device. Figure 5.15 shows a "normally open" contact in the de-energized position.

FIGURE 5.15 Symbol for a normally open pole of a relay or contactor

FIGURE 5.16 Symbols for relay or contactor coil: either symbol may be used for each device.

FIGURE 5.17 Symbol for a normally closed pole of a relay or contactor

The term **normally** refers to the position of a set of contacts when the device is de-energized. Figure 5.15 shows a **normally open** set of contacts and Figure 5.16 shows a **normally closed** set of contacts. Normally open contacts close and normally closed contacts open when the relay or contactor is energized. In Figure 5.18, a relay is shown with normally open and closed contacts.

The terms "normally open," "normally closed," "energized," and "de-energized" are important in understanding relays and contactors on wiring diagrams. Figure 5.19(a) shows a relay with two normally open contacts and one normally closed contact in the de-energized position (with no voltage to the coil). Figure 5.19(b) shows the same contacts in the energized position (with voltage to the coil). In the de-energized position, the current will not flow through contacts 1 and 2, but current will flow through contact 3. In the energized position, the current flow is through 1 and 2 but not through 3.

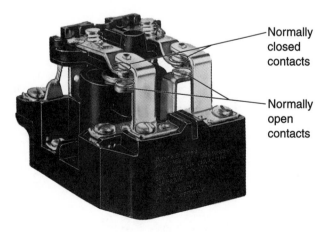

Normally closed contacts

Normally open contacts

FIGURE 5.18 Normally closed and open set of contacts *(Courtesy of Siemens Electromechanical Components, Inc.)*

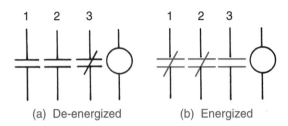

1   2   3   1   2   3

(a) De-energized   (b) Energized

FIGURE 5.19 Symbols showing de-energized and energized relays

5.3 MAGNETIC STARTERS

A **magnetic starter** is the same type of device as a contactor in terms of the ampere rating of the device. But the magnetic starter has a means of overload protection in it, whereas the contactor has none. Figure 5.20 shows a picture of the magnetic starter and its symbol. The principle of operation of the magnetic starter will be covered in Chapter 10.

5.4 SWITCHES

An electric **switch** is a device that opens and closes to control some load in an electric circuit. Electric switches can be opened and closed by temperature, pressure, humidity, flow, or by some manual means. You must become familiar with the symbols used for switches because in most cases they control the loads in the system. The symbol will also indicate what is initiating the action of the switch.

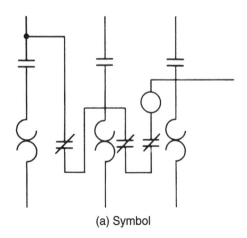

(a) Symbol

(b) Magnetic starter

FIGURE 5.20 Magnetic starter *(Courtesy of Furnas Electric Company)*

FIGURE 5.21 Single-pole–single-throw manual switch

A manually operated switch is a switch that is opened and closed by manual force. Figure 5.21 shows a simple manually operated switch. The poles of a manual switch are the number of contacts that are included in the switch. The **throw** indicates how the switch may be operated. For example, a single-pole—single-throw switch has one set of contacts and two positions: an open and a closed position, as shown in Figure 5.21. A double-pole—double-throw switch has two sets of contacts and three positions, as shown in Figure 5.22. Symbols for these two switches and for two other basic types of manual switches are shown in Figure 5.23.

FIGURE 5.22 Double-pole–single-throw manual switch

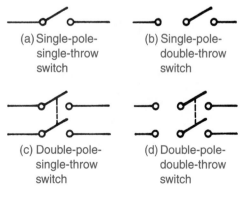

(a) Single-pole-single-throw switch

(b) Single-pole-double-throw switch

(c) Double-pole-single-throw switch

(d) Double-pole-double-throw switch

FIGURE 5.23 Symbols for manual switches

(a) Switch

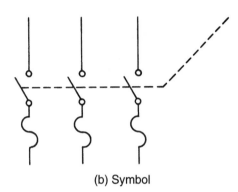

(b) Symbol

FIGURE 5.24 Three-pole fusible disconnect

There are other types of manual switches used in the industry. The **disconnect switch** is used to open and close the main power source to a piece of equipment or load. Figure 5.24 shows a three-pole disconnect switch and its symbol. The **push-button switch**, as shown in Figure 5.25, is a switch used to open and close a set of contacts by pressing a button. The symbols for the normally closed and the normally open push-button switches are also shown in Figure 5.25.

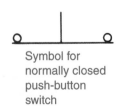

Symbol for
normally closed
push-button
switch

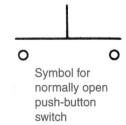

Symbol for
normally open
push-button
switch

FIGURE 5.25 Push-button switch

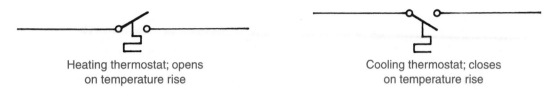

Heating thermostat; opens
on temperature rise

Cooling thermostat; closes
on temperature rise

FIGURE 5.26 Symbols for heating and cooling thermostats

The most important type of switch in a control system is the mechanically operated switch. **Thermostats** are mechanically operated switches used in most control systems. Thermostats are said to be mechanically operated because the temperature-sensing element moves a set of contacts by a mechanical linkage. Thermostats are designed for heating, cooling, or both. The cooling thermostat is designed to close on a temperature rise and open on a temperature fall. The heating thermostat is designed to open on a temperature rise and close on a temperature fall. The symbols for these two types of thermostats, shown in Figure 5.26, indicate their function. Figure 5.27 shows a modern thermostat.

Pressure switches are used for different functions in modern control circuits. The purpose of the pressure switch determines whether it opens or closes on a rise or fall in pressure. The pressure range of the switch is not part of the symbolic representation. Figure 5.28 shows the symbols for pressure switches. Letter designations in the symbols often denote the pressure ranges and purposes of the switches. Figure 5.29 shows some common pressure switches used in the industry.

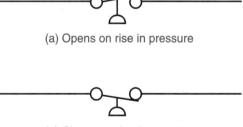

(a) Opens on rise in pressure

(a) Closes on rise in pressure

FIGURE 5.27 Thermostat *(Courtesy of Honeywell, Inc.)*

FIGURE 5.28 Symbols for pressure switches

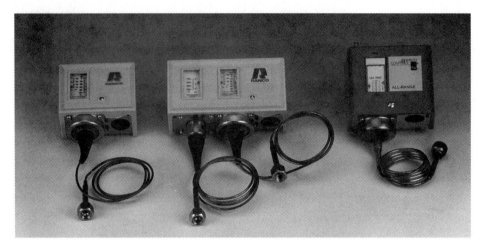

FIGURE 5.29 Some common pressure switches

5.5 SAFETY DEVICES

Safety devices are important in today's modern systems. Components are becoming more expensive each year. Thus, it is vital that these components be protected from adverse conditions such as low voltage, high ampere draw, and overheating. It is for this reason that you should become familiar with symbols for safety devices. Overloads and safety devices are sometimes a combination of a load and a switch. They differ from the relay in their purpose and overall design.

All motors are designed to operate on a certain current draw. If for some reason this rating is exceeded, the motor must be cut off immediately to prevent damage and possible destruction of the component. A burned-out motor is often caused by a malfunction in the safety devices.

The **fuse** is the simplest type of overload device. The fuse is effective against a large overload, but it is less effective against small overloads. The fuse is nothing more than a piece of metal designed to carry a certain load. Any higher load will cause the fuse to break the circuit. Figure 5.30 shows two symbols for a fuse. Figure 5.31 shows some common fuses in use today.

The second type of overload device is designed to protect the motor against small and large overloads. This type is divided into two categories: thermal and magnetic. The **thermal overload** is operated by heat, and the **magnetic overload** is operated by magnetism, which is directly proportional to the current draw.

The thermal overload can be a **pilot duty device**, which breaks the control circuit and locks the motor out. The pilot duty types of overloads are

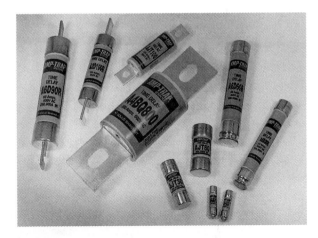

FIGURE 5.31 Some common fuses *(Courtesy of Gould Shawmut, Newburyport, MA)*

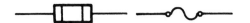

FIGURE 5.30 Symbols for fuses

most common on motors larger than 3 horsepower. The thermal overload can also be a line voltage device, which breaks the power line to the component being protected.

The bimetal element is the simplest of the thermal overloads. When it gets warm, it warps to open the circuit, as shown symbolically in Figure 5.32. Some bimetal elements are furnished with heaters, as shown symbolically in Figure 5.33. The heater allows the bimetal disc to react to an overload more quickly because the current flow is proportional to heat.

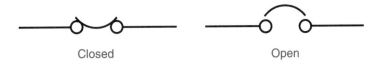

Closed   Open

FIGURE 5.32 Symbols for bimetal overload (closed and open)

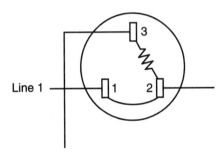

FIGURE 5.33 Symbol for three-wire bimetal overload

FIGURE 5.34 Symbol for thermal overload relay

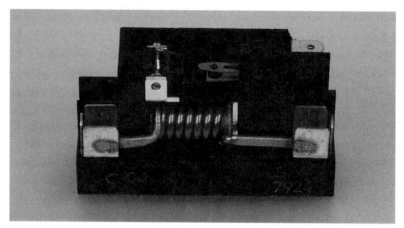

Magnetic overload device

Symbol

FIGURE 5.35 Magnetic overload *(Courtesy of Bill Johnson)*

The thermal overload relay, whose symbol is shown in Figure 5.34, is a simple device with a thermal element and a switch that opens on a rise in temperature.

The magnetic overload symbol is the same as the symbol for a relay with one normally closed contact. The current flow is relayed to the overload coil. Since current flow is proportional to the strength of the magnetic field, the relay can be designed to energize only on a high current draw. Figure 5.35 shows a magnetic overload and its symbol. The letter designation of this device will distinguish between the magnetic overload and the common relay.

5.6 TRANSFORMERS

The **transformer** decreases or increases the incoming voltage to a desired voltage. In most air-conditioning control circuits, it is not practical to pull large wires for a long distance. Therefore, a 24-volt control circuit, which

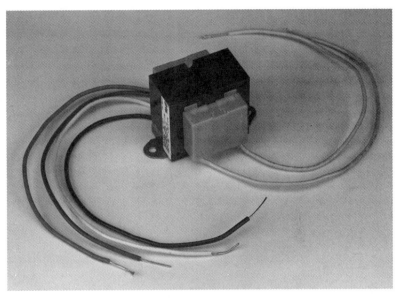

(a) Symbol   (b) Transformer

FIGURE 5.36 Transformer

is safer, less expensive, and a better method of control, is used. Figure 5.36 shows a transformer and its symbol. The voltage is also given with the symbol in some cases.

5.7 SCHEMATIC DIAGRAMS

Most modern heating, cooling, and refrigeration systems are becoming more complex with more controls and safety devices. Advances in controls and control systems require you to be able to read schematic diagrams. If you are able to read **schematic diagrams**, you will know what the unit should be doing.

The schematic diagram is the most useful and easiest to follow of any electric diagram. The schematic diagram tells how, when, and why a system works as it does. In most cases, service technicians use schematic diagrams to troubleshoot control systems. The schematic wiring diagram includes the symbols and the line representations so the user can easily identify loads and switches along with the circuits.

All electric circuits contain a source of electrons, a device that uses electron flow, and a path for the electrons to follow. In most cases, the source of electrons is an alternating current voltage supply. The device using the

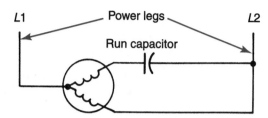

FIGURE 5.37 Schematic diagram showing power supply

electron flow is a motor, heater, relay coil, or any other load device. The path for the electrons to follow is a wire or any type of conductor.

In the schematic diagram, the source of electrons, the power supply, is represented by two lines drawn downward and listed as $L1$ and $L2$, as shown in Figure 5.37. There is a potential difference of 240 volts between $L1$ and $L2$. If a path is created between $L1$ and $L2$, current will flow.

All electrical loads in the unit are placed between $L1$ and $L2$, along with the switches controlling the load. Figure 5.38 shows a complete circuit in schematic form with a compressor and the switch (thermostat) that controls it. When the switch is closed, the compressor will run. In Figure 5.38, the source of electrons is from $L1$ and $L2$, the path is the connecting wire, and the device using the electron flow is the compressor. The compressor operates when the thermostat is closed.

Figure 5.39 shows a full schematic diagram similar to the diagrams you will be using on the job. All schematic diagrams are broken down into a circuit-by-circuit arrangement. Most schematic diagrams contain a legend that cross-references the components and their letter designation to the name of the component. Look at the legend in Figure 5.39.

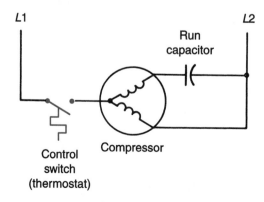

FIGURE 5.38 Schematic diagram of a complete circuit

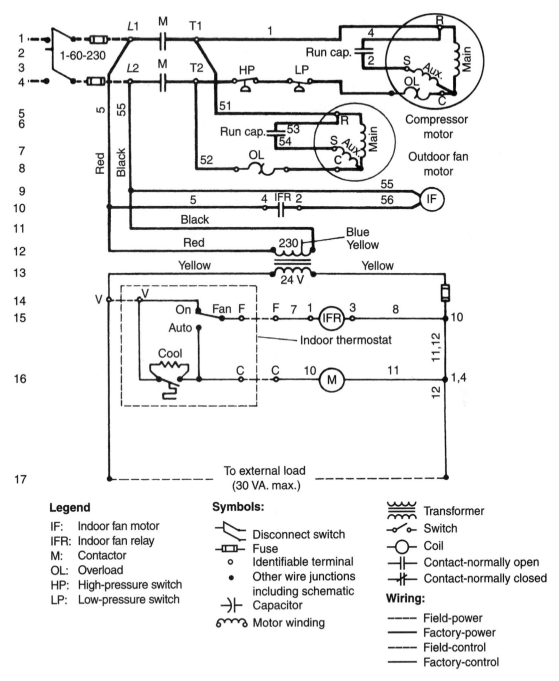

FIGURE 5.39 Complete schematic diagram for small packaged unit *(Courtesy of Westinghouse Electric Corp.)*

5.8 PICTORIAL DIAGRAMS

The **pictorial diagram**, also referred to as a label or line diagram, is intended to show the actual internal wiring of the unit. The pictorial diagram shows all the components of the control panel as a blueprint, including all the interconnecting wiring. It does not show the unit to scale, however. Components that are not shown in the control panel itself are shown outside the panel and labeled. The pictorial diagram is used to locate specific components or wires when troubleshooting from a schematic diagram. A typical pictorial diagram used in the industry is shown in Figure 5.40.

It is difficult to determine from a pictorial diagram how a system operates, and only an experienced mechanic can follow a complex pictorial diagram. Thus, most air-conditioning technicians use the schematic diagram to find the cause of the problem. Then they use the pictorial diagram to locate the position of the component at fault. In cases where the wiring is simple, however, a pictorial diagram may be the only diagram furnished with the equipment.

The **factual diagram** consists of a pictorial diagram along with a schematic diagram. Many air-conditioner manufacturers supply factual diagrams so service technicians can locate the relay or component in the control panel.

5.9 INSTALLATION DIAGRAMS

The **installation diagram** is used to help the installation electrician to wire the unit properly. The diagram gives specific information about terminals, wire sizes, color coding, and breaker or fuse sizes. The diagram does not provide details about equipment operation because the electrician has no need for this information. Figure 5.41 shows an installation diagram. The installation wiring diagram shows little internal wiring and is therefore almost useless to industry technicians.

SUMMARY

Loads are devices that use electricity to do useful work. Figure 5.42 gives a review of the symbols used for solenoids, motors, and heaters, the typical loads found in the industry. Most symbols have some type of letter designation to identify more clearly the component referred to.

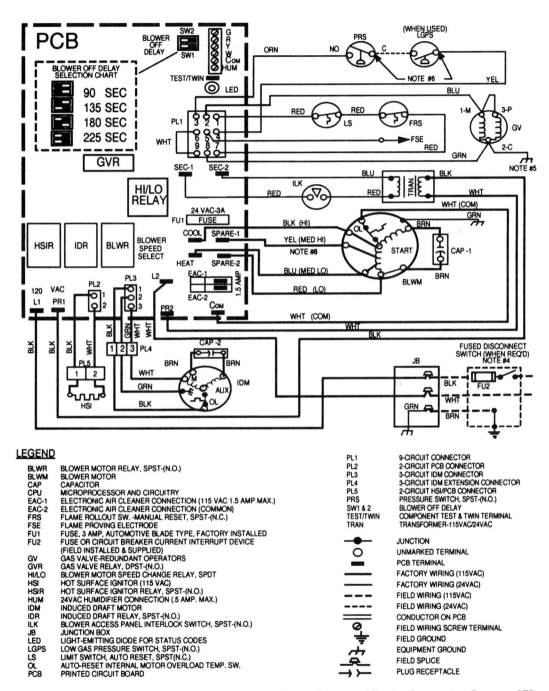

LEGEND

BLWR	BLOWER MOTOR RELAY, SPST-(N.O.)	PL1	9-CIRCUIT CONNECTOR
BLWM	BLOWER MOTOR	PL2	2-CIRCUIT PCB CONNECTOR
CAP	CAPACITOR	PL3	3-CIRCUIT IDM CONNECTOR
CPU	MICROPROCESSOR AND CIRCUITRY	PL4	3-CIRCUIT IDM EXTENSION CONNECTOR
EAC-1	ELECTRONIC AIR CLEANER CONNECTION (115 VAC 1.5 AMP MAX.)	PL5	2-CIRCUIT HSI/PCB CONNECTOR
EAC-2	ELECTRONIC AIR CLEANER CONNECTION (COMMON)	PRS	PRESSURE SWITCH, SPST-(N.O.)
FRS	FLAME ROLLOUT SW. -MANUAL RESET, SPST-(N.C.)	SW1 & 2	BLOWER OFF DELAY
FSE	FLAME PROVING ELECTRODE	TEST/TWIN	COMPONENT TEST & TWIN TERMINAL
FU1	FUSE, 3 AMP, AUTOMOTIVE BLADE TYPE, FACTORY INSTALLED	TRAN	TRANSFORMER-115VAC/24VAC
FU2	FUSE OR CIRCUIT BREAKER CURRENT INTERRUPT DEVICE (FIELD INSTALLED & SUPPLIED)		

LEGEND (right column symbols):

- JUNCTION
- UNMARKED TERMINAL
- PCB TERMINAL
- FACTORY WIRING (115VAC)
- FACTORY WIRING (24VAC)
- FIELD WIRING (115VAC)
- FIELD WIRING (24VAC)
- CONDUCTOR ON PCB
- FIELD WIRING SCREW TERMINAL
- FIELD GROUND
- EQUIPMENT GROUND
- FIELD SPLICE
- PLUG RECEPTACLE

GV	GAS VALVE-REDUNDANT OPERATORS
GVR	GAS VALVE RELAY, DPST-(N.O.)
HI/LO	BLOWER MOTOR SPEED CHANGE RELAY, SPDT
HSI	HOT SURFACE IGNITOR (115 VAC)
HSIR	HOT SURFACE IGNITOR RELAY, SPST-(N.O.)
HUM	24VAC HUMIDIFIER CONNECTION (.5 AMP. MAX.)
IDM	INDUCED DRAFT MOTOR
IDR	INDUCED DRAFT RELAY, SPST-(N.O.)
ILK	BLOWER ACCESS PANEL INTERLOCK SWITCH, SPST-(N.O.)
JB	JUNCTION BOX
LED	LIGHT-EMITTING DIODE FOR STATUS CODES
LGPS	LOW GAS PRESSURE SWITCH, SPST-(N.O.)
LS	LIMIT SWITCH, AUTO RESET, SPST(N.C.)
OL	AUTO-RESET INTERNAL MOTOR OVERLOAD TEMP. SW.
PCB	PRINTED CIRCUIT BOARD

FIGURE 5.40 A typical pictorial diagram used in the industry *(Courtesy of Carrier Corporation, Syracuse, NY)*

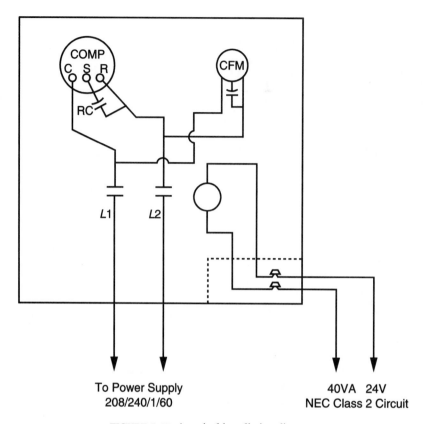

To Power Supply
208/240/1/60

40VA 24V
NEC Class 2 Circuit

FIGURE 5.41 A typical installation diagram

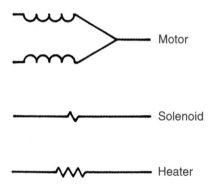

FIGURE 5.42 Review of symbols for loads

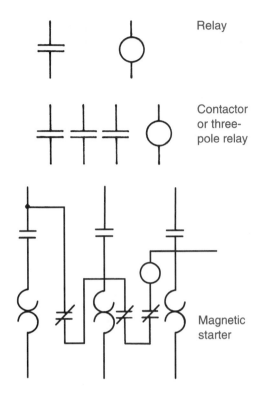

FIGURE 5.43 Review of symbols used for contactors and relays

Loads are controlled by relays and contactors, which share the same symbol and perform similar tasks. The major difference between relays and contactors is the amount of current each can carry. If a compressor is being operated by a device, you can assume the device is a contactor. If a small fan motor is being operated by a device, you can assume the device is a relay. A relay is used for small loads, and a contactor is used for large loads. Figure 5.43 reviews the symbols for some of these devices.

Relays and contactors are controlled by switches. Some of the switches used in the industry are manual, push-button, thermostat, and pressure. Thermostats are made for two purposes: to operate either a heating or a cooling system. The symbols for thermostats denote whether they are used for heating or cooling. Pressure switches are much the same as thermostats; their symbols also denote which way they open or close and under what condition. Pressure switches can be used for low or high pressure and are usually denoted by letter designations.

In any system using motors, protective devices are important to prevent damage to the motors or to larger components of the system. The most important type of safety device is for motor protection. A fuse, magnetic overload, thermal overload line break, thermal overload pilot duty, or a thermal overload relay could be used. Many overloads are built directly into the larger components.

Transformers are devices that increase or decrease the incoming voltage to some desired voltage. Transformers are used in the industry mainly in control circuits.

Schematic diagrams tell air-conditioning, heating, or refrigeration technicians when and why a system works as it does. Schematic diagrams show the symbols for devices and the interconnecting wiring of a unit in a circuit-by-circuit arrangement. Schematic diagrams are used most frequently by service technicians to troubleshoot equipment and systems.

Pictorial diagrams show an exact layout of the control panel with the external components shown outside the panel and labeled. The pictorial diagram can be used as a troubleshooting diagram on a simple system, such as a window air conditioner. In most cases, pictorial diagrams are used to find the placement of a component in the panel. Factual diagrams are a combination of the schematic and pictorial, with each shown separately.

Installation diagrams are used to help the installation electrician correctly connect the wiring to the unit. Appendix 1 shows most of the electrical symbols used by major refrigeration, heating, and air-conditioning manufacturers.

REVIEW QUESTIONS

1. What are the three types of electrical diagrams used in the heating, cooling, and refrigeration industry?

2. A load is an electrical device that _____.
   a. produces electricity
   b. directs the flow of electricity
   c. assists in the starting of motors
   d. uses electricity to do useful work

3. What is the major load of an air-conditioning system?

4. Identify the following symbols for loads:

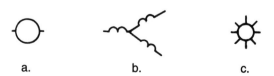

a.

b.

c.

d.

e.

5. What is the major difference between a contactor and a relay?

6. What do the terms "normally open" and "normally closed" refer to with regard to a switch or set of contacts? Draw a normally open and normally closed contacts.

7. What is the difference between a magnetic starter and a contactor?

8. Identify the following symbols for relays and contactors:

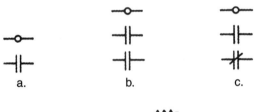

a.   b.   c.

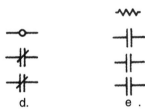

d.   e.

9. Draw a heating and a cooling thermostat and explain the difference between them.

10. A three-pole contactor would allow how many paths for current flow?

11. A disconnect switch is used to _____.

    a. open and close the main power source to a piece of equipment
    b. stop and start a compressor
    c. control the operation of an electric heater
    d. open when an unsafe condition occurs

12. What determines whether a pressure switch opens and closes on a rise of pressure?

13. Identify the following symbols for switches:

a.   b.   c.

d.   e.

14. What is the difference between a thermal overload and a magnetic overload? Draw the symbol for each.

15. What is the purpose of a transformer? Draw the symbols for a transformer.

16. Which of the following is not a requirement for an electric circuit?

    a. a source
    b. a path
    c. a load
    d. a signal light

17. What is the purpose of a legend on a schematic diagram?

18. A factual diagram is _____.

    a. a pictorial diagram with wire colors denoted
    b. a combination of pictorial and installation wiring diagrams
    c. a combination of schematic and pictorial wiring diagrams
    d. none of the above

19. Identify the following symbols for safety devices.

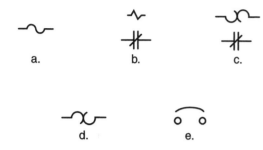

20. Which of the following are *not* components of a contactor or relay?

    a. a solenoid
    b. contacts
    c. thermal element
    d. mechanical linkage

21. True or False: The schematic diagram tells service technicians how to wire a system.

22. What is the difference between a pilot duty and a line break overload?

23. What type of switch would be used to open or close a set of contacts at a certain pressure?

24. What is the purpose of a fuse in an electrical system?

25. True or False: A solenoid valve is a device that opens or closes to control the flow of some element in the system.

26. What is a signal light used for in a control system?

27. Change the following normally open elements from the de-energized position to the energized position.

28. Draw the symbols for the following electrical devices.

    a. heating thermostat
    b. pressure switch (closes on rise)
    c. heater
    d. motor
    e. solenoid coil
    f. normally open push-button switch

29. Add letter designations to the symbols to indicate the following:

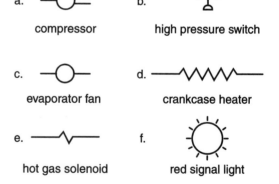

30. Draw a symbol for a magnetic starter.

LAB MANUAL REFERENCE

For experiments and activities dealing with material covered in this chapter, refer to Chapter 5 in the Lab Manual.

6

Reading Schematic Diagrams

OBJECTIVES

After completing this chapter, you should be able to

- Read and interpret the schematic of dehumidifier.
- Read and interpret the schematic of a window air conditioner.
- Read and interpret the schematic of a walk-in cooler.
- Read and interpret the schematic of a commercial freezer.
- Read and interpret the schematic of a gas furnace with a standing pilot.
- Read and interpret the schematic of a small packaged residential air conditioner.
- Read and interpret the schematics of light commercial air-conditioning systems with control relays.
- Read and interpret the schematics of light commercial air-conditioning systems with lockout relays.
- Read and interpret the schematics of two-stage heating and two-stage cooling systems.
- Read and interpret the schematics of heat pumps with defrost boards and with defrost timers.
- Read and interpret the schematic of a commercial refrigeration system with pump down.
- Read and interpret most diagrams found in the refrigeration, heating, and air-conditioning industry.

KEY TERMS

Balance point   Defrost cycle
Combustion chamber   Dehumidifier
Control relay   Gas furnace

Heat pump

Light commercial air-conditioning system

Limit switch

Line voltage control system

Lockout relay

Low-voltage control system

Multistage thermostat

Pump-down control system

Reversing valve

Set point

Short cycle

INTRODUCTION

Many types of wiring diagrams are used in the refrigeration, heating, and air-conditioning industry. Most of these diagrams can be used for multiple purposes, such as installation, locating electrical components in a control panel, and troubleshooting. One limitation, however, is that most diagrams are drawn to emphasize the part of the industry for which they are being used. For example, an installation or connection diagram shows the installation technician how to make the correct electrical connections when installing the equipment, but it does not show enough detail to enable technicians to troubleshoot the electrical system. The pictorial diagram can be used for troubleshooting and installation, but it is extremely cluttered and hard to follow for tasks other than locating electrical components in a control panel. However, if you had to completely rewire a control panel, a pictorial diagram would certainly be useful. The schematic diagram is the best diagram overall for the service technician to use when troubleshooting control circuits.

One of the most important tools available to the refrigeration, heating, and air-conditioning technician in the field is the schematic wiring diagram, sometimes called the ladder diagram. Technicians must understand the operation of the equipment they are assigned to install or troubleshoot. A proficient technician should be able to read and interpret schematic diagrams in order to:

• Understand how a piece of equipment operates.
• Know what the unit should be doing when it is in a specific mode of operation.
• Be able to make installation connections.
• Be able to troubleshoot and repair refrigeration, heating, and air-conditioning systems.

Refrigeration, heating, and air-conditioning equipment ranges from the simple to the complex. A schematic diagram shows the technician the electrical components that are used in the system. Diagrams can usually be found attached to one of the service panels of the equipment as shown in Figure 6.1 or attached to electrical panel covers. These diagrams can be schematic, pictorial, a combination of the two, or both. The schematic dia-

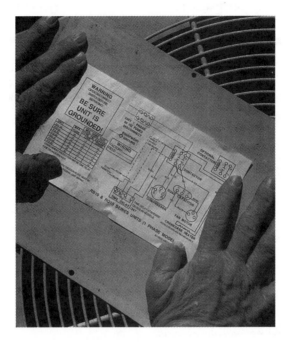

FIGURE 6.1 Diagram on service panel of air conditioner

gram not only identifies the electrical components that are in the unit but also illustrates how the unit works and electrical connections that need to be made. Technicians should refer to schematic diagrams when they want to know what electrical connections are needed to install the equipment. A schematic diagram makes troubleshooting easier because it breaks the control system down into a circuit-by-circuit arrangement. It is the easiest type of diagram to follow. The schematic diagram tells how, when, and why a system works as it does. A schematic diagram is shown in Figure 6.2.

6.1 SCHEMATIC DIAGRAM DESIGN

The schematic diagram resembles a ladder in that it is made up of two vertical lines representing the incoming electrical source (Figure 6.3). The electrical source is electrical energy being supplied to the equipment. Most electrical circuits shown in a schematic diagram are arranged between these lines, as shown in Figure 6.4. The schematic wiring diagram uses symbols to represent electrical components in the system. The symbols used by most equipment manufacturers are similar, but they are not alike in many cases. The symbol used for switches is a straight line with the controlling element connected to the switch as shown in Figure 6.5(a). Manual switches show no controlling element, as shown in

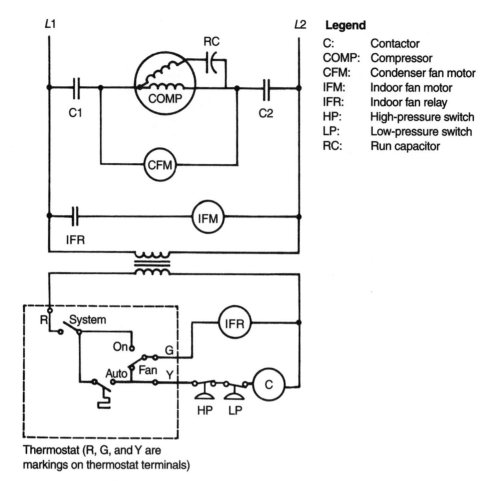

Legend

C:	Contactor
COMP:	Compressor
CFM:	Condenser fan motor
IFM:	Indoor fan motor
IFR:	Indoor fan relay
HP:	High-pressure switch
LP:	Low-pressure switch
RC:	Run capacitor

Thermostat (R, G, and Y are markings on thermostat terminals)

FIGURE 6.2 Schematic diagram

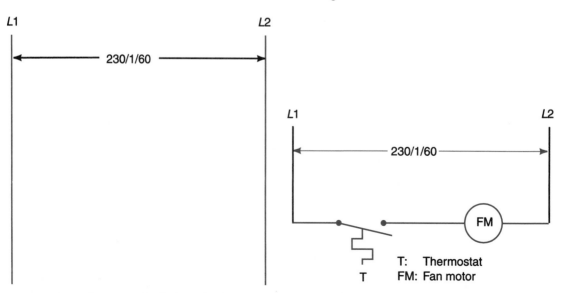

FIGURE 6.3 Power supply of schematic diagram

FIGURE 6.4 Circuits of schematic diagram

	Single pole single throw switch	AGS	Hot gas solenoid
	Single pole double throw switch		Relay coil
	Double pole single throw switch	R	Red signal light
	Double pole double throw switch	Comp	Compressor (single phase)
	Thermostat opens on rise (heating)	Comp	Compressor (three phase)
	Thermostat closes on rise (cooling)	Htr	Heater
	Pressure switch opens on rise	CH	Crankcase heater
	Pressure switch closes on rise		Defrost heater
	Normally closed push button		Fuse
	Normally open push button		Circuit breaker
H	Humidistat		In-line thermal overload
	Time delay relay		Magnetic overload
	Relay contacts		Pilot duty thermal overload
Comp	Compressor		Relay or contactor with two NO contacts
EFM	Evaporator fan motor		Relay with one NO and one NC contact
CFM	Condensor fan motor		Three-pole contactor
RVS	Reversing valve solenoid		Transformer

A B C D E

FIGURE 6.5 (A) Symbols for switches (B) Symbols for loads (C) Symbols for safety controls (D) Symbols for contactors and relays (E) Symbols for transformer

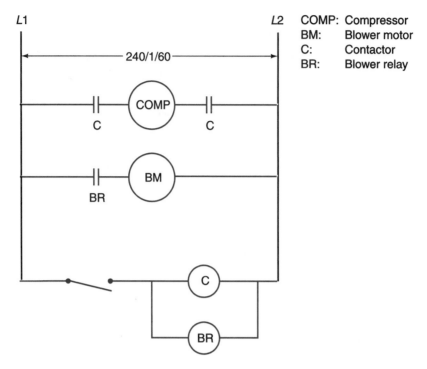

FIGURE 6.6 Parallel circuits in a schematic

Figure 6.5(a). Loads are represented by a variety of symbols like circles or zigzagged lines, (Figure 6.5[b]). Motors, relay coils, and signal lights are generally shown as circles, while solenoids and heaters are shown using zigzagged lines. Motors can be shown with their windings inside the circle. For most safety elements, a curved line represents their symbols, as shown in Figure 6.5(c). Figure 6.5(d) shows some contactors and relays commonly used on equipment today. Figure 6.5(e) shows a symbol for transformers. A transformer is used in most low-voltage control systems to reduce the voltage from line voltage to control voltage, usually 24 volts. A **low-voltage control system** is a control system that is operated by less voltage than that applied to the equipment. Most residential control systems are 24-volt control systems.

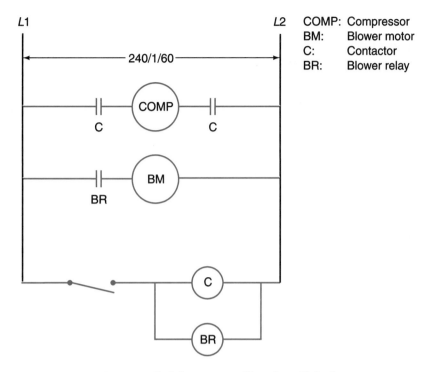

FIGURE 6.7 Switches connected in series with loads

A schematic diagram is made up of series and parallel circuits. Each circuit that is connected between the vertical lines or power supply is in parallel, as shown highlighted in Figure 6.6. A parallel circuit has more than one path of current flow; the loads shown in Figure 6.6 are connected parallel to each other. Switches that control loads are connected in series with each load, as shown highlighted in Figure 6.7. A series circuit is an electrical circuit with only one path of current flow. Series circuits are used when safety devices and operating controls are used to control a load. The entire schematic diagram shown in Figure 6.8 is a series-parallel circuit.

The circuit-by-circuit arrangement of a schematic diagram has important benefits. Technicians can be overwhelmed by the complexity of the equipment or control circuits they are trying to troubleshoot. It is much simpler

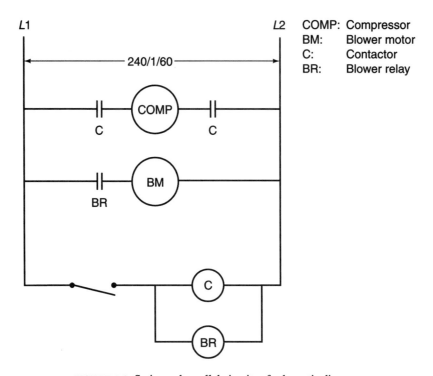

FIGURE 6.8 Series and parallel circuits of schematic diagram

to focus on a small section of a diagram than on the entire diagram. For example, examine the diagram in Figure 6.9(a) and then look at one circuit of the same diagram highlighted in Figure 6.9(b). Isolating the circuit that is shown in Figure 6.9(b) does not change how the unit or control system operates, but it does allow the technician to concentrate on the part of the system that he suspects is causing the problem. This is much simpler than trying to troubleshoot by focusing on the entire schematic diagram. A technician can never lose sight of the fact that the control circuit operates as a unit, but one can certainly make troubleshooting easier by focusing on the circuit that is not operating correctly.

Technicians must be able to read and interpret schematic diagrams so they can perform the tasks that will be assigned to them in the industry.

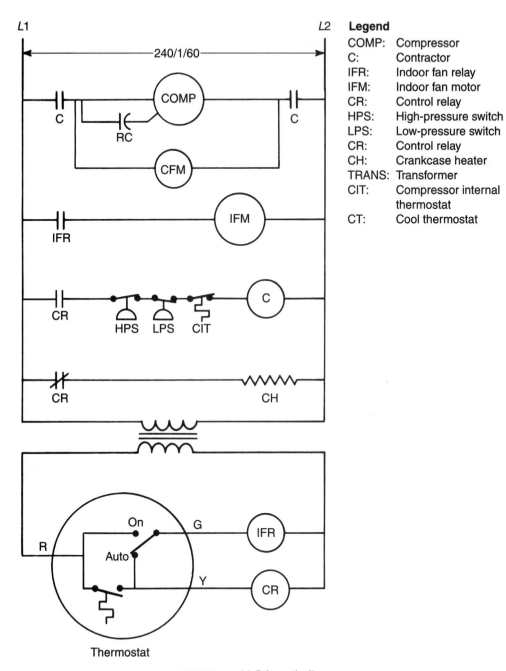

FIGURE 6.9 (a) Schematic diagram

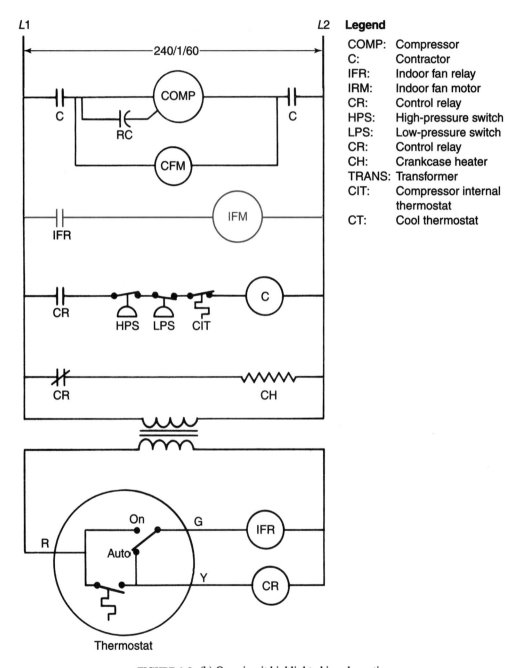

Legend

COMP: Compressor
C: Contractor
IFR: Indoor fan relay
IRM: Indoor fan motor
CR: Control relay
HPS: High-pressure switch
LPS: Low-pressure switch
CR: Control relay
CH: Crankcase heater
TRANS: Transformer
CIT: Compressor internal thermostat
CT: Cool thermostat

FIGURE 6.9 (b) One circuit highlighted in schematic

6.2 READING BASIC SCHEMATIC DIAGRAMS

In this section, six basic schematic diagrams will be explained in detail:

- Figure 6.11 Dehumidifier
- Figure 6.12 Simple window air conditioner
- Figure 6.13 Walk-in cooler
- Figure 6.14 Commercial freezer
- Figure 6.15 Gas furnace with standing pilot
- Figure 6.16 Packaged air-conditioning unit

Dehumidifier

A **dehumidifier** takes the humidity out of air in a specific area. Dehumidifiers are most commonly used to decrease moisture problems in basements, computer rooms, and other rooms where moisture is excessive. They come in many forms and sizes, but their function is the same no matter the size. The dehumidifier discussed will be the type that is used in a residence; it is very simple in design and construction and is shown in Figure 6.10. The dehumidifier uses a **line voltage control system**. A line voltage control system uses line voltage to operate the system.

The electrical components of a dehumidifier are the (1) humidistat, (2) compressor, and (3) fan motor. The schematic diagram for a simple residential dehumidifier is shown in Figure 6.11. A humidistat is a device that

FIGURE 6.10 Dehumidifier

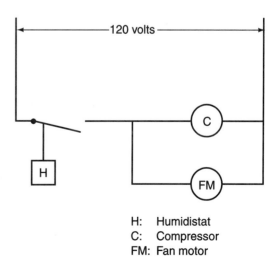

H: Humidistat
C: Compressor
FM: Fan motor

FIGURE 6.11 Schematic diagram of dehumidifier

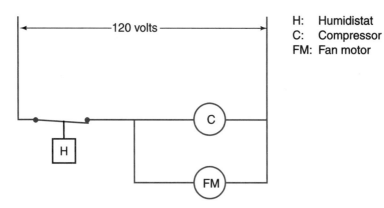

FIGURE 6.12 Schematic diagram of an operating dehumidifier with the active circuits highlighted

opens and closes a switch in reference to a set humidity. The compressor moves refrigerant throughout the refrigeration system, and the fan motor forces air across the evaporator to remove moisture and across the condenser to remove heat from the refrigerant. If the humidity increases to the **set point** of the humidistat, the humidistat will close and start the compressor and fan motor as shown highlighted in Figure 6.12. The set point of a control is the point at which the contacts open or close, depending on the function of the control. When the humidity falls below the set point, the humidistat will open, stopping the compressor and fan motor (see Figure 6.11).

Simple Window Air Conditioner

Window air-conditioning units (see Figure 6.13) are common throughout the United States. They are used to cool small areas and, in some cases, whole houses, depending on the capacity of the window air conditioner.

FIGURE 6.13 Window air conditioner *(Courtesy of Friedrich Air Conditioning Co.)*

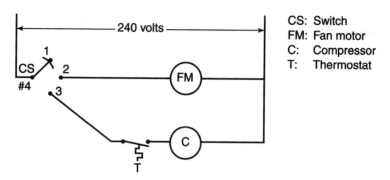

FIGURE 6.14 Schematic diagram of simple window air conditioner

Window air conditioners range from simple to complex, depending on the many options that are available to consumers today. Many window units are equipped with multiple-speed fan motors, options that allow the fan motor to cycle on and off with the compressor, and heating capabilities.

The window air conditioner that will be discussed is a simple design that has a single-speed fan motor operated by the control switch and a compressor that is operated by the control switch and thermostat, as shown schematically in Figure 6.14. The control switch has three positions: off, fan only, and cool, which operates the fan and compressor. When the control switch is in the off position, it connects the common terminal 4 to terminal 1, which makes no electrical connection to system components (see Figure 6.14). The fan-only position operates the fan only for air circulation and connects the common terminal 4 of the control switch to terminal 2 of the switch that will start the fan motor; this circuit is highlighted in Figure 6.15. When the control switch is in the cool position, an electrical connection is made between terminal 4 of the control switch and terminals 2

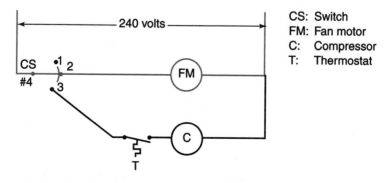

FIGURE 6.15 Schematic diagram of simple window air conditioner with fan motor circuits highlighted

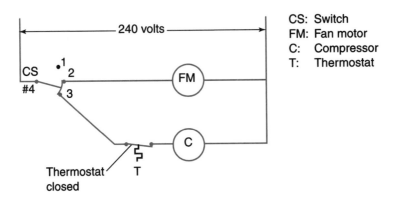

FIGURE 6.16 Schematic diagram of simple window air conditioner in the cooling mode of operation

and 3; these circuits are highlighted in Figure 6.16. The connection that is made between terminals 4 and 2 will operate the fan. The connection that is made between terminals 4 and 3 will operate the compressor if the temperature of the conditioned space is above the set point of the thermostat. When the temperature of the conditioned space reaches the set point, the thermostat opens, interrupting the power supply to the compressor, while the fan motor continues to run as shown in Figure 6.17. When the occupant shuts the window unit off, all components are stopped (see Figure 6.14).

Walk-In Cooler

A walk-in cooler provides the refrigeration for a refrigerated area. These units vary in size from small, like those used in a small grocery store, to extremely large, like those in manufacturing facilities. The indoor section

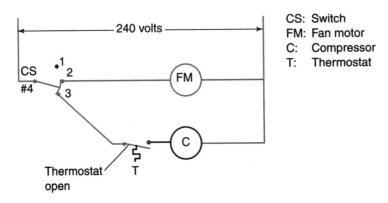

FIGURE 6.17 Schematic diagram of simple window air conditioner in the cooling mode of operation when the set point is reached

of the walk-in cooler includes an evaporator and a fan motor to move air across the evaporator, which is connected to a condensing unit. The condensing unit includes a compressor and a condenser fan motor that are usually located outside and connected to the evaporator section electrically and mechanically (by refrigerant lines).

The electrical components of a walk-in cooler include an evaporator fan motor, a compressor, a condenser fan motor, and electrical controls. The schematic diagram for a walk-in cooler is shown in Figure 6.18. The defrost timer consists of a timer motor that rotates and a set of contacts that opens for a certain period of time at selected intervals. Defrosting is accomplished by shutting the compressor and condenser fan motor off and operating the evaporator fan motor. In walk-in cooler applications this is usually enough to keep the evaporator coil defrosted. The walk-in cooler also has a high-pressure switch, which is a safety control that will interrupt supply voltage to the compressor if the discharge pressure reaches an unsafe condition. (Discharge pressure reaches an unsafe condition when the condenser is unable to adequately reject the heat that has been absorbed by the evaporator.) The low-pressure switch is used as the operating control for the walk-in cooler. Low-pressure switches are often used as operating controls because the pressure corresponds to the temperature in a refrigeration system. All of these electrical components can be seen in Figure 6.18.

Figure 6.19 shows the schematic diagram of a walk-in cooler control system with the system operating. The active circuits are highlighted. The defrost timer motor operates continuously whenever power is supplied to

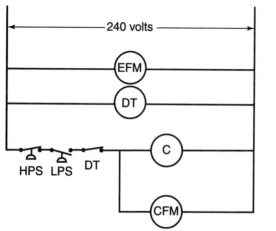

FIGURE 6.18 Schematic diagram of walk-in cooler

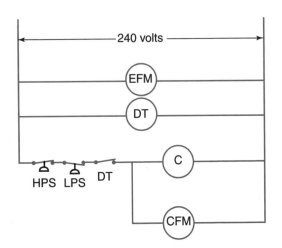

FIGURE 6.19 Schematic diagram of operating walk-in cooler with active circuits highlighted

the control system; this allows the defrost timer to initiate a defrost cycle at preset time intervals. The circuit controlling the compressor and condenser fan motor has three electrical switches in series with these components; the high-pressure switch, the low-pressure switch, and the defrost timer contacts. In normal operation, the low-pressure switch opens and closes to maintain the desired temperature in the cooler. The defrost timer contacts open and close at designated periods of time to defrost the evaporator coil, and the high-pressure switch will only open if a high-pressure condition occurred in the refrigeration system. The opening of the high-pressure switch would stop the compressor and condenser fan motor to prevent damage to the compressor.

Commercial Freezer

A commercial freezer is a refrigeration system that is designed to maintain a temperature lower than 32°F in a refrigerated case. When a temperature lower than 32°F must be maintained in a refrigerated space, the evaporator must be defrosted often enough to remove the frost from the evaporator coil. Frost removal from the evaporator coil is accomplished by using a source of heat. The most popular heat source is electric heaters, but in some cases hot gas from the compressor is transferred to the evaporator. A schematic diagram of a commercial freezer is shown in Figure 6.20.

The loads in the commercial freezer are compressor, condenser fan motor, evaporator fan motor, defrost heater, defrost timer motor. The controls needed for the freezer to work automatically are the thermostat, defrost thermostat, and defrost timer. The defrost timer is used to start and stop the

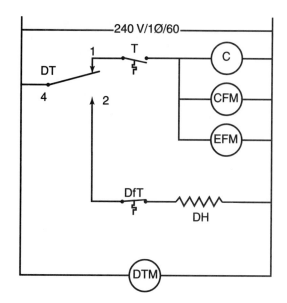

DT: Defrost timer
DTM: Defrost timer motor
T: Thermostat
C: Compressor
CFM: Condenser fan motor
EFM: Evaporator fan motor
DfT: Defrost thermostat
DH: Defrost heater

FIGURE 6.20 Schematic diagram of commercial freezer

defrost cycle along with a defrost thermostat that terminates the defrost cycle if the temperature rises too high. The defrost timer used in this application is a single-pole–double-throw switch controlled by a timer motor. The defrost timer used in commercial freezers is shown in Figure 6.21. This type of defrost timer can be set to go into defrost for a specific number of

FIGURE 6.21 Commercial freezer defrost timer

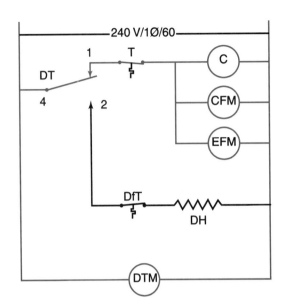

FIGURE 6.22 Commercial freezer schematic diagram in freezing mode of operation with active circuits highlighted

defrost cycles in a 24-hour period, and the defrost cycle time can be set from 30 to 120 minutes. The defrost timer motor operates whenever electrical energy is supplied to the control system. The thermostat is the operating control of the freezer and is used to maintain freezer temperature.

Figure 6.22 highlights the circuits that are active when the freezer is in the *freezer mode of operation*. The defrost timer contacts are closed between terminal 4 and terminal 1. The thermostat is the operating control in this mode of operation, it opens and closes to maintain the desired temperature in the freezer. The compressor, condenser fan motor, and evaporator fan motor are connected electrically in parallel and will operate together when the thermostat is closed.

The *defrost mode of operation* is highlighted in Figure 6.23. The defrost timer contacts have changed from the normal operating position of terminal 4 to terminal 1 to terminal 4 to terminal 2. When the defrost timer contacts are in this position, the compressor, condenser fan motor, and evaporator fan motor are de-energized. The defrost heater is energized through the defrost timer contacts 4 and 2 and the defrost thermostat. The defrost thermostat is a safety control that is normally closed but will open if the evaporator coil reaches a temperature that would be detrimental to the frozen product. The defrost heater remains energized until the defrost timer contacts change from 4 and 2 to 4 and 1, returning the system to normal operation. If the defrost thermostat opens, the defrost heater will be

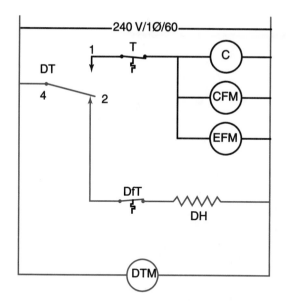

DT:	Defrost timer
DTM:	Defrost timer motor
T:	Thermostat
C:	Compressor
CFM:	Condenser fan motor
EFM:	Evaporator fan motor
DfT:	Defrost thermostat
DH:	Defrost heater

FIGURE 6.23 Commercial freezer schematic diagram in defrost mode of operation with active circuits highlighted

shut off. Some defrost timers automatically return the system to the frozen food mode of operation, but the type shown in this diagram will only shut the heater off.

Gas Furnace with Standing Pilot

The **gas furnace** shown in Figure 6.24 burns natural or LP gas to heat a structure. A standing pilot burns continuously until extinguished at the end of the heating season. The furnace has a low-voltage control system, which means it is 24 volts in most cases. In low-voltage control systems, the electrical system is divided into two sections by using a transformer. The transformer decreases line voltage—in this case 120 volts—to 24 volts (see Figure 6.25). The schematic diagram of a gas furnace with a standing pilot is shown in Figure 6.25. The fan motor is the only line voltage load in the electrical circuit and is controlled by the fan switch. The gas valve is a 24-volt load; it is used to stop and start the flow of gas to the furnace burners and is controlled by the heating thermostat. The pilot and gas burners are not shown in the schematic diagram because they are not electrical components.

When the temperature in the structure falls below the set point, the heating thermostat will close and provide 24 volts to the gas valve if the **limit switch** is closed (see circuit highlighted in Figure 6.26). The burners of the

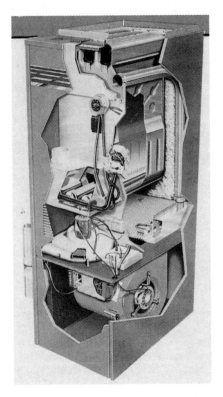

FIGURE 6.24 Gas furnace *(Courtesy of Heil-Quaker Corp.)*

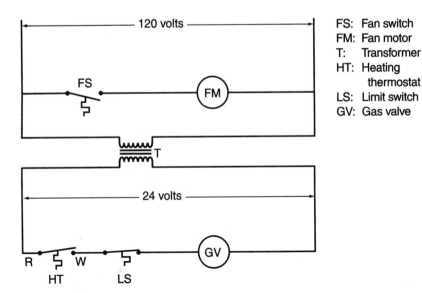

FS: Fan switch
FM: Fan motor
T: Transformer
HT: Heating thermostat
LS: Limit switch
GV: Gas valve

FIGURE 6.25 Schematic diagram of gas furnace with standing pilot

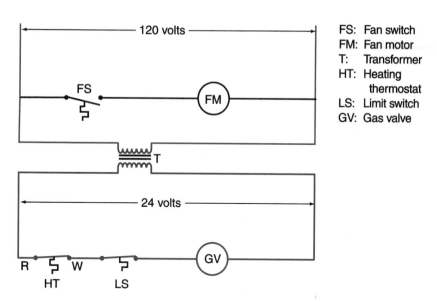

FIGURE 6.26 Schematic diagram of gas furnace when initially put into the heating mode with the active circuits highlighted

furnace will ignite, supplying heat to the **combustion chamber**, the section of the furnace that encloses the gas flame. The set point of a limit switch is the point at which the switch opens. If the temperature of the heat exchanger reaches an unsafe level, the limit switch will open, interrupting the voltage supply to the gas valve. As the combustion chamber is heated, the temperature of the fan switch begins to rise, and at the set point the fan switch will close, supplying heated air to the structure as shown highlighted in Figure 6.27. When the set point of the heating thermostat is reached, the 24-volt power supply is interrupted to the gas valve, extinguishing the flame. The fan motor will continue to operate until the combustion chamber cools and the fan switch opens.

Packaged Air-Conditioning Unit

A simple air-cooled packaged unit with a low-voltage control system is used to provide conditioned air to cool structures (see Figure 6.28). A packaged unit is manufactured in one piece with all components included, with the exception of the thermostat. Common components in an air-cooled packaged unit are the compressor, condenser fan motor, evaporator fan motor, and the controls necessary to provide automatic operation. The control system will include a transformer, which provides the 24 volts for the control system, along with an operating control and the necessary

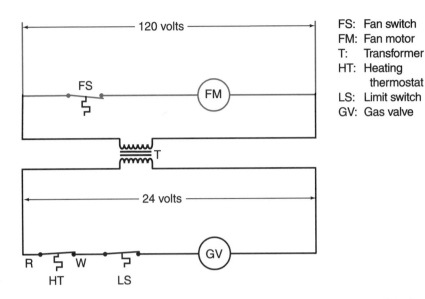

FS: Fan switch
FM: Fan motor
T: Transformer
HT: Heating thermostat
LS: Limit switch
GV: Gas valve

FIGURE 6.27 Schematic diagram of gas furnace after combustion chamber has warmed up and the fan motor is operating with active circuits highlighted

FIGURE 6.28 Packaged air-conditioning unit

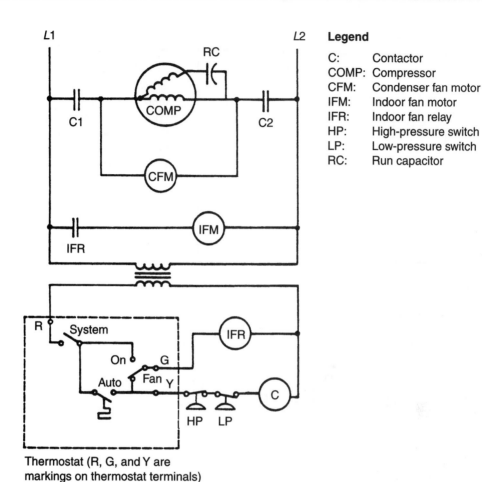

Legend

C: Contactor
COMP: Compressor
CFM: Condenser fan motor
IFM: Indoor fan motor
IFR: Indoor fan relay
HP: High-pressure switch
LP: Low-pressure switch
RC: Run capacitor

Thermostat (R, G, and Y are markings on thermostat terminals)

FIGURE 6.29 Schematic diagram of packaged air-conditioning unit

safety controls. Most residential air-conditioning systems have a 24-volt control system as shown in Figure 6.29.

The schematic diagram shown in Figure 6.29 is divided into two sections. The portion above the transformer is the line voltage section and is highlighted in Figure 6.30. The portion below the transformer is the 24-volt control circuit and is highlighted in Figure 6.31. The only connection between the line and low-voltage sections is the transformer.

Contactors and relays have coils that close the contacts when energized. These coils and the contacts themselves are not connected electrically. Often the coil will have a different voltage from the current passing through the contacts. In such cases the schematic diagram will show the

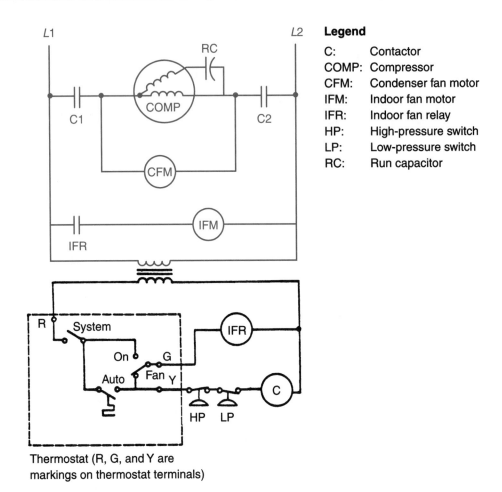

FIGURE 6.30 Schematic diagram with line voltage section highlighted

coils of relays and contactors in the low-voltage section of the diagram. The contacts of the relays or contactors will be in the line voltage section of the diagram in series with the loads they are controlling.

The schematic for the low-voltage thermostat used on the unit is shown in Figure 6.32. The thermostat is a combination of several switches along with the actual temperature-sensing portion of the control, which in this case is a cooling thermostat. Most low-voltage thermostats are also equipped with a heating function because most comfort cooling systems also include a heating function. The switch in the upper left of the thermostat diagram is a system switch. It cuts the total control system off; that is, it stops the entire

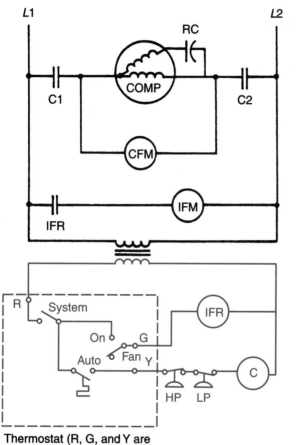

Thermostat (R, G, and Y are
markings on thermostat terminals)

FIGURE 6.31 Schematic diagram with low-voltage section highlighted

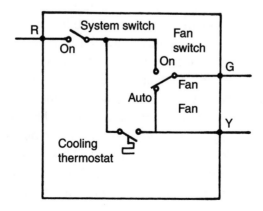

FIGURE 6.32 Schematic diagram of the 24-volt thermostat with all switches

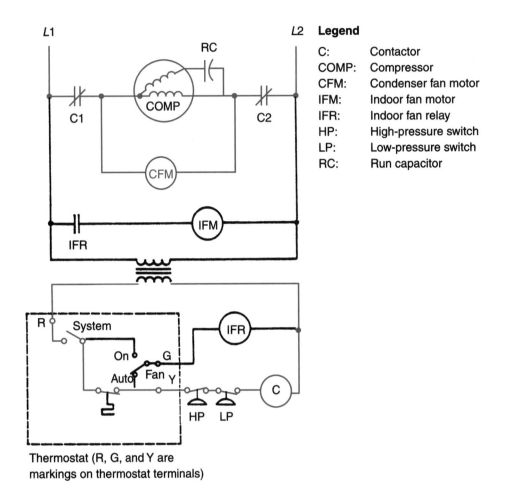

FIGURE 6.33 Thermostat and contactor coil circuits of a packaged unit

system from operating. The fan switch has both "on" and "automatic" positions. The "on" position permits fan-only operation. In the "automatic" position, the fan operates only when the cooling thermostat is closed.

The low-voltage control portion (see Figure 6.31) of the control system opens and closes the contacts in relays and contactors in the line voltage portion of the circuit (see Figure 6.30), which operates the condenser fan motor, indoor fan motor, and compressor. The contactor coil in the low-voltage control system is energized by the closing of the cooling thermostat if the system switch is closed. This circuit is highlighted in the low-voltage section of Figure 6.33. When the contactor is energized, its contacts (C1 and C2) close, starting the compressor and condenser fan

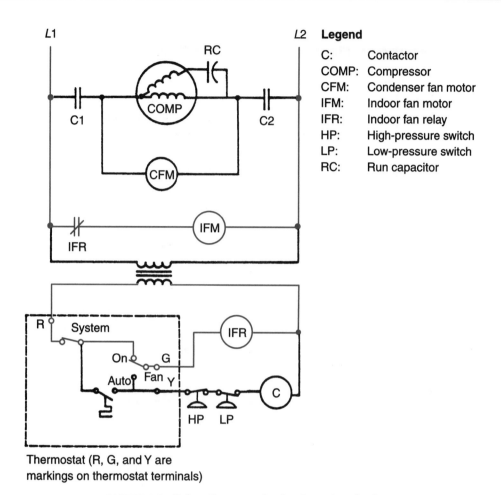

Legend

C:	Contactor
COMP:	Compressor
CFM:	Condenser fan motor
IFM:	Indoor fan motor
IFR:	Indoor fan relay
HP:	High-pressure switch
LP:	Low-pressure switch
RC:	Run capacitor

Thermostat (R, G, and Y are markings on thermostat terminals)

FIGURE 6.34 Indoor fan motor circuits of a packaged unit

motor shown highlighted in the line voltage section of Figure 6.33. The high-pressure and low-pressure switches in the contactor coil circuit will de-energize the contactor coil by opening if the refrigeration system reaches an unsafe operating condition, be it high or low pressure.

Figure 6.34 shows the low- and line voltage circuits that operate the indoor fan motor. The indoor fan coil is energized through the thermostat if the fan switch is in the "auto" position or by the fan switch if it is set in the "on" position (see the highlighted area in the low-voltage section of Figure 6.34). The indoor fan motor is started when the indoor fan relay contacts are closed; it is highlighted in Figure 6.34 in the line voltage section of the diagram.

The following steps give the sequence of operation of the air-conditioning unit. Refer to Figure 6.29 as you read each step.

1. The contactor coil is energized by the closing of the cooling thermostat when the system switch is in the "on" position. When the contactor closes, the compressor and condenser fan motor start.
2. The indoor fan motor operates continually with the fan switch in the "on" position because the IFR coil is energized, closing the contacts. If the fan switch on the thermostat is in the "automatic" position, the fan operates only when the contactor coil is energized.
3. The high-pressure and low-pressure switches are safety devices that de-energize the contactor if either opens. The opening of the high-pressure and low-pressure switches will not affect the operation of the indoor fan motor.

The schematic diagrams for residential and small commercial systems are usually simple and easy to understand. It is important for service technicians to be able to interpret wiring diagrams so they can service the equipment. A schematic diagram tells a technician how and when a piece of equipment operates as it does.

6.3 READING ADVANCED SCHEMATIC DIAGRAMS

In Section 6.2, six basic schematic diagrams were covered in detail. However, in the industry many control systems go far beyond the basic diagrams that were discussed. As the size and complexity of control systems and equipment grows, more electrical components are required to adequately protect and operate the equipment. The following six advanced schematic diagrams will be discussed.

- Figure 6.35: Light commercial air-conditioning control system with a control relay
- Figure 6.38: Light commercial air-conditioning control system with a lockout relay
- Figure 6.41: Two-stage heating, two-stage cooling control system
- Figure 6.46: Heat pump with defrost timer
- Figure 6.60: Heat pump with defrost board
- Figure 6.66: Commercial refrigeration system using a pump-down control system

Each circuit being discussed will be highlighted and explained in detail.

Light Commercial Air-Conditioning Control System with a Control Relay

A **light commercial air-conditioning system** is used on smaller commercial installations, usually from 5 to 25 tons. These systems generally have larger loads (electrical devices that have a higher current draw) like compressors, condenser fan motors, and evaporator fan motors that must be controlled by the control system. These larger loads require larger contactors because of their current requirements, and therefore the contactor coils must be larger to adequately close the contacts. The larger contactors oftentimes have difficulty closing if a 24-volt control system is used. Larger contactors operate better with line voltage because of their size. When it becomes necessary to use larger contactors, a control relay is used for control. The control relay is a 24-volt relay controlled by an operating control, usually a thermostat. The contacts of the control relay are placed in series with the contactor coil, which has a higher voltage and easily closes the larger contactor. A schematic diagram of a light commercial packaged unit is shown in Figure 6.35 with the control relay highlighted.

The light commercial air-conditioning packaged unit operates much like the smaller residential packaged unit. On a call for cooling, the thermostat closes, making electrical connections between R (power supply to thermostat), Y (cool function), and G (fan function) and energizing both the control relay and the indoor fan relay as shown highlighted in Figure 6.36. When the control relay is energized, the contacts close, sending line voltage to the contactor coil if the safety controls (high-pressure switch, low-pressure switch, and compressor internal thermostat) in the circuit are closed. The high-pressure switch would open if an unsafe high discharge pressure existed in the refrigeration system. The low-pressure switch would open if an unsafe low suction pressure existed in the refrigeration system. The compressors internal thermostat would open if the motor windings of the compressor's were at an unsafe temperature. Once the control relay and the indoor fan relay are energized, their contacts close, energizing the contactor and the indoor fan relay. When the contactor is energized, the contactor contacts close, starting the compressor and condenser fan motor. At the same time the indoor fan relay contacts close, starting the indoor fan motor. The line voltage circuits are highlighted in Figure 6.37.

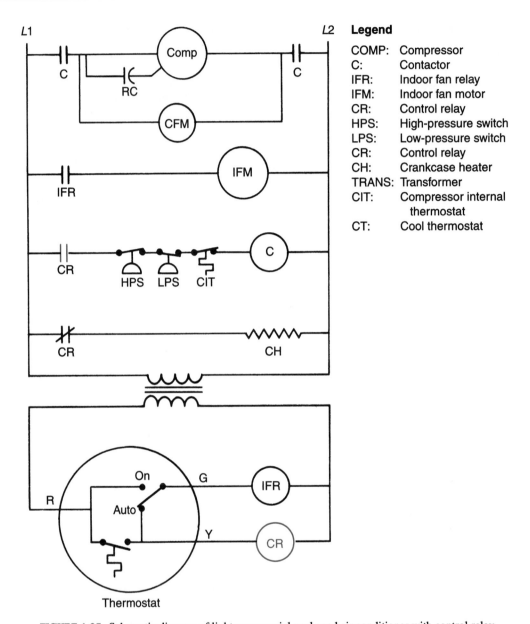

FIGURE 6.35 Schematic diagram of light commercial packaged air conditioner with control relay

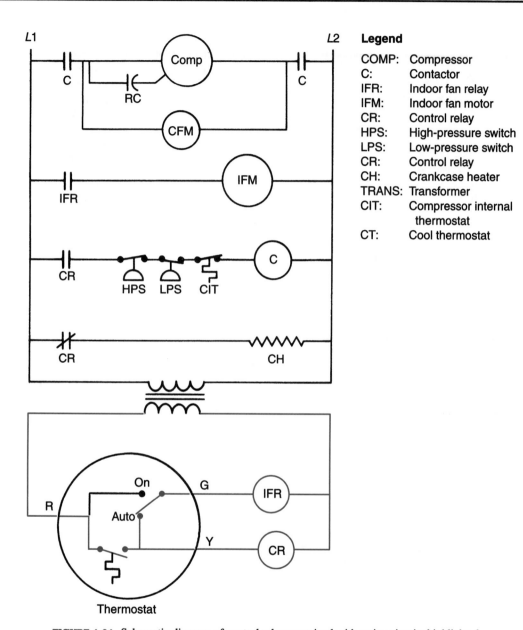

Legend

COMP:	Compressor
C:	Contactor
IFR:	Indoor fan relay
IFM:	Indoor fan motor
CR:	Control relay
HPS:	High-pressure switch
LPS:	Low-pressure switch
CR:	Control relay
CH:	Crankcase heater
TRANS:	Transformer
CIT:	Compressor internal thermostat
CT:	Cool thermostat

FIGURE 6.36 Schematic diagram of control relay energized with active circuits highlighted

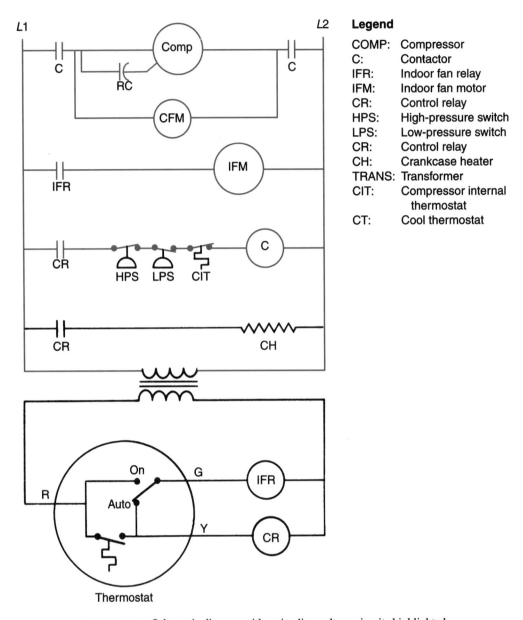

FIGURE 6.37 Schematic diagram with active line voltage circuits highlighted

Air-Conditioning System with Lockout Relay

Many air-conditioning systems have some type of lockout built into the control system that keeps the system off until the control system can be reset by turning the system off and back on. A **lockout relay** is a relay that can open a control circuit in an air-conditioning system and remain open until the occupant or the service technician resets the thermostat or the power supply, depending on whether the lockout relay is low- or line voltage. The lockout relay used in the past was an impedance relay, which is a relay with a high-resistance coil. Lockout relays are used in control systems to make the occupant aware that there is a problem and to prevent the equipment from short cycling. A compressor or other load **short cycles** when it continually stops and starts in rapid succession. The lockout relay looks identical to a normal relay and can only be distinguished by the number on the outside of the relay case. The old style lockout relay will be covered in order to familiarize you with the electrical circuitry. The newer lockout relays are solid-state devices that are much simpler to install and service; they will be discussed in the basic electronics chapter (chapter 18) of this text.

Figure 6.38 shows a schematic diagram of an air conditioner with a lockout relay. The lockout relay coil and contacts are shown highlighted in Figure 6.38. The lockout circuit of the diagram will be the only circuit covered. The remainder of the diagram is similar to those that have been discussed previously. The lockout relay in this schematic protects the compressor only; the indoor fan motor is not affected and will not be locked out by the action of the lockout relay. On a call for cooling, the thermostat closes and makes an electrical path between R, G, and Y. If all safety components (HPS, LPS, CT, and the normally closed HR contacts) are closed in the circuit to the contactor coil, closing the contacts of the contactor will start the compressor. At the same time, the indoor fan relay coil will be energized, closing the indoor fan relay contacts and starting the indoor fan motor. These circuits are highlighted in Figure 6.39.

Assume that the unit is operating and the high-pressure switch opens due to a dirty condenser. At that time the contactor is de-energized because the 24-volt supply has been broken to the contactor coil. At the same time, an electrical circuit has been made from the power source going to the circuit through the lockout relay coil. The lockout relay is energized because the contactor coil becomes a conductor. The contactor coil becomes a conductor because its resistance is less than that of the holding relay coil, and current will always take the path of least resistance. This circuit is shown highlighted in Figure 6.40, along with other circuits in the schematic that are energized. The circuit can be reset by opening and closing the thermostat or the 24-volt system switch.

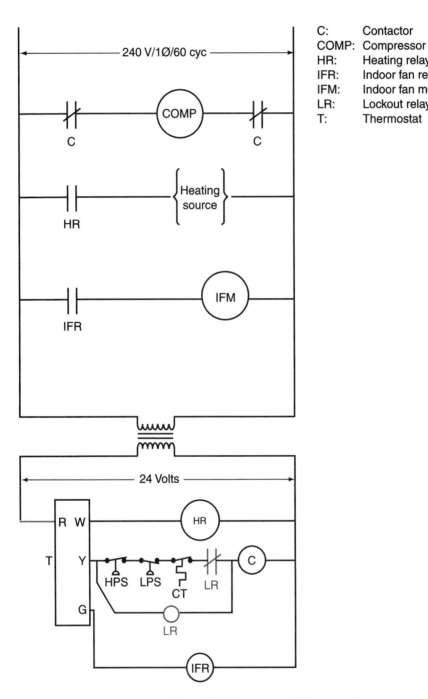

C: Contactor
COMP: Compressor
HR: Heating relay
IFR: Indoor fan relay
IFM: Indoor fan motor
LR: Lockout relay
T: Thermostat

FIGURE 6.38 Air-conditioning system with lockout relay

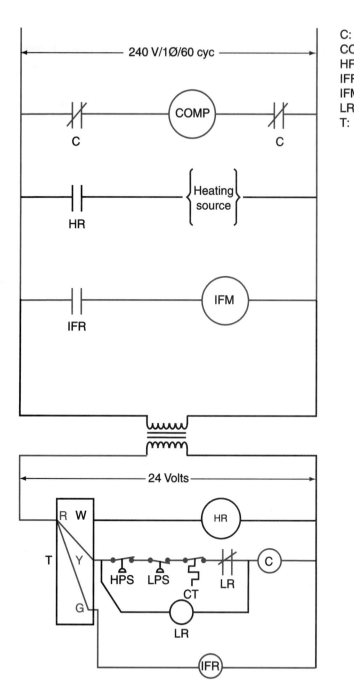

C: Contactor
COMP: Compressor
HR: Heating relay
IFR: Indoor fan relay
IFM: Indoor fan motor
LR: Lockout relay
T: Thermostat

FIGURE 6.39 Air-conditioning system with lockout relay in the operating mode

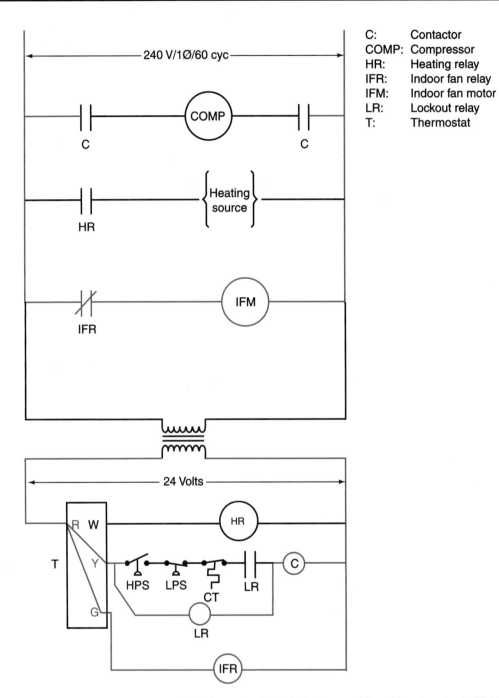

FIGURE 6.40 Air-conditioning system with lockout relay in the locked-out position with active circuits highlighted

Air-Conditioning System Controlled by a Multistage Thermostat

Many air-conditioning and heating systems incorporate multistage thermostats to more efficiently control the temperature of a structure. Multistage thermostats are used in light commercial air-conditioning systems that have more than one section of heating or cooling units and on heat pumps that have supplementary electric resistance heat. A multistage thermostat operates heating or cooling units or portions of heating and cooling equipment to maintain the desired temperature in a structure. A multistage thermostat could be used in a structure that has two air-conditioning condensing units or two heating units. In this case the control system would start the first cooling or heating units or stages; the other units or stages would not start unless the thermostat determined that one unit or stage could not adequately maintain the desired temperature. Multistage thermostats have a temperature range between the closing of the first and second stages. For example, in a two-stage cooling thermostat the first stage closes at the set point and the next stage closes only if the building temperature rises to the difference between stages, usually two to three degrees.

The operation of a two-stage cooling, two-stage heating control system will be covered in this section. The schematic diagram for the system is shown in Figure 6.41. The fan circuit operates like previously discussed fan circuits.

Heating Operation

If the control system is set for heating and the temperature in the structure drops below the set point, the first stage of the heating thermostat closes, making an electrical connection between R, W1, and G. The G terminal operates the fan function, which closes on both the heating and cooling cycles. The W1 terminal operates the first-stage heating cycle. The connection from G allows 24 volts to pass to the IFR coil, closing the contacts and starting the indoor fan motor. The connection from W1 allows 24 volts to pass to the HC1 coil, closing the contacts of HC1 and allowing voltage to pass through them, energizing Heater 1. The electrical connections made when the first stage of the heating thermostat closes is shown in Figure 6.42. The electric heater circuits are equipped with a safety thermostat ($L1$ and $L2$) that would open if the temperature around the electric heater(s) reached an unsafe condition.

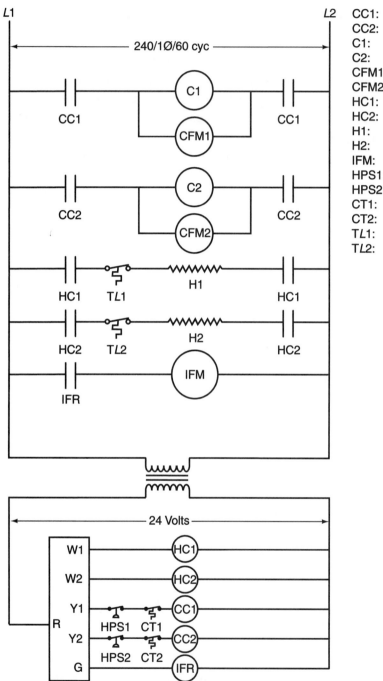

CC1:   Cooling contactor 1
CC2:   Cooling contactor 2
C1:   Compressor 1
C2:   Compressor 2
CFM1: Condenser fan motor 1
CFM2: Condenser fan motor 2
HC1:   Heat contactor 1
HC2:   Heat contactor 2
H1:   Heater 1
H2:   Heater 2
IFM: Indoor fan motor
HPS1: High pressure switch 1
HPS2: High pressure switch 2
CT1:   Compressor thermostat 1
CT2:   Compressor thermostat 2
TL1:   Temperature limit 1
TL2:   Temperature limit 2

FIGURE 6.41 Schematic diagram of two-stage cooling, two-stage heating control system

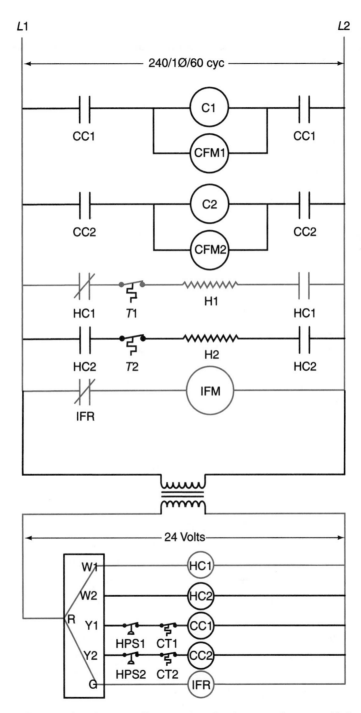

FIGURE 6.42 Schematic diagram of two-stage cooling, two-stage heating control system with the first stage in operation and the active circuits highlighted

If the temperature in the structure continues to drop, the second stage of the thermostat will close and make an electrical connection between R and W2. The first-stage heating system and the fan remain energized when the second stage closes. When this connection is made, the HC2 coil will be energized, closing the contacts and allowing voltage to pass through them to energize Heater 2. This and the circuits that are energized when the first-stage heating thermostat closes are highlighted in Figure 6.43.

Cooling Operation

On a call for cooling, an electrical connection is made between R, Y1, and G in the thermostat. The indoor fan motor is energized with the closing of the indoor fan relay contacts, starting the fan motor. The connection between R and Y1 allows voltage to pass through the safety controls (high-pressure switch 1 and compressor thermostat 1) to energize the compressor contactor 1 coil. This closes the contacts and starts compressor 1 and condenser fan motor 1 as shown highlighted in Figure 6.44. If the temperature continues to rise, then the second-stage cooling thermostat will close, making an electrical connection between R and Y2. This will energize compressor contactor 2 and start compressor 2 and condenser fan motor 2 as shown highlighted in Figure 6.45. The indoor fan motor, compressor 1, and condenser fan motor1 continue to operate.

Heat Pump with Defrost Timer

A **heat pump** is an air-conditioning system that uses a **reversing valve** to reverse the refrigeration cycle in order to provide heating. A reversing valve is an electrically controlled valve that allows for the reversing of the refrigeration cycle on a heat pump.

The refrigeration cycle of a heat pump is basically the same as that of any air-conditioning system except for the reversing valve and other refrigeration system components. When a heat pump is operating in the air-conditioning mode, the indoor coil is the evaporator and the outdoor coil is the condenser. In the heating mode, the refrigerant flows in the reverse direction so that the outdoor coil becomes the evaporator and the indoor coil becomes the condenser. In summer, the heat pump removes the heat from the inside air and expels it to the outside air, but in the winter the heat is absorbed from the outside air and released inside.

When the heat pump is operating in the heating mode, there must be a way to defrost the outside coil. This is done by temporarily reversing the cycle and using the cooling mode of operation to defrost the coil. This is

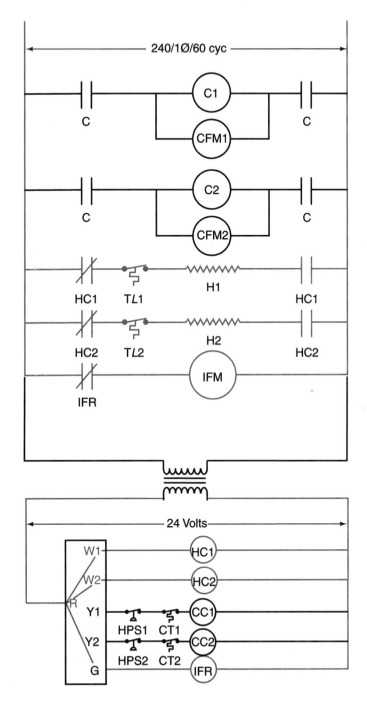

FIGURE 6.43 Schematic diagram of two-stage cooling, two-stage heating control system with the second stage in operation and the active circuits highlighted

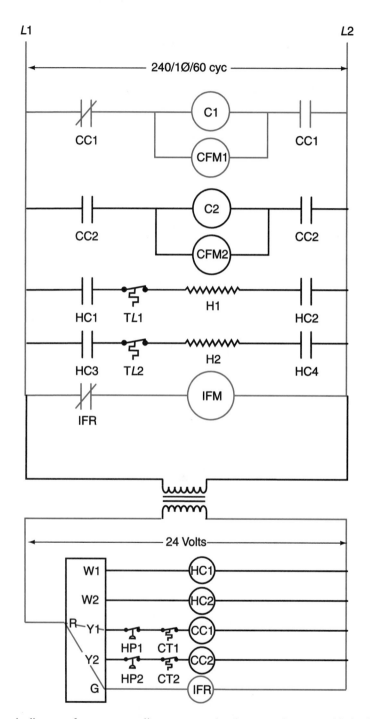

FIGURE 6.44 Schematic diagram of two-stage cooling, two-stage heating control system with the first-stage cooling mode in operation and the active circuits highlighted

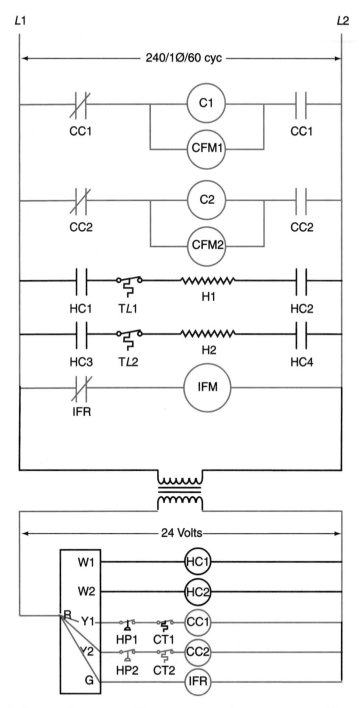

FIGURE 6.45 Schematic diagram of two-stage cooling, two-stage heating control system with the second-stage cooling mode in operation and the active circuits highlighted

referred to as the **defrost cycle**. In the defrost mode, a set of resistance heaters is energized to keep the indoor fan from moving cold air and causing drafts in the structure. Many methods are used in the industry to defrost a heat pump.

The efficiency of a heat pump decreases as the outdoor temperature drops, and at a certain point, called the **balance point**, the heat pump can no longer supply enough heat to properly condition the structure. Supplementary electric resistance heaters are added to the circuit to provide the additional heat needed to properly condition the space.

A two-stage heating, one-stage cooling or two-stage heating, two-stage cooling thermostat is used to operate most heat pumps, depending on the control circuitry designed by the manufacturer. The heat pump diagram in this text uses a two-stage heating, two-stage cooling thermostat.

The heat pump discussed here is a split-system heat pump, which means the unit is divided into an outdoor section and an indoor section. The thermostat is mounted in the indoor space.

The heat pump schematic will be covered in four stages: (1) cooling cycle, (2) heating cycle, (3) defrost cycle, and (4) resistance heating cycle. The schematic will be discussed according to the system's sequence of operation.

A heat pump schematic is shown in Figure 6.46 with all the electrical components identified by reference numbers and letters. The legend for the heat pump schematic is shown in Figure 6.47; it includes the names of the parts, abbreviations, and the reference letters or number in parentheses. These reference numbers and letters are locator points for electrical components in the diagram. The same legend will be used for all four explanations.

The Cooling Cycle

The cooling cycle schematic of the heat pump is shown in Figure 6.48. Only the electrical components that are functional in the cooling mode of operation are shown on the diagram.

Disconnect 1 (1) and disconnect 2 (20) are in the "on" position with the thermostat system switch (A) in the "off" position. Disconnect 1 (1) supplies power to the outdoor unit, thus supplying power to the trickle circuit through the start winding of the compressor motor (5), with the run capacitor (4) acting as a crankcase heater for the compressor. Disconnect 2 (20) supplies power to the indoor unit, thus supplying power to the transformer (16), which provides 24 volts to the control circuitry. The outdoor sensor

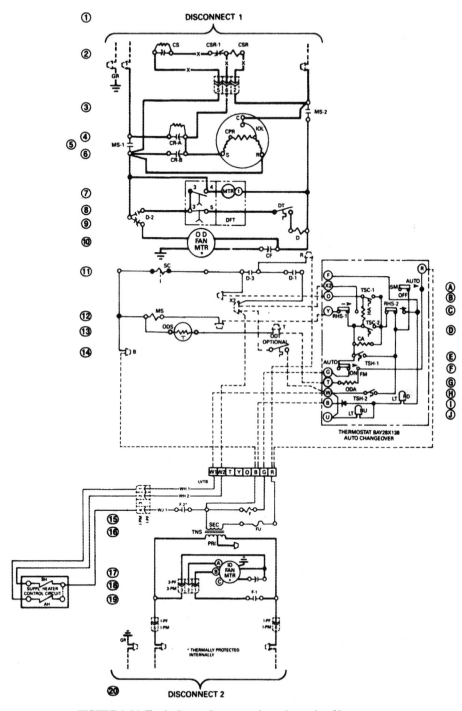

FIGURE 6.46 Typical complete operation schematic of heat pump

Legend

AH:	Supplementary Heat Contactor	(19)
BH:	Supplementary Heat Contactor	(18)
CA:	Cooling Anticipator	
CR:	Run Capacitor	(4 & 6)
CPR:	Compressor	
D:	Defrost Relay	(9)
DFT:	Defrost Timer	(7)
DT:	Defrost Termination Thermostat	(8)
F:	Indoor Fan Relay	(15)
FM:	Manual Fan Switch	(F)
HA:	Heating Anticipator	
HTR:	Heater	
IOL:	Internal Overload Protection	
LT:	Light	
LVTB:	Low-voltage Terminal Board	
MS:	Compressor Motor Contactor	(5, 3, & 12)
MTR:	Motor	
ODA:	Outdoor Temperature Anticipator	(G)
ODS:	Outdoor Temperature Sensor	(13)
ODT:	Outdoor Thermostat	
RHS:	Resistance Heat Switch	(C)
SC:	Switchover Valve Solenoid	(11)
SM:	System Switch	(A)
TNS:	Transformer	(16)
TSC:	Cooling Thermostat	(B & D)
TSH:	Heating Thermostat	(E & H)

FIGURE 6.47 Legend of diagrams in figures 6.45–6.51 *(Courtesy of The Trane Company)*

(13) is energized through the outdoor temperature anticipator (G) in the thermostat whenever the disconnect for the indoor unit is turned on. These circuits are highlighted in Figure 6.49.

If the thermostat system switch (A) is moved to the "auto" position, the cooling anticipator is energized and provides heat to the thermostat bimetal during the "off" cycle. Due to the voltage drop across the cooling anticipator, the motor starter (12) will not energize. The first-stage cooling thermostat (B) closes and completes the 24-volt circuit from terminal R to terminal O of the thermostat, thus energizing the switchover (reversing) valve solenoid (11). These circuits are highlighted in Figure 6.50 along with those from the previous explanation.

The second-stage cooling thermostat (D) closes and completes the 24-volt circuit from R to Y, thus energizing the motor starter (12), and from R to G,

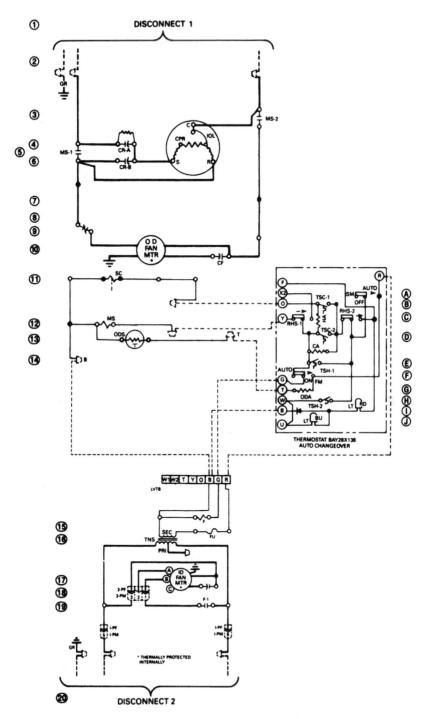

FIGURE 6.48 Typical cooling cycle schematic of heat pump *(Courtesy of The Trane Company)*

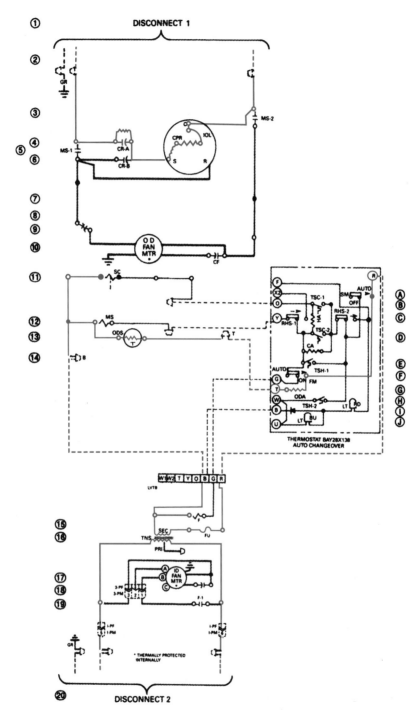

FIGURE 6.49 Typical cooling cycle schematic of heat pump with active circuits highlighted (*Courtesy of The Trane Company*)

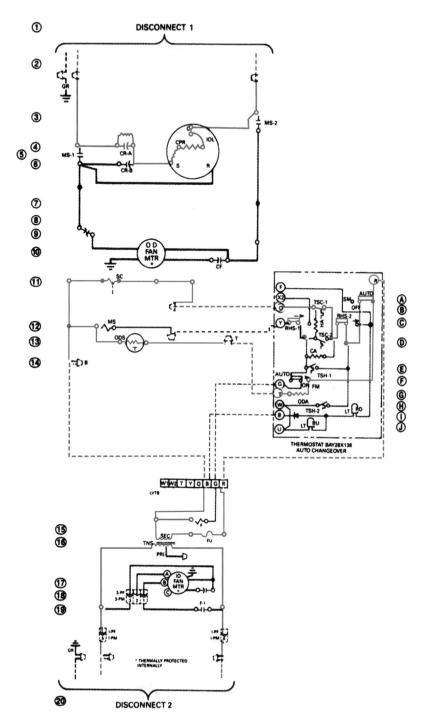

FIGURE 6.50 Typical cooling cycle schematic of heat pump with active circuits highlighted *(Courtesy of The Trane Company)*

thus energizing the fan relay (15) after the temperature rises 0.7 to 1.5°F above the temperature of the first-stage cooling thermostat. Simultaneously, the contacts (3 and 5) of the motor starter close, energizing the compressor and the outdoor fan motor; the indoor fan motor is energized through the indoor fan relay contacts (19). Once the second-stage cooling thermostat closes, the cooling anticipator is bypassed. These circuits are highlighted along with the previous two explanations in Figure 6.51. The heat pump is now operating in the cooling mode.

The Heating Cycle

The heating cycle schematic of the heat pump is shown in Figure 6.52. Only the electrical components that are functional in the heating mode of operation are shown in the diagram.

Disconnect 1 (1) and disconnect 2 (20) are in the "on" position with the thermostat system switch (A) in the "off" position. Disconnect 1 (1) supplies power to the outdoor unit, thus supplying power to the trickle circuit through the start winding of the compressor motor (5), with the run capacitor (4) acting as a crankcase heater for the compressor. Disconnect 2 (20) supplies power to the transformer (16) and provides 24 volts to the control circuitry. The outdoor thermostat sensor (13) is energized through the outdoor anticipator (G) in the thermostat whenever the disconnect for the indoor unit is turned on. These circuits are highlighted in Figure 6.53.

If the thermostat system switch (A) is moved to the "auto" position, the first-stage heating thermostat (E), when closed, completes the 24-volt circuit from terminal R to terminal Y. This energizes the motor starter (12). It also complete the 24-volt circuit from terminal R to terminal G, thus energizing the indoor fan relay (15). The heating anticipator is energized when the first-stage heating thermostat closes. Simultaneously, the contacts (3 and 5) of the motor starter close, energizing the compressor and the outdoor fan motor; the indoor fan motor is energized through the indoor fan relay contacts (19). The second set of contacts (15) of the indoor fan relay provides an interlock for the heater control circuit. These circuits are highlighted along with those previously discussed in the heating cycle and illustrated in Figure 6.54.

The second-stage heating thermostat closes after a temperature drop of 0.7°F to 1.5°F. When the second-stage heating thermostat closes, the 24-volt circuit from terminal R to terminals W and U energizes the supplementary heat contact coil (19) and the blue light on the thermostat. If the outdoor thermostat (14) is closed, the W circuit can also energize the supplementary contactor (18). These circuits are shown highlighted, along with the two previously discussed, in Figure 6.55.

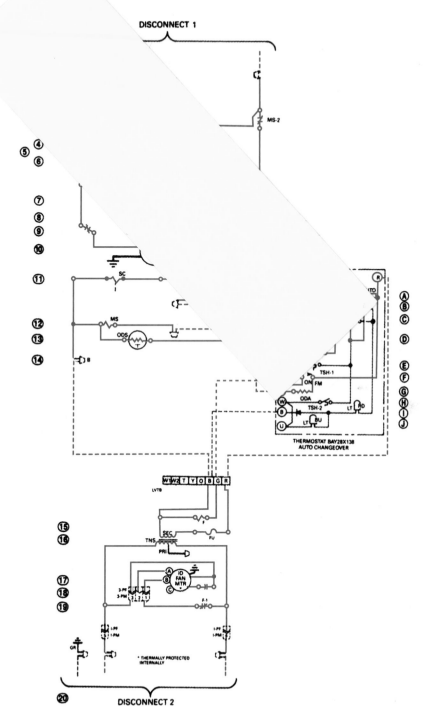

FIGURE 6.51 Typical cooling cycle schematic of heat pump with active circuits highlighted *(Courtesy of The Trane Company)*

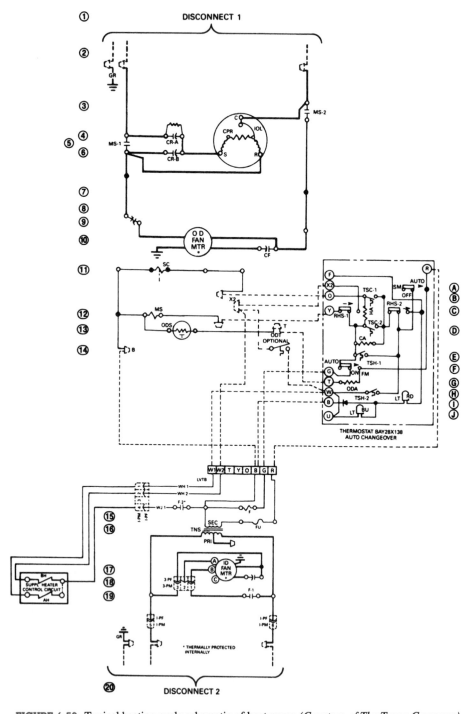

FIGURE 6.52 Typical heating cycle schematic of heat pump *(Courtesy of The Trane Company)*

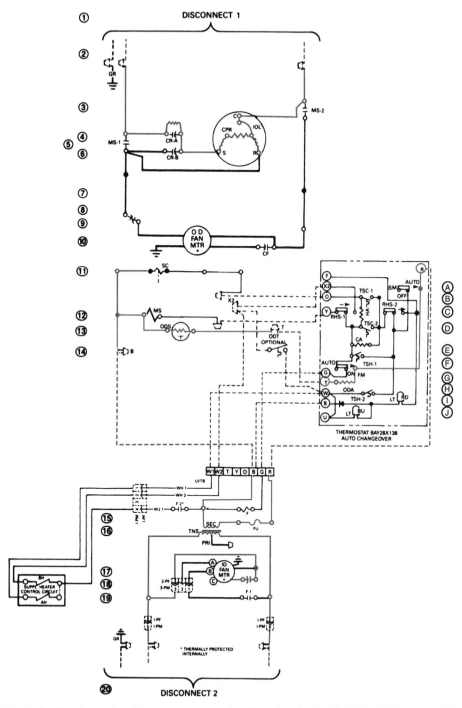

FIGURE 6.53 Typical heating cycle of heat pump schematic with active circuits highlighted *(Courtesy of The Trane Company)*

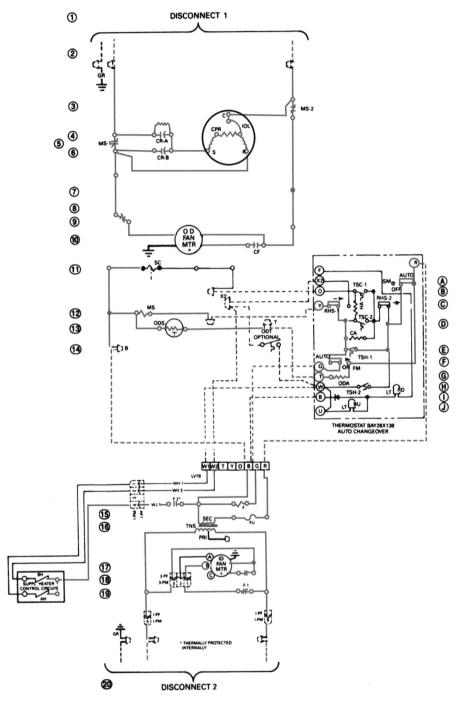

FIGURE 6.54 Typical heating cycle of heat pump schematic with active circuits highlighted *(Courtesy of The Trane Company)*

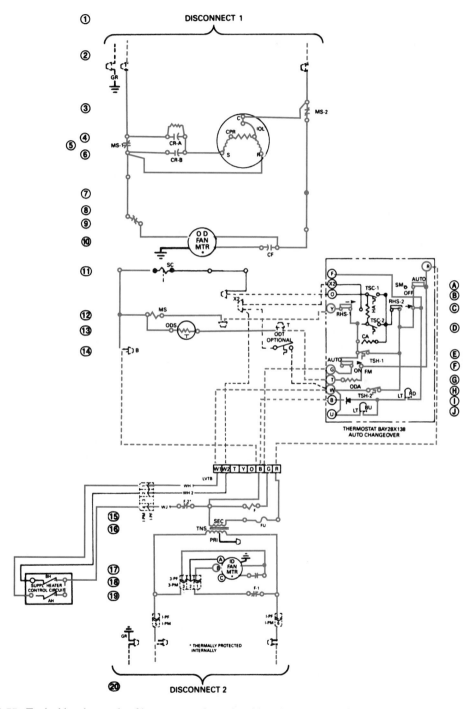

FIGURE 6.55 Typical heating cycle of heat pump schematic with active circuits highlighted *(Courtesy of The Trane Company)*

The Defrost Cycle

When operating in the heating mode at temperatures below 40°F outdoors, the heat pump's outdoor coil will accumulate frost or ice. The defrost cycle schematic of the heat pump is shown in Figure 6.56. Only the electrical components that are functional in the defrost mode of operation are shown. Defrosting is accomplished by putting the heat pump in the cooling mode so that the outdoor coil becomes the condenser without outdoor fan operation long enough to defrost the coil.

The heat pump is operating in the heating mode. Initiation of the defrost cycle requires that two conditions be met: (1) the lower circuit of the outdoor coil must be below 26°F, as detected by the defrost termination thermostat (8), and (2) the defrost timer (7 and 8) must have completed the required of 45- or 90-minute compressor run time. If these two conditions are met, the timer cam switch (7) will close and complete a circuit through the closed defrost termination switch (8) and the defrost timer switch (8) to the defrost relay coil (9). When the defrost relay is energized, defrost relay contacts 1 (11) and 3 (11) complete the circuitry to the supplementary heater contactor (18) and the switchover valve solenoid coil (11). Contact 2 of the defrost relay opens the circuit to the outdoor fan motor (10) and provides a holding circuit for the defrost relay coil.

Termination of the defrost cycle normally occurs when the defrost termination thermostat (8) opens after the liquid temperature leaving the outdoor coil reaches approximately 52°F. If the defrost termination thermostat does not terminate the defrost cycle within 10 minutes, the override switch in the defrost timer will open and de-energize the defrost relay and thus terminate the defrost cycle.

The Resistance Heating Cycle

Most heat pumps are equipped with an emergency heat mode that allows the customer to isolate the pump part of the heat mode and heat the residence by supplementary electric resistance heat. The emergency heat mode provides heat when the normal heating mode suffers a mechanical failure. The emergency heat cycle of the heat pump is shown in Figure 6.57. (Only the electrical components that are functional are shown.)

The emergency heat switches 1 and 2 (C) are mechanically linked to each other within the thermostat. When emergency heat switch 1 is moved from the "normal" position to the "emergency" position, terminal Y is removed from the circuit. This isolates the motor starter and closes terminal X2, which energizes the supplementary heat contactor coils (18 and 19)

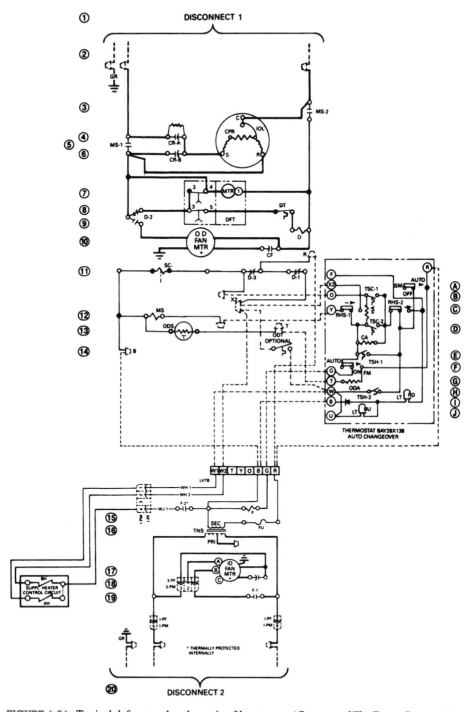

FIGURE 6.56 Typical defrost cycle schematic of heat pump *(Courtesy of The Trane Company)*

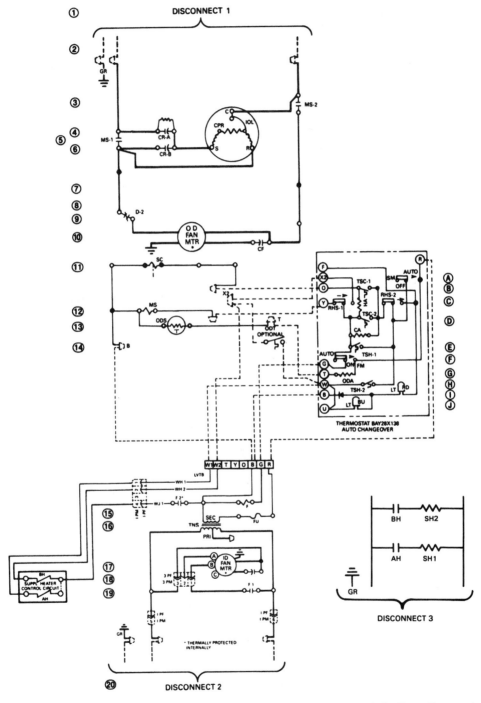

FIGURE 6.57 Typical resistance heat cycle schematic of heat pump *(Courtesy of The Trane Company)*

if the outdoor thermostat (14) is closed. If the outdoor thermostat is open, the supplementary heat relays will be energized by the second-stage heating thermostat when needed. These circuits are highlighted in Figure 6.58.

The entire heat pump schematic is shown in Figure 6.59; it can be understood when the technician looks at each mode of operation and then breaks it down into a circuit-by-circuit arrangement.

Heat Pump with Solid-State Defrost Board

The heat pump discussed in this section has the operating characteristics of the previous heat pump except for the method of defrost initiation and termination. The defrost boards on many modern-day heat pumps are nothing more than a solid-state timer that closes for a specified period of time on a particular time cycle, plus other electrical devices such as relays that facilitate an effective defrost cycle. In some cases, the technician can set the defrost cycle time and the time that the heat pump remains in defrost by making certain electrical connections on the solid-state defrost board. This option is useful because of the severity of the winters in some areas where heat pumps are currently being installed. Many heat pumps are being manufactured today that are equipped with a demand defrost system, which defrosts the outdoor coil on the heating cycle only when there is a definite need for the coil to be defrosted. These types of electronic defrost controls are more complex and expensive than the normal solid-state defrost board that will be discussed in this section.

The schematic diagram for a heat pump with a solid-state defrost board is shown in Figure 6.60. The main operating control discussed in this section is the thermostat. The letter designations used on heat pump thermostats vary from one manufacturer to another. One of the common letter designations is used in this section. The R terminal on the thermostat receives 24 volts from the transformer. The thermostat, depending on the setting, will distribute 24 volts to the desired terminals to cause the system to accomplish the purpose for which the thermostat has been set. The outdoor unit includes the compressor, condenser fan motor, reversing valve, and defrost board with the defrost thermostat.

The indoor and outdoor units are supplied with 240 V/single phase/ 60 cycles. The active circuits that are supplied with electrical energy and the blower motor circuitry are highlighted in Figure 6.61. The transformer primary is supplied with 240 volts from the indoor unit power supply. The control voltage from the transformer secondary is 24 volts, which is supplied to the defrost board and the R terminal on the thermostat. When the

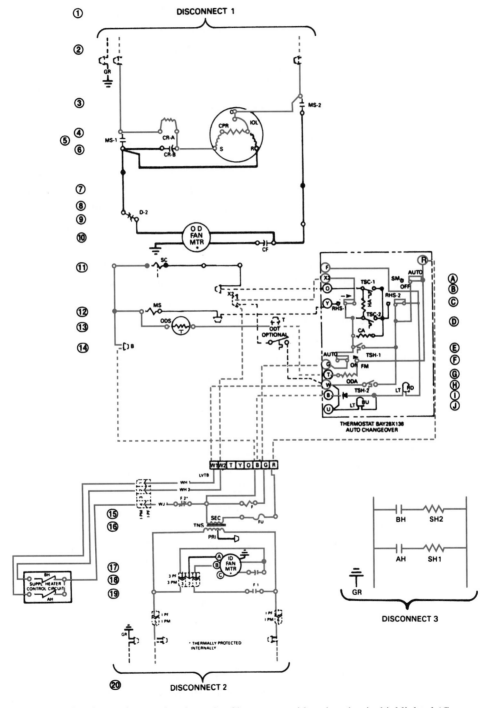

FIGURE 6.58 Typical resistance heat cycle schematic of heat pump with active circuits highlighted *(Courtesy of The Trane Company)*

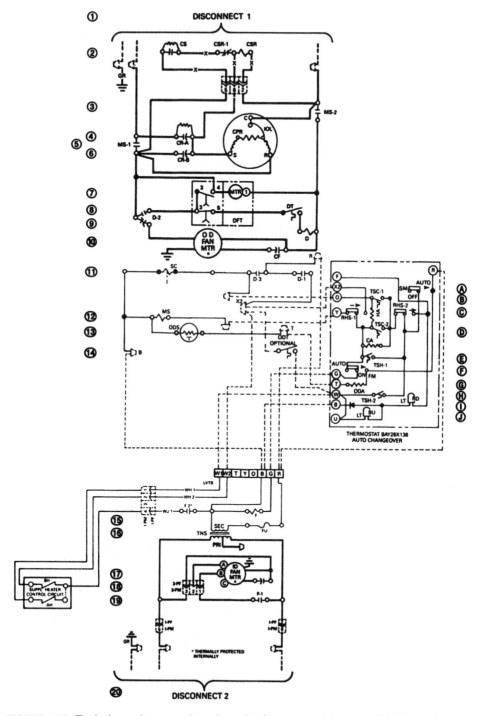

FIGURE 6.59 Typical complete operation schematic of heat pump *(Courtesy of The Trane Company)*

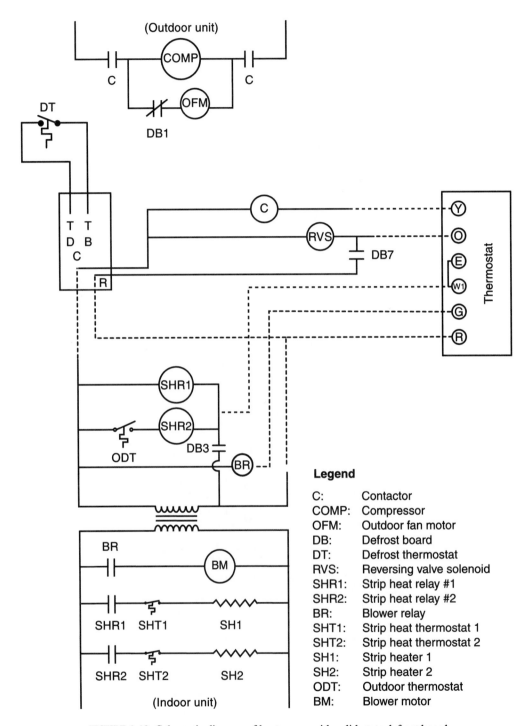

FIGURE 6.60 Schematic diagram of heat pump with solid-state defrost board

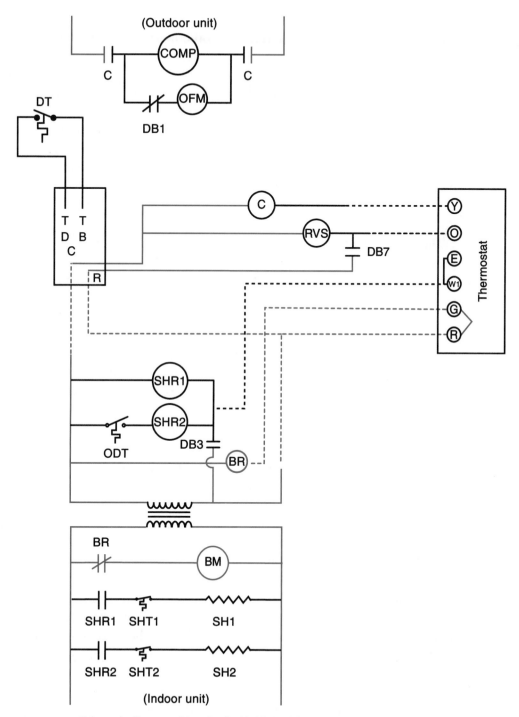

FIGURE 6.61 Schematic diagram of fan circuits highlighted for a heat pump with solid-state defrost board

thermostat fan switch is set in the "on" position, an electrical connection is made between the R and G terminals of the thermostat, supplying 24 volts to the blower relay and starting the blower motor.

When the fan switch is set in the "automatic" position, the fan will operate when the unit is in either cooling or heating mode. The thermostat is the controlling element for the blower motor in the "automatic" or continuous position. In the "automatic" position, the fan runs when the thermostat is calling for heating or cooling. When the thermostat fan switch is set in the continuous position, the indoor fan will run continuously. The thermostat calling for heat makes the electrical connection between R and Y, which sends electrical energy to the contactor of the outdoor unit and starts the compressor and outdoor fan motor. These circuits are highlighted in Figure 6.62.

The supplementary heat cycle of the heat pump is used when the compressor cannot supply enough heat to adequately heat the structure. When this occurs, the indoor temperature will continue to fall, even though the compressor is operating. The second-stage heating thermostat will then close, supplying electrical energy from the R terminal to W1. This supplies 24 volts to the supplementary heat relay 1 (SHR1) and supplementary heat relay 2 (SHR2) if the outdoor thermostat (ODT) is closed (Figure 6.63.). Many thermostats are equipped with emergency heat switches that will prevent the compressor from operating and will allow the supplementary heaters to operate when the thermostat is set for emergency heat.

The reversing valve was previously described as the electrical device that reverses the refrigerant flow in a heat pump. Heat pumps can operate with the reversing valve energized on either heating or cooling, but the reversing valve is energized only on the cooling cycle of the heat pump that is being discussed. One advantage of not energizing the reversing valve on the heating cycle is that if the valve fails, the system will still operate in the heating mode during cold weather.

The operation of the heat pump in the cooling cycle is similar to its operation in the heating cycle, except that the reversing valve is energized. If the thermostat is set in the cooling mode and the temperature is above the set point of the thermostat, the reversing valve is energized from the R terminal on the thermostat to the O terminal. The thermostat also makes an electrical connection between R and Y, energizing the contactor and starting the compressor and outdoor fan motor. The schematic for the cooling cycle is shown in Figure 6.64 with the cooling circuits highlighted.

In the heating cycle, the outdoor coil becomes the evaporator and the indoor coil becomes the condenser. When the heat pump is operating in the heating cycle in a cold ambient condition, the outdoor coil, being the

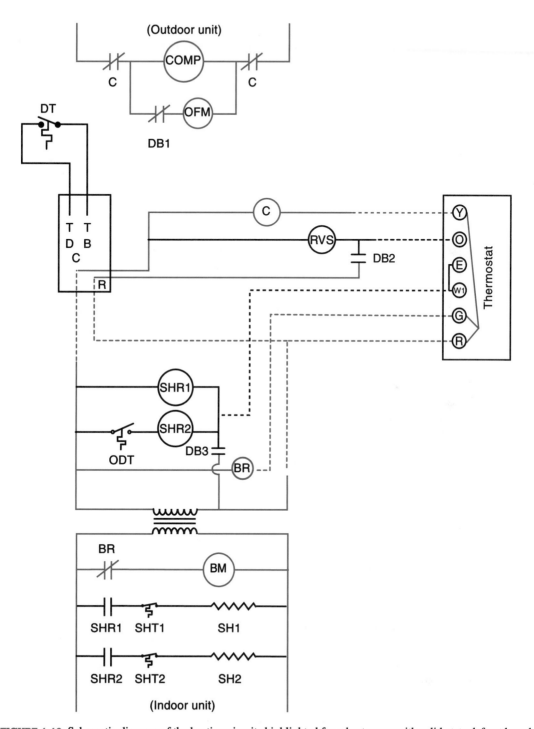

FIGURE 6.62 Schematic diagram of the heating circuits highlighted for a heat pump with solid-state defrost board

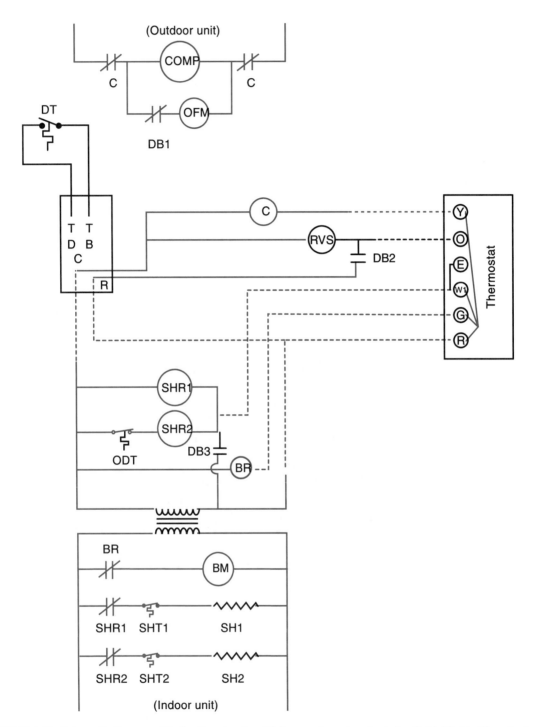

FIGURE 6.63 Schematic diagram of the heating circuits with the supplementary heat circuits highlighted for a heat pump with solid-state defrost board

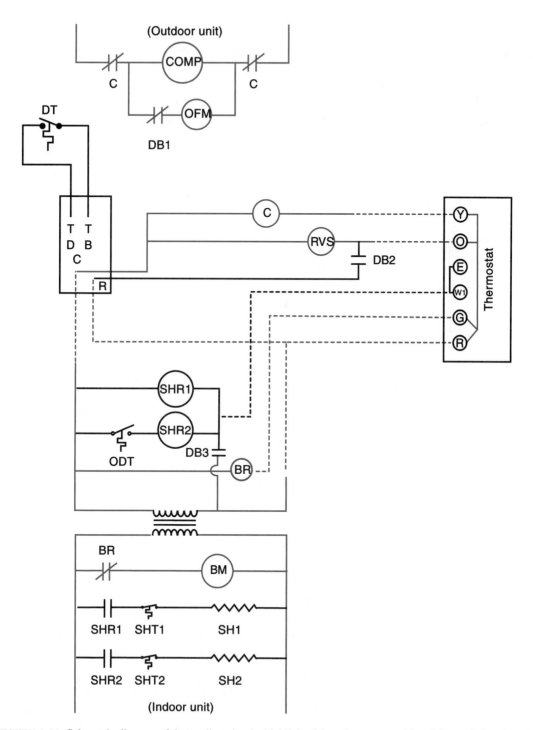

FIGURE 6.64 Schematic diagram of the cooling circuits highlighted for a heat pump with solid-state defrost board

evaporator, will frost. If the outdoor coil is not cleaned of the frost, the system will dramatically lose efficiency and heating capacity. To defrost the outdoor coil in the heating cycle of operation, the heat pump must be temporarily put in the cooling mode in order to supply the outdoor coil with hot discharge gas, which defrosts the coil. However, the heat pump will be cooling, so the supplementary heat must be started to temper the air going into the structure.

Figure 6.65 shows the heat pump diagram with the electrical circuits that are active in the defrost cycle highlighted. The defrost circuits contain the defrost relay, which contains DB1, DB2, and DB3 contacts. These contacts will remain closed until the timer in the defrost board cycles out or the defrost thermostat opens, at which time the defrost cycle will be terminated. The DB1 contacts are normally closed and they now open, stopping the outdoor fan motor during the defrost cycle to prevent the flow of cold air across the outdoor coil. The DB2 contacts are normally open and they now close, energizing the reversing valve. The supplementary heater or heaters, depending on the outdoor temperature, are energized through the normally open DB3 contacts.

Commercial Refrigeration System with Pump-Down Cycle

A **pump-down control system** stops the liquid flow of refrigerant with a solenoid valve in the liquid line and allows the compressor to transfer most of the refrigerant from the low side to the high side of the refrigeration system. This type of cycle is used on many commercial refrigeration systems and some commercial air-conditioning systems. The advantage of using a pump-down cycle is that when the system cycles off, the refrigerant cannot migrate back to the crankcase of the compressor to cause a flooded start situation. The schematic diagram of a commercial freezer with electric defrost is shown in Figure 6.66.

Figure 6.67 shows a commercial freezer in the normal operating cycle with the active circuits highlighted. If the system is off and the thermostat closes with the defrost timer contacts in the 1 to 4 position, the liquid line solenoid (LLS) valve is energized. The opening of the liquid line solenoid valve allows liquid refrigerant to pass through the liquid line to the metering device. When the pressure in the system due to the opening of the liquid line solenoid valve increases enough to close the low-pressure switch, the contactor coil is energized, closing the contacts of the contactor and starting the compressor and the condenser fan motor if the safety controls (LPS and CMT) in the circuit are closed. The condenser fan

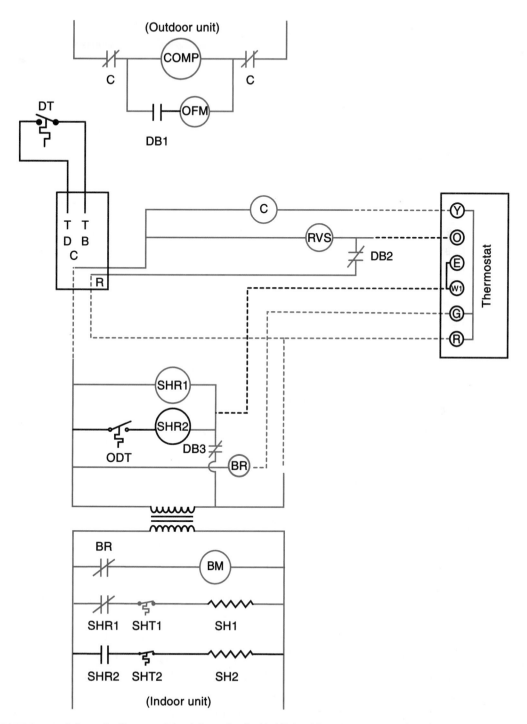

FIGURE 6.65 Schematic diagram of the defrost circuits highlighted for a heat pump with solid-state defrost board

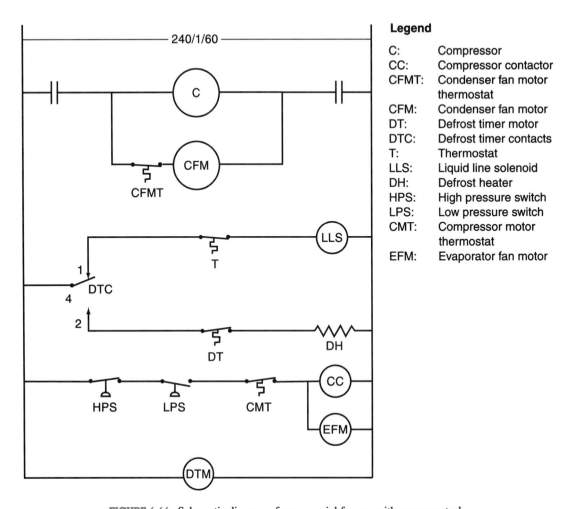

FIGURE 6.66 Schematic diagram of commercial freezer with pump control

motor is controlled by the condenser fan motor thermostat; it operates when the ambient temperature of the condenser reaches the set point of the condenser fan motor thermostat. The evaporator fan motor is started at the same time. The defrost timer motor is energized whenever power is supplied to the unit.

The defrost cycle of the system is initiated when the defrost timer contacts change to the 4 and 2 position. The defrost cycle is shown in Figure 6.68 with the active circuits highlighted. This action breaks the electrical connection to the liquid line solenoid valve, stopping the flow of refrigerant to the metering device. The compressor will continue to operate until most of the

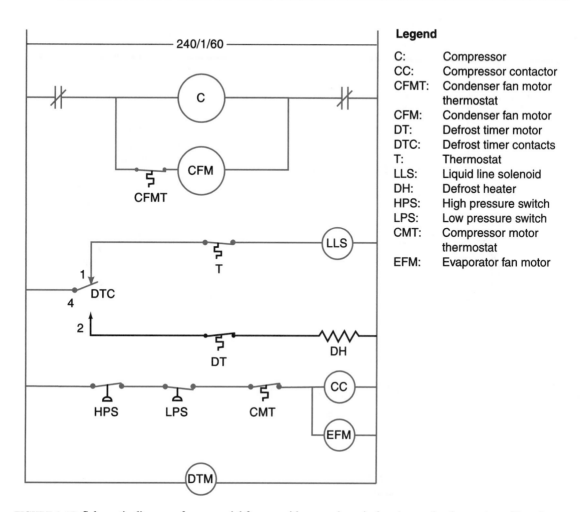

Legend

C:	Compressor
CC:	Compressor contactor
CFMT:	Condenser fan motor thermostat
CFM:	Condenser fan motor
DT:	Defrost timer motor
DTC:	Defrost timer contacts
T:	Thermostat
LLS:	Liquid line solenoid
DH:	Defrost heater
HPS:	High pressure switch
LPS:	Low pressure switch
CMT:	Compressor motor thermostat
EFM:	Evaporator fan motor

FIGURE 6.67 Schematic diagram of commercial freezer with pump down in freezing mode of operation with active circuits highlighted

refrigerant is captured in the high side of the system. When the defrost timer contacts are in the 4 and 2 position and the defrost thermostat is closed, the defrost heater will be energized to defrost the evaporator coil. The defrost cycle can be terminated when the defrost timer contacts (4 and 2) open. If the defrost thermostat opens, indicating a temperature increase in the evaporator, the defrost heater will be de-energized. However, this action will not put the system back into the freezing mode of operation. The defrost timer will continue to run until contacts 4 and 2 open and contacts 4 and 1 close, returning the system to normal operation.

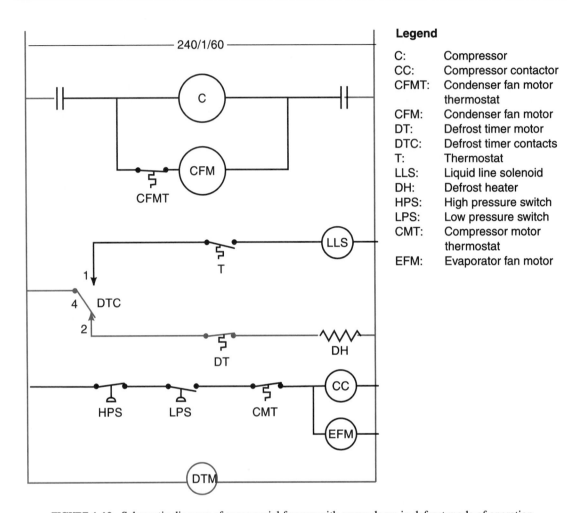

Legend

C: Compressor
CC: Compressor contactor
CFMT: Condenser fan motor
        thermostat
CFM: Condenser fan motor
DT: Defrost timer motor
DTC: Defrost timer contacts
T: Thermostat
LLS: Liquid line solenoid
DH: Defrost heater
HPS: High pressure switch
LPS: Low pressure switch
CMT: Compressor motor
        thermostat
EFM: Evaporator fan motor

FIGURE 6.68 Schematic diagram of commercial freezer with pump down in defrost mode of operation

SUMMARY

In this chapter, the schematics of several different applications were covered in detail from the simple electrical circuits that are used in window air conditioners and dehumidifiers to more advanced schematics such as those of heat pumps and light commercial air-conditioning systems. On most occasions schematic diagrams are not simple, but they are easy to understand if the technician looks at them circuit by circuit. The technician only needs to determine the part of the circuit that is not operating properly and then to direct attention to that circuit. It is essential that the technician develop the skills required to read schematic diagrams in order to adequately troubleshoot heating, air-conditioning, and refrigeration systems.

Schematic diagrams are laid out in a circuit-by-circuit arrangement. Each of these individual circuits is connected in parallel across two lines that represent the power supply. Most of these parallel circuits include a major load with the switches that control these loads connected in series. Technicians can easily troubleshoot a control system by utilizing a schematic diagram if they focus on the load and circuit that is at fault.

Small appliances that operate on 120 or 240 volts with line voltage control systems tend to be simple and easy to understand. Examples of these systems are dehumidifiers, window air conditioners, walk-in coolers, and commercial reach-in freezers. Not all 120- and 240-volt appliances are simple. Simplicity depends upon the number of loads and switches, and some of these appliances have many loads and switches.

Equipment that uses a low-voltage control system and requires a transformer and low-voltage components to control the loads in the system tends to be complex. The major difference is that line voltage, usually 240/1/60 or 208/240/3/60, supplies power to the equipment. The loads are supplied with the line voltage, and, in some cases, part of the control system also uses line voltage. The line voltage is supplied to a transformer that reduces the voltage for the primary control system. Low-voltage control systems are used because of their accuracy and safety.

No matter what type of control system is used in a heating, air-conditioning, or refrigeration system, it is essential that the technician be able to read schematic diagrams in order to efficiently troubleshoot the system.

REVIEW QUESTIONS

1. Why should an air-conditioning technician learn to read schematic diagrams?

2. Briefly explain the layout of a schematic diagram.

3. Most schematic diagrams are made up of _____.
   a. series circuits
   b. parallel circuits
   c. series-parallel circuits
   d. none of the above

4. Schematic diagrams break the wiring of control systems down into a(n) _____ arrangement.
   a. circuit-by-circuit
   b. series circuit
   c. individual circuit
   d. none of the above

5. Draw the symbols for the following components.

   a. heating thermostat
   b. cooling thermostat
   c. pressure switch (opens on rise)
   d. thermal line break overload
   e. fuse
   f. magnetic overload
   g. crankcase heater
   h. three-pole contactor
   i. relay with one NO and one NC contact
   j. red signal light
   k. three-phase compressor motor

6. What is the name of an appliance that is used to remove moisture?

   a. humidifier
   b. dehumidifier
   c. air conditioner
   d. both a and c

7. What is the position of the control switch if the window unit in Figure 6.14 is to operate the fan and compressor if the thermostat is closed?

   a. 4 and 1
   b. 4, 1, and 2
   c. 4, 2, and 3
   d. 4 and 3

8. True or False: The defrost timer motor operates continuously in Figure 6.18.

9. True or False: The evaporator fan motor is controlled by the defrost timer in Figure 6.18.

10. What electrical component initiates the defrost cycle of the commercial freezer shown in Figure 6.20?

    a. thermostat
    b. defrost timer
    c. defrost thermostat
    d. defrost heater

11. The thermostat of the commercial freezer in Figure 6.20 is connected in series with which of the following electrical devices?

    a. compressor
    b. condenser fan motor
    c. evaporator fan motor
    d. all of the above

12. What is the purpose of the fan switch in Figure 6.25?

13. If the temperature in the furnace schematic shown in Figure 6.25 were to increase to an unsafe condition, what electrical device would interrupt the power supply to the gas valve?

    a. fan switch
    b. heating thermostat
    c. limit switch
    d. all of the above

14. What is the purpose of the transformer in Figure 6.25?

15. True or False: The system in Figure 6.29 is using a line voltage control system.

Use Figure 6.29 to answer questions 16–24:

16. The HP and LP switches are connected in _____ with the contactor coil.
    a. series
    b. parallel

17. The contactor is controlled by _____.
    a. the disconnect switch
    b. the high-pressure switch
    c. the cooling thermostat
    d. none of the above

18. With the fan switch in the "on" position, the indoor fan will _____.
    a. run only when the thermostat is closed
    b. run when the temperature exceeds 95°F
    c. run continously
    d. none of the above

19. The compressor and condenser fan motor start _____.
    a. at separate times
    b. at the same time

20. The compressor and condenser fan motor are connected in _____.
    a. series
    b. parallel

21. Under what conditions would a control relay be used in a control system?

22. What is the purpose of a lockout relay?

23. Explain why the lockout relay coil energizes instead of the contactor coil.

24. True or False: The first- and second-stage cooling will be energized at the same time when a two-stage cooling thermostat closes on a call for cooling.

25. True or False: A two-stage cooling thermostat would be used on a system with one condensing unit.

Use Figure 6.46 to answer questions 26–29:

26. What is the purpose of the defrost cycle of a heat pump?

27. What two factors initiate the defrost mode in the diagram?

28. What type of thermostat is used with the heat pump in the diagram?

29. Explain the operation of the switchover valve and its purpose.

30. The reversing-valve solenoid is energized on the _____ cycle.
    a. heating
    b. cooling

31. The defrost cycle of the heat pump schematic in Figure 6.54 is initiated by _____.
    a. time
    b. temperature
    c. pressure
    d. both b and c

32. What is a pump down system and why is it used in commercial refrigeration system?

LAB MANUAL REFERENCE

For activities dealing with material covered in this chapter, refer to Chapter 6 in the Lab Manual.

7

Alternating Current, Power Distribution, and Voltage Systems

OBJECTIVES

After completing this chapter, you should be able to

- Explain the basic difference between direct and alternating current.
- Briefly explain how alternating current is produced.
- Explain the difference between single-phase and three-phase power distribution systems.
- Explain inductance, reactance, and impedance.
- Explain a basic power distribution system.
- Explain the common voltage systems.
- Identify the common voltage systems.

KEY TERMS

Alternator
Capacitive reactance
Delta system
Effective voltage
Frequency
Impedance
Inductance
Inductive reactance

Peak voltage
Phase
Power factor
Reactance
Sine wave
Single phase
Three phase
Wye system

INTRODUCTION

Two types of current are used in the heating, cooling, and refrigeration industry today: direct current (DC) and alternating current (AC). Current is the flow of electrons. Direct current is an electron flow in only one direction. Alternating current is an alternating (back and forth) flow of electrons; that is, the electrons reverse their direction of flow at regular intervals. Direct current will not be discussed in this chapter because it is used in the industry only for special applications.

Most current produced by electric utilities is alternating current. In the rare instances that direct current is needed, it can be produced by a direct current generator or a rectifier. It is usually produced by the consumer. Direct current has limited use in the industry; it is used mostly for refrigeration transportation equipment, electronic air cleaners, and electronic control components.

Alternating current is used in most heating, cooling, and refrigeration equipment. Alternating current equipment is cheaper and more trouble free than direct current equipment. In addition, alternating current is easier to produce than direct current.

Because of the popularity and wide use of alternating current, it is important for industry technicians to be familiar with its theory. In addition, technicians should be familiar with the way power is distributed by electric utilities and the many types of voltage-current systems available.

We begin our study with a discussion of some basic ideas of alternating current.

7.1 BASIC CONCEPTS OF ALTERNATING CURRENT

Alternating current, or AC, is an electron flow that alternates, flowing in one direction and then in the opposite direction at regular intervals. It is produced by cutting a magnetic field with a conductor.

The **sine wave** is often used as a graphical representation of alternating current. Figure 7.1 shows a sine wave, the graph of alternating current. The letter X represents a conductor in a particular position as it is rotated in the magnetic field. At point A, X is at a potential of 0. However, the potential (voltage) increases when the conductor is rotated through the magnetic field until it reaches point B, where the potential peaks. When the conductor (X) is rotated from B to C, the potential decreases until at point C the potential is again 0. The direction of flow from point A to point C is termed positive. The direction is reversed on the lower half of the sine wave (point C back to point A) and is termed negative. From point C the potential increases until it reaches point D and peaks. The potential decreases from point D until the conductor reaches point A, where the potential is again 0. The curve shows the conductor from point A through a complete clockwise revolution back to point A.

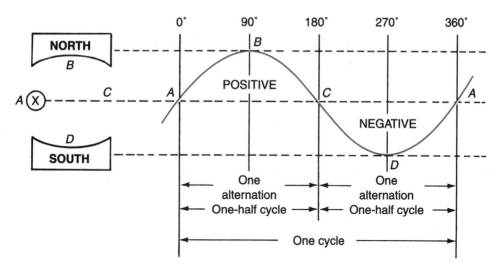

FIGURE 7.1 Sine wave of alternating current

While the conductor is rotating from point A to point C, one positive alternation occurs. That is, the curve starts from 0, reaches the peak at point B, and returns to 0 at point C. The next alternation is negative as X goes from point C back to point A in the same manner as the first alternation.

Cycles and Frequency

When the conductor rotates through one complete revolution, it has generated two alternations, or flow reversals. Two alternations (changes in direction) equal one cycle (refer again to Figure 7.1). One cycle occurs when the rotor, or conductor, cuts the magnetic field of a north pole and a south pole.

The frequency of alternating current is the number of complete cycles that occur in a second. The frequency in known as *hertz* (Hz), but many times it is referred to as *cycles.* In most locations in the United States, the common frequency is 60 hertz. In some isolated cases in the United States and in some foreign countries, 25 hertz is the frequency used. The disadvantage to using 25 hertz is that the reversals can be detected by the human eye in light fixtures. The frequency of 60 hertz is considered standard frequency, but 50 hertz is also used in some cases.

Effective Voltage

Because alternating current starts at 0, reaches a peak, and then returns to 0, there is always a variation in voltage and an effective value has to be

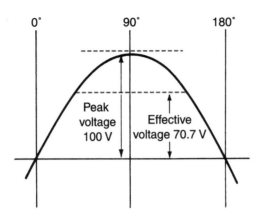

FIGURE 7.2 Effective voltage versus peak voltage

determined. Alternating current reaches a peak at 90 electrical degrees. This is known as **peak voltage.** The **effective voltage** of an alternating current circuit is 0.707 times its highest or peak voltage. This value is determined with respect to direct current so the effective voltage is equal to one direct current volt. Figure 7.2 shows a comparison of effective voltage versus peak voltage in an alternating current circuit. Alternating current amperage is 0.707 times the peak amperage. All electric meters are calibrated to read effective voltage and amperage.

Voltage-Current Systems

Alternating current is available to the consumer at several different voltages and with different current characteristics. The four basic voltage-current characteristics available are 240 volt–single phase–60 hertz, 240 volt–three phase–60 hertz, 208 volt–three phase–60 hertz, and 480 volt–three phase–60 hertz. These designations are often abbreviated; for example, 240 volt–single phase–60 hertz is abbreviated as 240 V/1ϕ-60 Hz. The V is the abbreviation for volts; ϕ (Greek letter phi) is the symbol for phase; and Hz is the abbreviation for hertz (which means the same as cycles).

Phase

The **phase** of an AC circuit is the number of currents alternating at different time intervals in the circuit. **Single-phase** current would have only a single current, while **three-phase** current would have three.

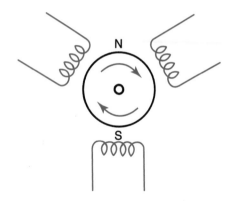

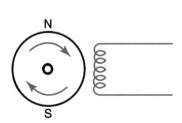

FIGURE 7.3 Winding arrangement of single-phase
alternator

FIGURE 7.4 Winding arrangement of three-phase alternator

Alternator

Alternating current is produced by an **alternator.** The alternator is made up
of a winding or set of windings called the stator and a rotating magnet
called the *rotor.* The number of windings used depends on the desired
phase characteristic of the current. When the rotor is rotated through the
field of the winding, an alternating electrical current is produced.

Single-phase current is produced by an alternator with one winding in the
stator as shown in Figure 7.3. Three-phase current is produced by an alternator with three windings in the stator that are wound 120° apart as shown
in Figure 7.4. The sine wave of three-phase current is shown in Figure 7.5.

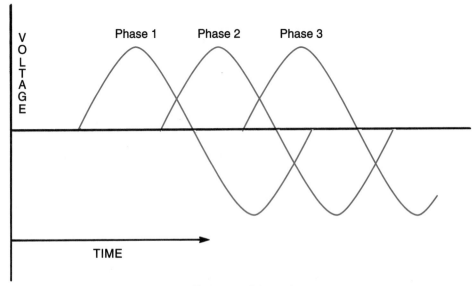

FIGURE 7.5 Sine wave of three-phase current

Inductance and Reactance

It is a common assumption that when the voltage is strongest, the amperage is also strongest. However, this is not always the case. The amperage of a circuit determines the strength of the magnetic field in the circuit. If the current increases, so will the magnetism. If the current decreases, so will the magnetism. The fluctuation of the magnetic strengths in an AC circuit, and in conductors cutting through more than one magnetic field, *induces* (causes) a voltage that counteracts the original voltage. This effect is called **inductance.**

The inductance in an AC circuit—that is, the effect of the magnetic fields—produces an out-of-phase condition between the voltage and amperage. The induction of the original voltage produces a second voltage due to the magnetic fields collapsing. This effect causes the voltage to lead the amperage, as shown in Figure 7.6.

Resistance in a direct current circuit is the only factor that affects the current flow. AC circuits are affected by resistance, but they are also affected by **reactance.** Reactance is the resistance that alternating current encounters when it changes flow. There are two types of reactance in AC circuits; inductive reactance and capacitive reactance. **Inductive reactance** is the opposition to the change in flow of alternating current, which produces an out-of-phase condition between voltage and amperage, as shown in Figure 7.6. **Capacitive reactance** is caused in AC circuits by

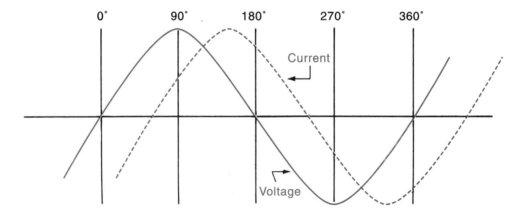

FIGURE 7.6 Sine wave of voltage and current when they are out of phase. Voltage leads the current in an AC circuit due to the effect of inductance.

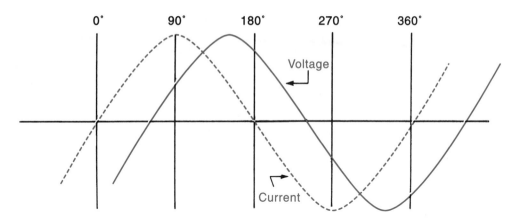

FIGURE 7.7 Sine wave of voltage and current when they are out of phase. Current leads the voltage in an AC circuit due to capacitive reactance.

using capacitors. When a capacitor is put in an AC circuit, it resists the change in voltage, causing the amperage to lead the voltage, as shown in Figure 7.7. The sum of the resistance and reactance in an AC circuit is called the impedance.

Power

As we have said, the voltage and current in an AC inductive circuit are out of phase with each other. When voltage and current are out of phase, they are not working together. Thus, the power (wattage) of the circuit must be calculated by using a voltmeter and an ammeter. This calculation gives the apparent wattage. A watt meter would measure the true power (wattage). The ratio between the true power and the apparent power is called the power factor and is usually expressed as a percentage. The following equation expresses this:

$$PF = \frac{\text{true power (measured)}}{\text{apparent power (calculated)}}$$

7.2 POWER DISTRIBUTION

Direct current was used in the beginning to supply consumers with their electrical needs, but it had many disadvantages. Transmission for a long distance without using generating stations to boost the direct current was

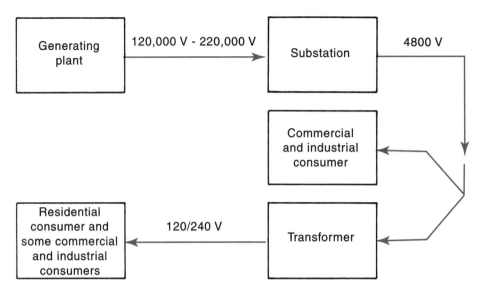

FIGURE 7.8 Layout of power distribution

impossible. The inability to raise and lower DC voltages and the need to use large transmission equipment were other problems. Alternating current can be transmitted with much less worry than direct current and has become the ideal power to supply to consumers because of its flexibility. Most equipment used in the industry incorporates alternating current as its main source of power.

Electric power is generated by rotating turbines through the use of gas, oil, coal, hydropower, or atomic energy. The rotating turbines have the effect of a rotating conductor in a magnetic field. The electricity is generated inside the plant and transferred outside the plant, where it is boosted to a large, easily transmitted voltage. This is frequently as high as 220,000 volts. The alternating current is then transmitted to a substation, where its voltage is reduced to around 4800 volts with the use of a step-down transformer. From the substation, the power is distributed to transformers that step down the voltage to a usable voltage. Or the power is supplied directly to consumers who use their own transformers to reduce the voltage. Figure 7.8 shows a diagram of electric power transmission from the generating plant to the customer.

In the following sections we will discuss the four common voltage systems available to consumers.

7.3 240 VOLT-SINGLE-PHASE-60 HERTZ SYSTEMS

Single-phase alternating current exists in most residences. Most small heating, cooling, and refrigeration equipment can be purchased for use on single phase. Any domestic appliance that operates on 120 volts is single-phase equipment. It is important to understand the makeup of the 240 V-1ϕ-60 Hz system because of its popularity in small commercial buildings and residences. Air-conditioning technicians must be able to determine the type of voltage system available to ensure proper installation and operation of the equipment.

In some older structures it is still possible to find a single-phase, two-wire system. This voltage system was used when few electric appliances were available and there was no need for 240-volt systems. The system uses two wires, supplying only 120 volts, one as a hot wire and one as a neutral. Hot wires are conductors that actually supply voltage to an appliance. The neutral is a wire connected to the ground and is identified by the color white. The ground wire is a safety circuit not intended to carry normal circuit current, but is provided to carry fault current to ground in the case of a short between the hot circuit conductor and the case or frame of a device. This conductor is either bare or colored green. In the voltage system shown in Figure 7.9, 120 volts is displaced through L1 but must be connected to a neutral to make a complete circuit.

The most common voltage system found today is the 240 V-1ϕ-60 Hz system. This voltage system consists of three wires, two hot wires and one ground. A schematic for the 240-volt system is shown in Figure 7.10. The

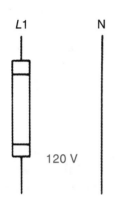

FIGURE 7.9 Schematic for 120 volt-single phase-60 hertz system using two wires

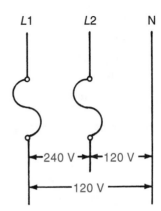

FIGURE 7.10 Schematic for 240 volt-single phase-60 hertz system using three wires

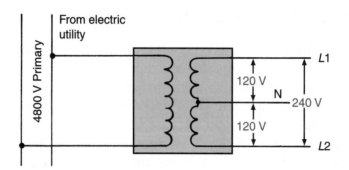

FIGURE 7.11 Transformer hookup of a 240 volt–single phase–60 hertz system

figure shows the system being used as a switch to break down the power being fed to a piece of equipment, such as a walk-in cooler, an electric furnace, or an air-conditioning unit. In this system, connecting either hot leg, $L1$ or $L2$, and the neutral directly to a load will supply 120 volts. Connecting $L1$ and $L2$ directly to a load will supply 240 volts.

The electric utility uses a transformer to produce the 240 V-1φ-60 Hz system. The transformer hookup of this system is shown in Figure 7.11. Pay close attention to the secondary side of the transformer. A transformer takes voltage at one value and by induction changes it to another value. Figure 7.11 shows a transformer with a 4800-volt primary winding and a 240-volt secondary winding. The primary winding, or input, is connected to the initial voltage, which is 4800 volts in Figure 7.11. The secondary winding produces the output, or new voltage, which is 240 volts in Figure 7.11.

Most equipment can be operated at plus or minus 10% of the rated voltage. For example, a piece of equipment rated at 240 volts could operate on a minimum voltage of 216 volts and a maximum voltage of 264 volts. The electric utility maintains a plus or minus 10% of the correct supply voltage. Air-conditioning equipment has a tendency to operate more satisfactorily on the maximum voltage than on the minimum voltage.

7.4 THREE-PHASE VOLTAGE SYSTEMS

Three-phase alternating current is common in most commercial and industrial structures. Three-phase electrical services supply three hot legs of power with one ground to the distribution equipment and then on to the equipment. Three-phase power supplies are more versatile than single-phase supplies because of the different voltage systems that are available. Thus, it is essential to understand the uses and advantages of three-phase voltage systems.

Three-phase electric power supplies have many advantages over the single-phase power supply when used in commercial and industrial applications. In most cases, residences do not use enough electric energy to warrant a three-phase power supply. The electric power consumer can buy three-phase electric power cheaper than single phase. Three-phase electric motors require no special starting apparatus, eliminating one trouble spot in the building and servicing of motors. Three-phase power offers better starting and running characteristics for motors than does single phase. Many large electric motors used in the industry are available for three-phase only, preventing many structures from using single-phase power supplies. Thus, most commercial and industrial structures are supplied with three-phase power.

The only disadvantage to three-phase systems is the higher cost of electric panels and distribution equipment.

There are basically two types of three-phase voltage systems used in commercial and industrial wiring systems: the delta transformer hookup, which will supply 240 V-3ϕ-60 Hz, and the wye transformer hookup, which will supply 208 V-3ϕ-60 Hz or 480 V-3ϕ-60 Hz.

7.5 240 VOLT-THREE-PHASE-60 HERTZ DELTA SYSTEM

The 240 V-3ϕ-60 Hz, or **delta system** is used in structures that require a large supply to motors and other three-phase equipment. The delta system is usually supplied to a structure with four wires, which include three hot legs and a neutral, but in some rare cases it is supplied with three hot wires only. Figure 7.12 shows the wiring layout of a 240 V-3ϕ-60 Hz system. This system is unique in that it contains a high leg, which in Figure 7.12 is $L1$. Connecting $L1$ to neutral, or ground, provides a range of 180 volts to 208 volts. Connecting $L1$ to $L2$, $L2$ to $L3$, or $L1$ to $L3$ provides 240 volts. Connecting $L2$ to N or $L3$ to N provides 120 volts. If one hot leg of a three-phase delta system is lost, only single-phase voltage is supplied and a single-phase condition exists. This condition can easily damage any three-phase equipment.

The transformer secondary hookup of a three-phase delta system is shown in Figure 7.13. The delta system takes its name from the Greek letter delta (Δ), which resembles the shape of the hookup, as can be seen in Figure 7.13. The high leg of the voltage occurs because the transformer winding between $L1$ and N is longer than the windings between $L2$ and N and between $L3$ and N.

The 240 V-3ϕ-60 Hz or delta voltage system is used primarily on systems that have many three-phase circuits of 240-volts and few 120-volt circuits.

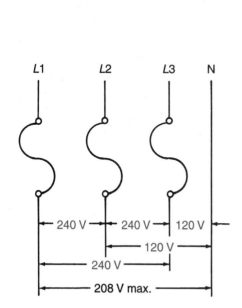

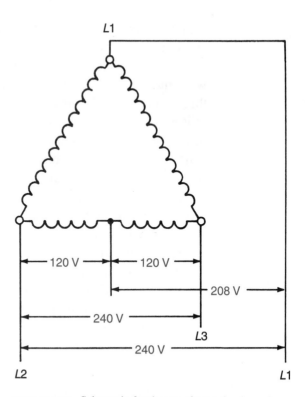

FIGURE 7.12 Schematic for 240 volt–three phase–60 hertz system using four wires

FIGURE 7.13 Schematic for the transformer hookup of a 240 volt–three phase–60 hertz system showing the delta transformer secondary hookup

The system consists of two available 120-volt legs when *L2* and *L3* are connected to a circuit with the neutral. If the high leg, or *L1*, is connected in a circuit with the neutral, it delivers between 180 volts and 208 volts. This will damage any 120-volt load or appliance.

Detection of the 240 V-3∅-60 Hz system can be accomplished simply and easily by testing across any two hot legs with a voltmeter. If 240 volts are read, the system is a delta system. The high leg should be identified by an orange marking. Another method of detection is to check the voltage between each hot leg and neutral with a voltmeter. If the voltage between any hot leg and neutral reads between 180 volts and 208 volts, the system is a delta system.

7.6 208 VOLT-THREE-PHASE-60 HERTZ WYE SYSTEM

The 208 V-3∅-60 Hz wye voltage system is common in structures that require a large number of 120-volt circuits, such as schools, hospitals, and office buildings. The **wye system** offers the versatility of using three-phase alternating current and the possibility of supplying many 120-volt circuits for lights, appliances, and other 120-volt equipment. The voltage ratings of some equipment are such that the equipment must be used with the right voltage system for proper operation and selection. Therefore, the detection of this system is important because of the lower-voltage requirements.

This system is supplied by four wires, one ground and three hot legs, as shown in schematic form in Figure 7.14. The 208-volt system is different from the 240-volt system in that it contains no high leg. In Figure 7.14, L1, L2, and L3 are the hot legs and N represents the ground. The voltage available between any two hot legs (L1, L2, or L3) is 208 volts. Connections between any hot leg (L1, L2, or L3) and the ground (N) provide 120 volts. As can be seen from the schematic, there are three available 120-volt power legs. This allows more 120-volt circuits than the 240-volt delta system.

The transformer secondary hookup of a 208 volt-three-phase wye system is shown in Figure 7.15. The wye system takes its name from the letter Y, which resembles the shape of the hookup, as can be seen in Figure 7.15. All three legs of the transformer windings are equal in winding length, as

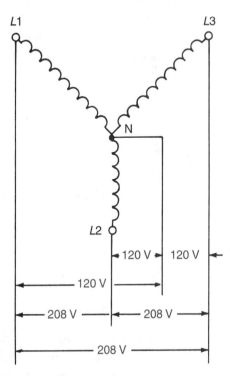

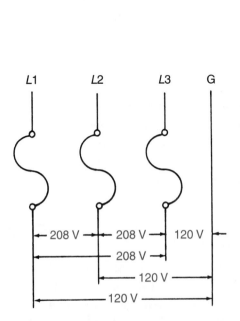

FIGURE 7.14 Schematic for 208 volt–three phase–60 hertz system using four wires

FIGURE 7.15 Schematic for the transformer hookup of a 208 volt–three phase–60 hertz system showing the wye transformer secondary hookup

shown in Figure 7.15, but in the delta system they are not. The wye transformer connection is also used on several higher-voltage systems.

Detection of the 208 V-3ϕ-60 Hz Y system is easy with the use of a voltmeter. If 208 volts are read across any two of the hot legs, or if 120 volts are read between all three hot legs and the ground, the system is a 208-volt wye system. The 208-volt wye system is a balanced system and contains no high leg that must be identified. But care should be taken to identify the ground when connections are being made.

7.7 HIGHER-VOLTAGE SYSTEMS

Higher-voltage systems are becoming increasingly popular because of their many advantages. The higher-voltage systems are used mostly in industrial structures, but in some cases they are used in commercial structures. It is necessary to understand and to be able to detect the higher-voltage system to ensure the safety of technicians and to prevent damage to the equipment.

The technician should take extra care when working around circuits with high voltage. Several high-voltage systems are available. There is a 240/480 volt-single phase system, a 240/416 volt–three phase system, and a 277/480 volt-single phase system. These systems can be identified with a voltmeter. For example, in the 277/480-volt system, the 277-volt reading is obtained between any hot leg and ground. The different systems are available with wye and delta hookups. The wye 277/480-volt system will be discussed in this section because of its popularity, but other high-voltage systems are available.

Advantages

Using higher-voltage systems has many advantages. There is little difference in the switches, relays, and other electric panels used in 208-volt and 480-volt systems. Motors may be wound differently for higher voltages, but they cost a little more. The service equipment and wiring may be smaller for a 480-volt system than for a 208-volt system. This might save the consumer a great deal of money.

Disadvantages

Disadvantages to using the higher-voltage systems stem mainly from problems that would be brought about by trying to implement a common high-voltage system in this country. If the consumer needs single-phase circuits with both 120 volts and 240 volts, additional transformers must be purchased. Single-phase equipment is more common than three phase.

Equipment manufacturers would like to work toward some common voltages so they could produce fewer types of equipment. Now they must manufacture many motors with different voltage ratings to meet consumers' needs.

The 277/480-Volt System

Figure 7.16 shows a schematic layout of the 277/480-volt wye system. This system is primarily used in industry, but occasionally there are commercial applications. The system has no means of supplying 120 volts or 240 volts at single phase without the use of a separate step-down transformer.

Between any hot leg ($L1$, $L2$, or $L3$) and neutral (N), 277-volt circuits are obtained. The 277-volt circuit is commonly used in commercial and industrial lighting systems to operate fluorescent lights. Between any two hot legs ($L1$ and $L2$, $L2$ and $L3$, or $L1$ and $L3$), 480 volts can be obtained. If three-phase is desirable, it can be obtained by connections to all three hot legs. Figure 7.17 shows the completely balanced transformer layout of a 277/480 V-3ϕ-60 Hz wye system.

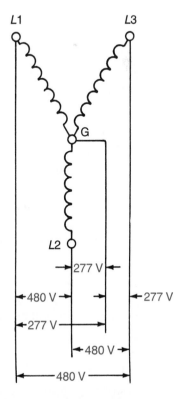

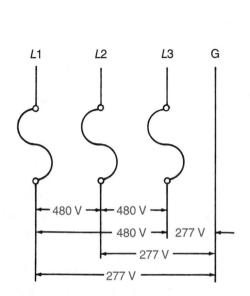

FIGURE 7.16 Schematic diagram for a 277/480 volt–three phase–60 hertz system

FIGURE 7.17 Schematic for the transformer hookup of a 277/480 volt–three phase–60 hertz wye system

Detection of the 480 V-3φ-60 Hz system is easy with the use of a voltmeter. If 480 volts are read across any two of the hot legs, or if 277 volts are read between the hot legs and the ground, the system is a 277/480-volt wye system. The 480-volt wye system is a balanced system and contains no high leg that must be identified. But care should be taken to identify the ground when connections are being made.

SUMMARY

Alternating current is used in most heating, cooling, and refrigeration equipment because its properties give it greater flexibility than direct current. It is easier to transmit over a long distance, and no expensive transmission equipment is required. It can be supplied at almost any voltage that a consumer wants.

Alternating current changes direction twice per cycle. The number of cycles per second is the frequency. Standard frequency is 60 hertz. The voltage leads the current in an AC circuit due to inductive reactance. The current leads the voltage due to capacitive reactance. The sum of the resistance and reactance in an AC circuit is called the *impedance.*

Voltage and current are out of phase in an AC circuit. Consequently, the power of the circuit must be calculated by using the effective wattage. The ratio between the true power and the effective power in an AC circuit is called the power factor.

An electric utility company can supply single-phase or three-phase current to structures at voltages from 208 volts to 480 volts. The generating plant supplies a high voltage that is stepped down so the consumer can use it. The common voltage characteristics are 240 V-1φ-60 Hz, 240 V-3φ-60 Hz, 208 V-3φ-60 Hz, and 480 V-3φ-60 Hz.

The 240 V-3φ-60 Hz, or delta, system is common throughout commercial and industrial structures. It adds efficiency to systems that have a large number of three-phase and 240-volt circuits. The system supplies three hot legs and one neutral, which deliver 240 volts between any two of the three hot legs. This system has a high leg. Caution should be taken not to connect any 120-volt circuit to the high leg. Between ground and the other two hot legs, 120 volts can be obtained. Detection of the delta system is essential to prevent damage to 120-volt equipment and to ensure proper equipment selection.

The 208-volt wye system is a well-balanced three-phase system that is commonly used on structures that need many 120-volt circuits as well as three-phase or 208-volt circuits. The system does not operate motors as efficiently as the 240-volt delta system, but it is more adaptable to the balancing of each leg of a service. Much equipment is made for either 208-volt or 240-volt systems, and some can be used on either system.

High-voltage systems are becoming increasingly common. In a high-voltage system smaller wire can be used, making the installation of equipment less expensive. While there are several high-voltage systems in use, the 277/480-volt wye system is most common. A large number of heating, cooling, and refrigeration systems are rated at higher voltages. Air-conditioning technicians should be able to identify the higher-voltage systems for safety and proper installation.

REVIEW QUESTIONS

1. Alternating current is _____.
   a. an alternating flow of electrons
   b. electrons reversing their flow at regular intervals
   c. the most commonly used current
   d. all of the above

2. What is the advantage of alternating current over direct current?

3. Direct current is used in which of the following applications in the industry?
   a. electronic control components
   b. magnetic starters
   c. three-phase motors
   d. none of the above

4. What is the common frequency of alternating current in the United States?
   a. 25 Hz
   b. 50 Hz
   c. 60 Hz
   d. none of the above

5. What is reactance?

6. What is impedance?

7. What is the sum of the resistance and reactance in an alternating current?
   a. induction
   b. alternation
   c. impedance
   d. none of the above

8. Give a brief description of the transmission of alternating current from the generating plant to the consumer.

9. Name the four voltage-current characteristics commonly used today.

10. The phase of alternating current is _____.

11. True or False: Inductance in an AC circuit is an effect that is due to the magnetic fields caused by the current flow.

12. What is the common electrical service to a residence?
    a. 240 volts-single phase-60 hertz
    b. 208 volts-three phase-60 hertz
    c. 240 volts-three phase-60 hertz
    d. 480 volts-three phase-60 hertz

13. What is the voltage range in which alternating current equipment can be operated?

14. The effective voltage of alternating current is _____ times its highest or peak voltage.

15. How can a 240 volt-single phase-60 hertz electrical system be detected?

16. What is the difference between single-phase and three-phase alternating current?

17. What is the voltage of a system with a wye transformer hookup?

18. What is the voltage of a system with a delta transformer hookup?

19. What are the advantages and disadvantages of a delta electric system?

20. How can the wye and delta systems be detected?

21. What is the advantage of the 480-volt wye system over the 208-volt wye system?

22. True or False: Single-phase alternating current is common in most commercial and industrial structures.

23. Sketch the delta and wye transformer hookup arrangements.

24. What are the advantages of the 208 volt–three phase–60 hertz system? In which types of structures is this system primarily used?

25. What is the 277 volt–single phase–60 hertz system used for?

LAB MANUAL REFERENCE

For experiments and activities dealing with material covered in this chapter, refer to Chapter 7 in the Lab Manual.

8

Installation of Heating, Cooling, and Refrigeration Systems

OBJECTIVES

After completing this chapter, you should be able to

- Understand the standard wire size as defined by the American Wire Gauge (AWG).
- Give the advantages and disadvantages of copper and aluminum conductors.
- Explain the factors that are considered when sizing an electrical circuit conductor.
- Correctly size and install electrical conductors for circuits used in the industry by the *National Electrical Code®* and manufacturers' instructions.
- Calculate the voltage drop in an electrical circuit.
- Explain the types of enclosures for disconnect switches that are available.
- Explain the types, sizes, and enclosures of disconnect switches that are used in the industry.
- Explain the types of electrical panels that are used to distribute electrical power to circuits in the structure.
- Install breakers in an electrical breaker panel.

KEY TERMS

American Wire Gauge
Breaker
Breaker panel
Disconnect switch
Distribution center

Fusible disconnect switch
Fusible load center
National Electrical Code®
Nonfusible disconnect switch

INTRODUCTION

The proper installation of heating, cooling, and refrigeration equipment is as important as any other phase of the industry. Installation covers a broad range of subjects, but one of the most important is the electric circuit servicing the equipment and its size. Thus, industry technicians should be familiar with the structure's circuitry and circuit components.

Once the electric utility delivers the power to the structure, the customer must bring it inside. Several kinds of electric panels are used in residences and many types are used in commercial and industrial structures to accomplish this. Hence, it is important for service and installation technicians to understand how electric power is distributed within the structure.

In this chapter we will discuss several components of the electric circuit servicing the heating, cooling, and refrigeration equipment. We will also discuss several types of electric panels that technicians may encounter on the job.

8.1 ■ SIZING WIRE

Manufacturers usually list in the installation instructions the correct wire and fuse size. But in many cases the person responsible for the installation must calculate the wire and fuse size. The *National Electrical Code®* governs the types and sizes of wire that can be used for a particular application and a certain amperage. The correct wire and fuse size is important to the life and efficiency of any equipment. Hence, the installation mechanic should know how to determine the size to use.

Copper is the most popular conductor in the industry. However, aluminum is used in some cases because of its low cost. Copper wire is a good conductor of electricity and has many other characteristics that contribute to its popularity. Copper wire bends easily, has good mechanical strength, resists corrosion, and can be easily joined together. Aluminum conductors, on the other hand, do not have all the good characteristics of copper. Aluminum conducts electricity well enough, but problems arise because aluminum corrodes easily. Thus, aluminum wire connections have a tendency to become loose after they have corroded.

Wire Size

Standard wire size is defined in the United States by the **American Wire Gauge** (AWG). The American Wire Gauge lists the largest wire, 0000 (4/0), down to number 50, which is the smallest wire. In the industry, wire sizes from number 20 to number 4/0 are the most common. The most popular sizes are from number 16 to number 4.

In some cases wire larger than 4/0 is needed. The circular mil system is used for this purpose. Circular mil sizing runs from 250 MCM (MCM is the abbreviation for 1000 circular mils), which is about 1/2 inch in diameter, to 750 MCM, which is about 1 inch in diameter. Circular mil sizing does exist in larger sizes. Figure 8.1 is a table that gives the data on round copper wire.

Factors to Consider in Wiring

The type of insulation surrounding the conductor usually determines its application and the amperage it can be used for. Insulation of different grades is used for different purposes. For example, insulation can be heat resistant, moisture resistant, heat and moisture resistant, or oil resistant. Figure 8.2 shows a partial table from the *NEC®* listing conductor application and insulations. *National Electrical Code®* and *NEC®* are registered trademarks of the National Fire Protection Association, Inc., Quincy, MA 02269.

Several factors should be considered when sizing circuit conductors. These factors are voltage drop, insulation type, enclosure, and safety. The voltage drop of a circuit, which takes into account the distance the conductors must be fed, must be calculated by the installation technician. The insulation type, the enclosure, and safety can be determined by using the tables of the *National Electrical Code®*. There are also wire-sizing tables in the *National Electrical Code®* that give the allowable amperage for both aluminum and copper conductors. The *National Electrical Code®* is considered a guide to safe wiring procedures. The code does not ensure that all systems following its procedures will be good systems, however. That is the responsibility of the designer.

Voltage Drop

Voltage drop in a conductor is of prime importance when sizing wire. Any voltage that drops between the supply and the equipment is lost to the equipment. If the voltage drop is large enough, it will seriously affect the operation of the equipment. But even a small voltage drop is detrimental to the equipment. The voltage drop can be easily measured when the equipment is operating by reading the voltage at the supply and subtracting from that the voltage read at the equipment. If we read 240 volts at the supply and 210 volts at the equipment when it is operating, then there is a voltage

American Wire Gauge (A.W.G.) Working Table (U.S. Bureau of Standards)*

| Gauge No. A.W.G. | Diameter in Mils | Cross Section | | Ohms per 1000 Ft. at 25 Deg. C. (77 Deg. F.) | Lb. per 1000 Ft. |
		Circular Mils	Square Inches		
0000	460	212000	.166	.0500	641
000	410	168000	.132	.0630	508
00	365	133000	.105	.0795	403
0	325	106000	.0829	.100	319
1	289	83700	.0657	.126	253
2	258	66400	.0521	.159	201
3	229	52600	.0413	.201	159
4	204	41700	.0328	.253	126
5	182	33100	.0260	.320	100
6	162	26300	.0206	.403	79.5
7	144	20800	.0164	.508	63.0
8	128	16500	.0130	.641	50.0
9	114	13100	.0103	.808	39.6
10	102	10400	.00815	1.02	31.4
11	91	8230	.00647	1.28	24.9
12	81	6530	.00513	1.62	19.8
13	72	5180	.00407	2.04	15.7
14	64	4110	.00323	2.58	12.4
15	57	3260	.00256	3.25	9.86
16	51	2580	.00203	4.09	7.82
17	45	2050	.00161	5.16	6.20
18	40	1620	.00128	6.51	4.92
19	36	1290	.00101	8.21	3.90
20	32	1020	.000802	10.4	3.09
21	28.5	810	.000636	13.1	2.45
22	25.3	642	.000505	16.5	1.94
23	22.6	509	.000400	20.8	1.54
24	20.1	404	.000317	26.2	1.22
25	17.9	320	.000252	33.0	0.970
26	15.9	254	.000200	41.6	0.769
27	14.2	202	.000158	52.5	0.610
28	12.6	160	.000126	66.2	0.484
29	11.3	127	.0000995	83.5	0.384
30	10.0	101.0	.0000789	105	0.304
31	8.9	79.7	.0000626	133	0.241
32	8.0	63.2	.0000496	167	0.191
33	7.1	50.1	.0000394	211	0.152
34	6.3	39.8	.0000312	266	0.120
35	5.6	31.5	.0000248	336	0.0954
36	5.0	25.0	.0000196	423	0.0757
37	4.5	19.8	.0000156	533	0.0600
38	4.0	15.7	.0000123	673	0.0476
39	3.5	12.5	.0000098	848	0.0377
40	3.1	9.9	.0000078	1070	0.0299

FIGURE 8.1 Data on round copper wire *(Courtesy of BICC Industrial Cable Company)*

Table 310.13 Conductor Application and Insulations (ROP 6-41,6-42, 6-43, 6-45) (*Continued*)

Trade Name	Type Letter	Maximum Operating Temperature	Application Provisions	Insulation	Thickness of Insulation			Outer Covering[1]
					AWG or kcmil	mm	Mils	
Extended polytetra-fluoro-ethylene	TFE	250°C 482°F	Dry locations only. Only for leads within apparatus or within raceways connected to apparatus, or as open wiring (nickel or nickel-coated copper only)	Extruded polytetra-fluoro-ethylene	14–10 8–2 1–4/0	0.51 0.76 1.14	20 30 45	None
Heat-resistant thermoplastic	THHN	90°C 194°F	Dry and damp locations	Flame-retardant, heat-resistant thermo-plastic	14–12 10 8–6 4–2 1–4/0 250–500 501–1000	0.38 0.51 0.76 1.02 1.27 1.52 1.78	15 20 30 40 50 60 70	Nylon jacket or equivalent
Moisture- and heat-resistant thermoplastic	THHW	75°C 167°F 90°C 194°F	Wet location Dry location	Flame-retardant, moisture- and heat-resistant thermo-plastic	14–10 8 6–2 1–4/0 213–500 501–1000	0.76 1.14 1.52 2.03 2.41 2.79	30 45 60 80 95 110	None
Moisture- and heat-resistant thermoplastic	THW[4]	75°C 167°F 90°C 194°F	Dry and wet locations Special applications within electric discharge lighting equipment. Limited to 1000 open-circuit volts or less. (size 14-8 only as permitted in 410.33)	Flame-retardant, moisture- and heat-resistant thermo-plastic	14–10 8 6–2 1–4/0 213–500 501–1000 1001–2000	0.76 1.14 1.52 2.03 2.41 2.79 3.18	30 45 60 80 95 110 125	None
Moisture- and heat-resistant thermoplastic	THWN[4]	75°C 167°F	Dry and wet locations	Flame-retardant, moisture- and heat-resistant thermo-plastic	14–12 10 8–6 4–2 1–4/0 250–500 501–1000	0.38 0.51 0.76 1.02 1.27 1.52 1.78	15 20 30 40 50 60 70	Nylon jacket or equivalent

FIGURE 8.2 Table of conductor application and insulation (*Reprinted with permission from NFPA 70-2002, the National Electrical Code® copyright © 2001, National Fire Protection Association, Quincy, MA 02269. This reprinted material is not the referenced subject which is represented only by the standard in its entirety.*)

Table 310.13 *Continued*

Trade Name	Type Letter	Maximum Operating Temperature	Application Provisions	Insulation	Thickness of Insulation			Outer Covering[1]
					AWG or kcmil	mm	Mils	
Moisture-resistant thermoplastic	TW	60°C 140°F	Dry and wet locations	Flame-retardant, moisture-resistant thermoplastic	14–10 8 6–2 1–4/0 213–500 501–1000 1001–2000	0.76 1.14 1.52 2.03 2.41 2.79 3.18	30 45 60 80 95 110 125	None
Underground feeder and branch-circuit cable — single conductor (For Type UF cable employing more than one conductor, *see* Articles 339, 340.)	UF	60°C 140°F 75°C 167°F [7]	See Article 340.	Moisture-resistant Moisture- and heat-resistant	14–10 8–2 1–4/0	1.52 2.03 2.41	60[6] 80[6] 95 [6]	Integral with insulation

FIGURE 8.2 *Continued*

drop of 30 volts in the circuit. The maximum recommended voltage drop for a branch circuit is 3%. The allowable voltage drop of most manufacturers is 10% below the nameplate rating.

Wire-Sizing Charts

Figure 8.3 shows the table from the *National Electrical Code®* for the allowable ampacities (amperages) of a conductor for copper and aluminum wire. The wire-sizing charts from the *NEC®* are usually accurate for sizing electric conductors unless the circuit is extremely long. For example, if a five-ton condensing unit draws 24 amperes, the service technician should add 25% of the full-load amperage to the total, which would be 6 amperes. This gives a total of 30 amperes for the wire-sizing data. From the table in Figure 8.3, we see that the circuit would require a copper conductor of No. 10 TW copper wire. Also from the table, we see the circuit would require a No. 8 TW aluminum conductor.

Figure 8.4 shows the manufacturer's electrical data for a piece of equipment. The wire sizes are given in the fourth column. Figure 8.5 shows the table of electrical characteristics of a specific model of equipment, but in

Table 310.16 Allowable Ampacities of Insulated Conductors Rated 0 Through 2000 Volts, 60°C Through 90°C (140°F Through 194°F), Not More Than Three Current-Carrying Conductors in Raceway, Cable, or Earth (Directly Buried), Based on Ambient Temperature of 30°C (86°F)

Size AWG or kcmil	Temperature Rating of Conductor (See Table 310.13.)						Size AWG or kcmil
	60°C (140°F)	75°C (167°F)	90°C (194°F)	60°C (140°F)	75°C (167°F)	90°C (194°F)	
	Types TW, UF	Types RHW, THHW, THW, THWN, XHHW, USE, ZW	Types TBS, SA, SIS, FEP, FEPB, MI, RHH, RHW-2, THHN, THHW, THW-2, THWN-2, USE-2, XHH, XHHW, XHHW-2, ZW-2	Types TW, UF	Types RHW, THHW, THW, THWN, XHHW, USE	Types TBS, SA, SIS, THHN, THHW, THW-2, THWN-2, RHH, RHW-2, USE-2, XHH, XHHW, XHHW-2, ZW-2	
	COPPER			ALUMINUM OR COPPER-CLAD ALUMINUM			
18	—	—	14	—	—	—	—
16	—	—	18	—	—	—	—
14*	20	20	25	—	—	—	—
12*	25	25	30	20	20	25	12*
10*	30	35	40	25	30	35	10*
8	40	50	55	30	40	45	8
6	55	65	75	40	50	60	6
4	70	85	95	55	65	75	4
3	85	100	110	65	75	85	3
2	95	115	130	75	90	100	2
1	110	130	150	85	100	115	1
1/0	125	150	170	100	120	135	1/0
2/0	145	175	195	115	135	150	2/0
3/0	165	200	225	130	155	175	3/0
4/0	195	230	260	150	180	205	4/0
250	215	255	290	170	205	230	250
300	240	285	320	190	230	255	300
350	260	310	350	210	250	280	350
400	280	335	380	225	270	305	400
500	320	380	430	260	310	350	500
600	355	420	475	285	340	385	600
700	385	460	520	310	375	420	700
750	400	475	535	320	385	435	750
800	410	490	555	330	395	450	800
900	435	520	585	355	425	480	900
1000	455	545	615	375	445	500	1000
1250	495	590	665	405	485	545	1250
1500	520	625	705	435	520	585	1500
1750	545	650	735	455	545	615	1750
2000	560	665	750	470	560	630	2000

CORRECTION FACTORS

Ambient Temp. (°C)	For ambient temperatures other than 30°C (86°F), multiply the allowable ampacities shown above by the appropriate factor shown below.						Ambient Temp. (°F)
21–25	1.08	1.05	1.04	1.08	1.05	1.04	70–77
26–30	1.00	1.00	1.00	1.00	1.00	1.00	78–86
31–35	0.91	0.94	0.96	0.91	0.94	0.96	87–95
36–40	0.82	0.88	0.91	0.82	0.88	0.91	96–104
41–45	0.71	0.82	0.87	0.71	0.82	0.87	105–113
46–50	0.58	0.75	0.82	0.58	0.75	0.82	114–122
51–55	0.41	0.67	0.76	0.41	0.67	0.76	123–131
56–60	—	0.58	0.71	—	0.58	0.71	132–140
61–70	—	0.33	0.58	—	0.33	0.58	141–158
71–80	—	—	0.41	—	—	0.41	159–176

* See 240.4(D).

FIGURE 8.3 Table for wire sizing (*Reprinted with permission from NFPA 70-2002, the National Electrical Code® copyright © 2001, National Fire Protection Association, Quincy, MA 02269. This reprinted material is not the referenced subject which is represented only by the standard in its entirety.*)

Recommended Wire Size for Air-Cooled Packaged Equipment

Unit Model Number	Voltage Characteristics	Minimum Circuit Ampacity	Minimum Wire Size (AWG)	Maximum Wire Length (ft)	Maximum Overcurrent Protection
ACP18	208/230-1-60	14	14	70	20
ACP24	208/230-1-60	20	12	80	30
ACP30	208/230-1-60	22	10	100	30
ACP36	208/230-1-60	28	10	90	45
ACP48	208/230-1-60	40	6	100	60
ACP48	208/230-3-60	25	10	85	40
ACP60	208/230-1-60	50	6	100	60
ACP60	208/230-3-60	35	8	95	50

FIGURE 8.4 Table of recommended wire size for an air conditioner

Electrical data

38AK007-012, 38AKS008-012

UNIT 38	NOMINAL VOLTAGE (V-Ph-Hz)	VOLTAGE RANGE*		COMPR		OFM	POWER SUPPLY	
		Min	Max	RLA	LRA	FLA	MCA	MOCP
AK007	208/230-3-60	187	254	19.0	142	1.9	25.6	35
	460-3-60	414	508	9.5	72	1.0	12.9	15
	575-3-60	518	632	7.6	58	1.9	11.4	15
AK008	208/230-3-60	187	254	25.0	185	3.1	34.4	45
	460-3-60	414	508	12.4	89	1.4	16.9	20
	575-3-60	518	632	10.4	78	1.4	14.4	15
AK012	208/230-3-60	187	254	34.5	239	3.1	46.2	60
	460-3-60	414	508	17.0	119	1.4	22.7	30
	575-3-60	518	632	14.3	90	1.4	19.3	25
AKS008	208/230-3-60	187	254	31.5	160	3.1	42.5	50
	380-3-60†	342	437	19.0	75	2.2	26.0	35
	460-3-60	414	508	15.7	80	1.4	21.0	25
	575-3-60	518	632	12.6	64	1.4	17.2	20
AKS009	208/230-3-60	187	254	39.7	198	3.1	52.7	70
	380-3-60†	342	437	24.0	93	2.2	32.2	40
	460-3-60	414	508	19.9	99	1.4	26.3	35
	575-3-60	518	632	15.9	79	1.4	21.3	25
AKS012	208/230-3-60	187	254	39.7	198	3.1	52.7	70
	380-3-60†	342	437	24.0	93	2.2	32.2	40
	460-3-60	414	508	19.9	99	1.4	26.3	35
	575-3-60	518	632	15.9	79	1.4	21.3	25

LEGEND

CSA	—	Canadian Standards Association
FLA	—	Full Load Amps
HACR	—	Heating, Air Conditioning and Refrigeration
LRA	—	Locked Rotor Amps
MCA	—	Minimum Circuit Amps
MOCP	—	Maximum Overcurrent Protection
OFM	—	Outdoor (Condenser) Fan Motor
RLA	—	Rated Load Amps

*Units are suitable for use on electrical systems where voltage supplied to the unit terminals is not below or above the listed limits.
†380-v units are export models not listed with UL or CSA.

NOTES:
1. In compliance with NEC (National Electrical Code) requirements for multimotor and combination load equipment (refer to NEC Articles 430 and 440), the overcurrent protective device for the unit shall be fuse or HACR breaker.
2. The MCA and MOCP values are calculated in accordance with the NEC, Article 440.
3. Motor RLA and LRA values are established in accordance with Underwriters' Laboratories (UL), Standard 1995.
4. The 575-v units are CSA-listed only.

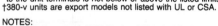

FIGURE 8.5 Table of electrical characteristics of a large condensing unit *(Courtesy of Carrier Corporation)*

this table the exact wire sizes are not given. However, the wire size amperages are given (the column headed MCA). The installation technician would use the wire size amperage column and the *NEC*® wire charts to size the wire.

Many items in the *NEC*® cover the sizing of electric conductors. Most technicians use the *NEC*® for reference. The preceding examples were given as a guide to show you how to use the *NEC*® charts.

CAUTION Use only electrical conductors that are the proper size for the load of the circuit according to the *NEC*® to avoid overheating and possible fire.

Calculating Voltage Drop

Figure 8.3 does not consider voltage drop, which must be calculated for long circuits. To calculate the voltage drop in a conductor, you must know the number of feet of wire that is used. The following formula is used to calculate voltage drop:

$$E = IR$$

For example, if No. 12 TW wire is to run 500 feet, what is the voltage drop over this distance when the supply voltage is 240 volts? First, we find the resistance of No. 12 wire from Figure 8.1. It is 1.6 ohms per 1000 feet. From Figure 8.3, we find the ampacity of No. 12 TW wire. It is 20 amperes. Substituting in the formula $E = IR$, we find $E = 20 \times .8$ (1.6 ohms per 1000 feet = .8 ohm per 500 feet). The voltage drop over 500 feet is 16 volts. Subtracted from the supply voltage of 240 volts, this gives 224 volts supplied to the equipment.

In most cases, the wiring that is run to heating, cooling, or refrigeration equipment will not exceed 75 feet to 100 feet. Then the wire size can be read directly from the table in Figure 8.3 because the voltage drop can be ignored on short-distance circuits. (There is a correction factor in the *NEC*® tables if the temperature exceeds 30°C.)

8.2 DISCONNECT SWITCHES

All heating, cooling, and refrigeration equipment should have some means for disconnecting the power supply at the equipment. The neutral conductor, if present, must not be disconnected by this device. Some equipment has a built-in method for disconnecting the power, such as a circuit breaker or fuse blocks. However, in most cases, a disconnecting device must be supplied and installed by an electrician or the installation technician. Disconnect switches are relatively simple and easy to install once the correct selection is made. Disconnect switches are used to provide a positive way to disconnect the power source to the equipment being serviced by the technician. The technician should lock the disconnect in the open position to prevent shock in case someone tries to close it.

A **disconnect switch** is a two- or three-pole switch mounted in an enclosure. The disconnect switch is basically a convenient, easy, and safe means of disconnecting power from the equipment for servicing and testing the equipment, or for safety purposes. The switch can be purchased with or without a space for fuses. Using fuses provides overcurrent protection for the conductors and equipment. In most cases, disconnect switches have a grounding lug mounted in the enclosure. The switches can have different arrangements. For example, a four-pole-three-fuse switch would be used on a three-phase circuit and would have a ground lug. A three-pole-three-fuse switch would be used for three-phase circuits and would not have a ground lug. A three-pole-two-fuse switch would be used on single-phase circuits and would have a ground lug. Figure 8.6 shows the schematics of two disconnect switches and Figure 8.7 shows these switches.

Disconnect switches can be purchased for general duty or heavy duty. The heavy-duty disconnect switch would be installed for equipment requiring frequent switching. The general-duty switch would be used for equipment requiring infrequent switching.

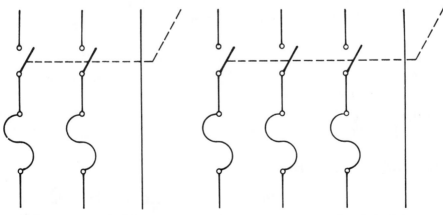

(a) 3-wire–2-fuse disconnect   (b) 4-wire–3-fuse disconnect

FIGURE 8.6 Schematic diagram of two fusible disconnect switches

(a) 3-wire–2-fuse disconnect   (b) 4-wire–3-fuse disconnect

FIGURE 8.7 Disconnect switches

Enclosures

The type of enclosure that the disconnect switch mounts in is determined by the conditions existing in the area of installation. A general type of enclosure could only be used where there were no problems of moisture, dust, or explosive fumes. A rain-tight disconnect enclosure could be used in areas of moisture but not where dust or explosive fumes exist. An explosion-proof enclosure could be used in any location, but it is much more expensive than other enclosures and hence is not used without reason.

Fusible and Nonfusible Switches

The purpose of a disconnect switch can be twofold. First, it can be used as a means of disconnecting the supply power going to the equipment. Second, it can be used as a safety device when fused correctly. If the only purpose of a disconnect switch is to break the power supply, then a **nonfusible disconnect switch** should be used. But if a means of protection for the wire or equipment is needed, a **fusible disconnect switch** should be used with the proper fuse sizes. Most equipment manufacturers will give the fuse sizes needed in the installation instructions. If fuse sizes are not given, the *National Electrical Code®* should be consulted.

The selection of a fusible disconnect switch is determined by duty, enclosure type, and size. Fuses are designed so that one size covers several different ampacities. The same size fuse can be purchased to cover from 1 to 30 amperes, from 30 to 60 amperes, from 70 to 100 amperes, and from 100 to 200 amperes. There are larger sizes available, but they are not used frequently. Disconnect switches are rated 30 amperes, 60 amperes, 100 amperes, 200 amperes, 400 amperes, and 600 amperes. A 30-ampere disconnect switch would be used for any load from 1 ampere to 30 amperes. A 200-ampere disconnect switch can be used with fuses from 100 amperes to 200 amperes. Other determining factors of the switches can easily be selected from a manufacturer's catalog.

Figure 8.8 shows a disconnect switch installed on a piece of equipment. In the installation of a disconnect switch, it is imperative that the wiring connections be made tightly. The line voltage wires should always be connected at the top of the enclosure. The load or wires supplying the unit

FIGURE 8.8 Disconnect switch installed on equipment

should be connected to the load side or bottom of the fuses. At no time should wires be installed in an enclosure without the proper connection.

 CAUTION The technician should make certain that all circuits are protected with properly sized fuses or breakers according to the *National Electrical Code.*®

8.3 FUSIBLE LOAD CENTERS

Fusible load centers, or **breaker panels,** are electric panels that supply the circuits in a structure with power and protect those circuits with fuses. Figure 8.9 shows a typical fusible load center and its schematic. Fusible load centers were popular until about 1968, when breaker panels gained

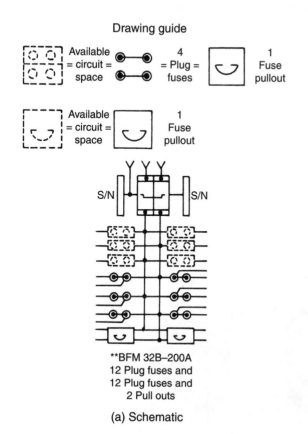

Drawing guide

(a) Schematic

(b) Load center

FIGURE 8.9 Fusible load center

prominence in the market. Air-conditioning technicians often find themselves working on fusible load centers when the owner of an older home decides to have air conditioning installed. Many fusible load centers are still in existence, so technicians must understand them and know how to make the correct connections.

The newest fusible load centers are built with the capacity of taking additional fuse blocks in the main lugs, which supply power to the entire panel, at any time if a space is available. This capacity allows the electrician to install the additional circuits required for any particular structure. It is relatively easy to add a circuit to the late-model panels by merely inserting the fuse blocks and screwing them in.

8.4 BREAKER PANELS

Breaker panels are usually installed in residences and industrial buildings today. **Breakers** are devices that detect any overload above their rating in a circuit and open the circuit automatically. The breaker must then be reset manually.

Breakers can be obtained for almost any application, no matter how large or small. Breakers are made with one, two, or three poles. The poles denote how many hot legs are being fed from the breaker to the appliance. A one-pole breaker supplies one hot leg and makes up a 120-volt–single-phase circuit. A two-pole breaker supplies two hot legs and makes a 240-volt–single-phase circuit. A three-pole breaker supplies three hot legs and makes up a three-phase circuit at the supplied voltage. Breakers are available with amperage ratings between one ampere and several hundred amperes, depending on the application. Figure 8.10 shows one-pole, two-pole, and three-pole breakers.

FIGURE 8.10 One-pole, two-pole, and three-pole breakers

Construction

Breaker panels are built in several different forms, but they all serve the same purpose. Different manufacturers build breaker panels that are similar in design, but the breakers from different manufacturers do not usually fit each others' panels. However, several manufacturers' equipment does interchange.

Breaker panels are built with or without main breakers. A main breaker installed in the breaker panel provides a main switch in the panel and adds a means of overload protection for the entire panel. A breaker panel can be obtained with main lugs and no main breaker, but the breaker should have some means of protection. Breaker panels are rated by how many amperes the main lugs can carry and by the rating of the main breaker. Figure 8.11 shows the main lugs of a breaker panel with a main breaker.

Breaker panels are built for use with single-phase or three-phase systems and for 250 volts or 600 volts. The breaker panel in the average residence is rated at 150 amperes or 200 amperes and is a general-duty type. The breakers snap into a residential panel (as they do in some commercial and industrial panels). Figure 8.12 shows a residential breaker panel and its schematic.

FIGURE 8.11 Breaker panel showing main lugs and main breaker *(Courtesy of Square D Company)*

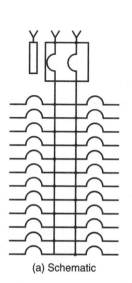

(a) Schematic

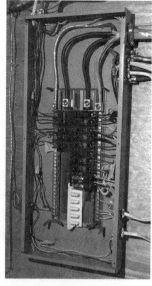

(b) Panel

FIGURE 8.12 Residential breaker panel (a) schematic (b) panel

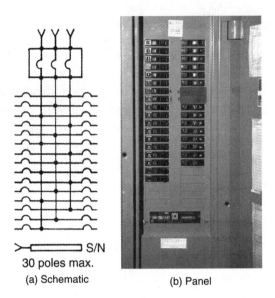

FIGURE 8.13 Industrial breaker panel

The commercial and industrial breaker panel can meet almost any specifications that the customer requires. The commercial and industrial panels are built for more rugged duty than are the usual residential panels. Most commercial and industrial panels use breakers that attach to main lugs with screws. Figure 8.13 shows a typical breaker panel, along with its schematic, used in commercial and industrial applications.

Installation

The installation of a breaker into a breaker panel causes little or no trouble. Breakers connect in some panels by attachment to the main lugs with clips. Other breakers are attached to the main lugs by screws. Technicians should be familiar with breakers and breaker panels to order the correct breaker for a particular use.

In most cases, breaker panels have some open spaces. If a situation arises where there is no spare opening, a breaker with two circuits can replace a standard one circuit breaker and is shown in Figure 8.14. Most manufacturers produce this type of breaker.

Figure 8.15 shows a breaker being installed in a breaker panel. Care should be taken to ensure the breaker is correctly aligned with the proper clip in the panel. When installing a breaker that is attached to the lug with a screw, make certain that the connection is tight. Before installing a breaker in a panel, you should cut off the power supply to the panel.

FIGURE 8.14 A two-circuit breaker used to replace a normal-sized breaker

FIGURE 8.15 Breaker being installed in a breaker panel

 CAUTION Wear shoes that have insulating soles and heels that are in good condition.

On rare occasions circuit breakers are faulty. Either they cannot be reset or they open the circuit on a lower amperage than the rating. If either condition exists, the breaker must be replaced. A breaker can be checked for resetting if a voltage reading is taken between the ground and the breaker. If voltage is available to the load side of the breaker under load, the breaker is good. In rare cases, a breaker could be stuck in the closed position, and if so, it should be replaced. The breaker that trips at a lower-than-rated amperage should be checked with an actual ampere reading of the circuit.

8.5 DISTRIBUTION CENTERS

Distribution centers are designed to distribute the electrical supply to several places in a large structure. Their use is largely confined to commercial and industrial applications. Figure 8.16 shows schematically the layout of a distribution system in a structure. In Figure 8.16, Panel A is the main distribution point between the electric service and other panels and the heating and air-conditioning equipment.

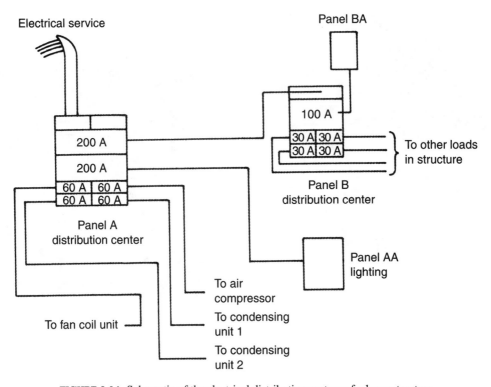

FIGURE 8.16 Schematic of the electrical distribution system of a large structure

FIGURE 8.17 Modern distribution panel
(Courtesy of Square D Company)

Figure 8.17 shows a modern distribution panel. In a very large structure, it would not be unusual to find several distribution centers. A distribution center often saves a great deal of money because many circuits are shortened and only the main circuit is lengthy. Distribution centers can be of the fusible or circuit-breaker design. The fusible design is the more popular mainly because of its moderate cost. Larger breakers are extremely expensive and hence are not used often.

8.6 INSTALLING ELECTRICAL CIRCUITS FOR REFRIGERATION, HEATING, AND AIR-CONDITIONING EQUIPMENT

The installation of refrigeration, heating, and air-conditioning equipment requires a power source being supplied to the equipment for proper operation. In almost all cases this power source will be alternating current being supplied through conductors from the power distribution equipment in the structure. Common power distribution equipment in structures today are fusible load centers, breaker panels, or large distribution centers. Fusible load centers and medium-duty breaker panels are used in most residences to distribute the electrical power to individual circuits within the structure. Distribution panels and heavy-duty breaker panels are used in most commercial and industrial applications for power distribution.

In addition to the power source there must be a path for the electrical energy to flow. It is the responsibility of the installation mechanic to determine the correct type and size of conductor that is to be used to supply the electrical energy to the equipment. The mechanic must also determine what type of conduit or covering, if any, is to be used to protect the conductors. Many different types of conduit are available to protect electrical conductors from the elements, including electrical metallic tubing, rigid conduit, plastic flexible conduit, polyvinylchloride (PVC) conduit, and many others.

The electrical circuit supplying power to the equipment must have a means to protect the wire and a method to disconnect the power when service is required. Overcurrent protection could come in the form of a circuit breaker or fuse. In order to protect the conductors supplying the electrical energy and the equipment, the installation technician will have to correctly size and install the chosen method of protection. The selection of the protective devices should meet the *National Electrical Code*® as well as local and state codes. In most applications, some form of a disconnect switch is used to interrupt the power supply to the equipment in the event that equipment needs service or needs to be shut down.

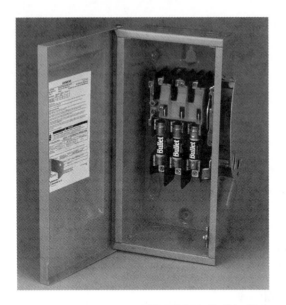 

FIGURE 8.18 Fusible disconnect switches used in the industry.

CAUTION Make sure that extension cords being used are in good condition.

CAUTION Use properly grounded tools connected to properly grounded circuits.

There are many different types of disconnect switches that are used in this industry. One of the most popular types is the fusible disconnect switches shown in Figure 8.18. Other types of disconnects are used in the industry and vary widely in style and design. All disconnects serve the same purpose and basically are wired in the same manner. The incoming energy source is connected to the $L1$ and $L2$ or line terminals in the disconnect, while the supply to the equipment is connected to the T_1 and T_2 or load terminals in the disconnect.

CAUTION Make sure all electrical connections are tight.

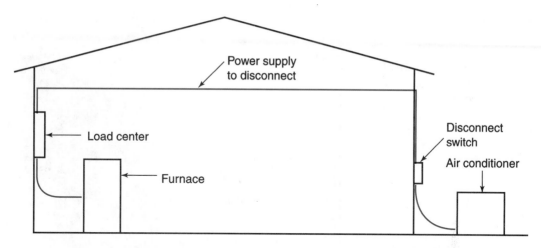

FIGURE 8.19 Electrical circuit between a circuit breaker load center and selected equipment

When installing a piece of air-conditioning equipment, the technician must determine the location of electrical energy that will be supplied to the selected equipment. In most residences this supply will be from a fusible load center or a circuit breaker load center as shown in Figure 8.12. In commercial and industrial applications, a circuit breaker panel or distribution panel will supply the electrical energy to the selected equipment. Once the source location has been determined the technician must determine how to get the electrical energy from the source to the equipment. Weather-resistant cable or conduit should be used from the disconnect to the outdoor unit. All codes that are applicable, the *National Electrical Code®*, and all state and local codes should be followed in the installation and sizing of the conductors supplying the load. The mechanic will have to make the proper connections at the power source whether it is service entrance equipment or a disconnect switch. From the source, conductors will have to be connected between the source and equipment with connections made at the load (equipment). Figure 8.19 shows a drawing of the electrical circuit between the power source—in this case, a circuit breaker load center—and selected equipment. Figure 8.20 shows a drawing of the electrical circuit between the power source—in this case, a disconnect switch—and selected equipment. Figure 8.21 shows an electrical connection between the disconnect and outdoor unit. Figure 8.22 shows the basic circuit requirements for the electrical connections of a residential air-conditioning or heat pump condensing unit.

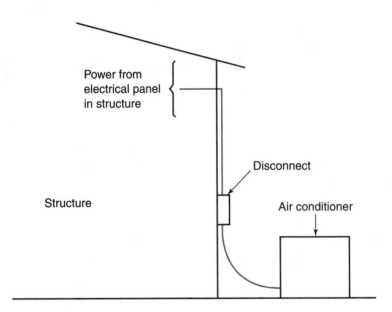

FIGURE 8.20 Electrical circuit between a disconnect and selected equipment

FIGURE 8.21 Photograph of an electrical connection between the disconnect and outdoor unit

> **CAUTION** Technicians should not stand in damp or wet areas when working with electric circuits with the power on.

Depending upon geographic location, in some cases the heating and air-conditioning technicians can only make connections to the equipment, and other electrical connections must be made by licensed electricians. This is determined by local and state codes.

The installation mechanic should follow all applicable codes as well as the installation instructions furnished by the manufacturer. Installation diagrams are usually furnished by the manufacturer and are located inside the equipment.

> **CAUTION** When working with ladders, make sure that they are not allowed to come in contact with power lines.

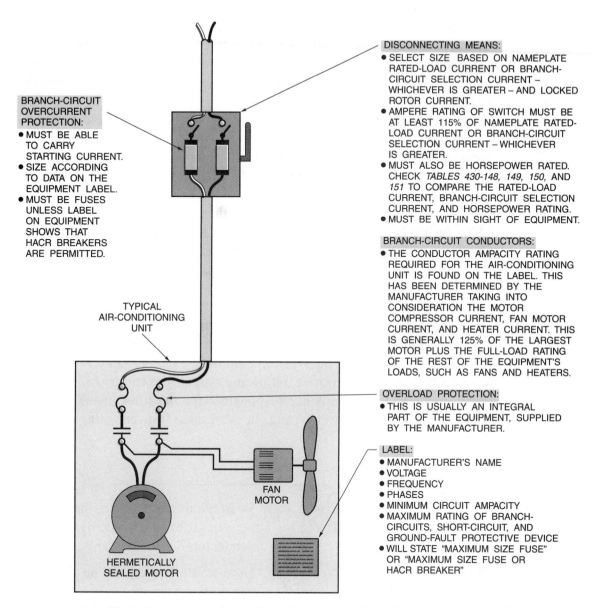

DISCONNECTING MEANS:
- SELECT SIZE BASED ON NAMEPLATE RATED-LOAD CURRENT OR BRANCH-CIRCUIT SELECTION CURRENT – WHICHEVER IS GREATER – AND LOCKED ROTOR CURRENT.
- AMPERE RATING OF SWITCH MUST BE AT LEAST 115% OF NAMEPLATE RATED-LOAD CURRENT OR BRANCH-CIRCUIT SELECTION CURRENT – WHICHEVER IS GREATER.
- MUST ALSO BE HORSEPOWER RATED. CHECK *TABLES 430-148, 149, 150,* AND *151* TO COMPARE THE RATED-LOAD CURRENT, BRANCH-CIRCUIT SELECTION CURRENT, AND HORSEPOWER RATING.
- MUST BE WITHIN SIGHT OF EQUIPMENT.

BRANCH-CIRCUIT OVERCURRENT PROTECTION:
- MUST BE ABLE TO CARRY STARTING CURRENT.
- SIZE ACCORDING TO DATA ON THE EQUIPMENT LABEL.
- MUST BE FUSES UNLESS LABEL ON EQUIPMENT SHOWS THAT HACR BREAKERS ARE PERMITTED.

BRANCH-CIRCUIT CONDUCTORS:
- THE CONDUCTOR AMPACITY RATING REQUIRED FOR THE AIR-CONDITIONING UNIT IS FOUND ON THE LABEL. THIS HAS BEEN DETERMINED BY THE MANUFACTURER TAKING INTO CONSIDERATION THE MOTOR COMPRESSOR CURRENT, FAN MOTOR CURRENT, AND HEATER CURRENT. THIS IS GENERALLY 125% OF THE LARGEST MOTOR PLUS THE FULL-LOAD RATING OF THE REST OF THE EQUIPMENT'S LOADS, SUCH AS FANS AND HEATERS.

TYPICAL AIR-CONDITIONING UNIT

OVERLOAD PROTECTION:
- THIS IS USUALLY AN INTEGRAL PART OF THE EQUIPMENT, SUPPLIED BY THE MANUFACTURER.

LABEL:
- MANUFACTURER'S NAME
- VOLTAGE
- FREQUENCY
- PHASES
- MINIMUM CIRCUIT AMPACITY
- MAXIMUM RATING OF BRANCH-CIRCUITS, SHORT-CIRCUIT, AND GROUND-FAULT PROTECTIVE DEVICE
- WILL STATE "MAXIMUM SIZE FUSE" OR "MAXIMUM SIZE FUSE OR HACR BREAKER"

FAN MOTOR

HERMETICALLY SEALED MOTOR

FIGURE 8.22 The basic requirements for the electrical connection of a residential air conditioner or heat pump

SUMMARY

For proper installation and maintenance of heating, cooling, and refrigeration equipment, industry technicians should be familiar with the electric circuits and circuit components of a structure. The life and safety of the equipment depend on the use of the correct size of wiring. Voltage drop, insulation type, enclosure, and safety are the determining factors in wire sizing. Manufacturers usually provide instructions on the size of wire to use with their equipment, but in some cases the technician must calculate the size. The *National Electrical Code®,* is a guide that should be used in sizing wire properly.

Various types of electric panels are used in the industry. The disconnect switch, sometimes called the safety switch, is commonly used on equipment, along with another electric panel. In most structures, the electric panels are not located close enough to the equipment to be considered safe for disconnecting the equipment. Therefore, in most cases disconnect switches should be installed on or close to the equipment.

The fusible load center or breaker panel is used in most residences. The breaker panel is also popular in commercial and industrial panels. Breakers are designed to trip or break the circuit when an overload occurs. Some breakers clip to the main lugs; others screw to them.

The distribution center is used in commercial and industrial plants as a means of distributing power to other electric panels in the structure. Service technicians should be familiar with power distribution because of the many voltage ranges of modern equipment. The mechanic or installer must be able to pick up power out of any type of electric panel. In many areas, industry technicians are responsible for the total installation of equipment, including power wiring. In other areas, power wiring must be done by an electrician.

REVIEW QUESTIONS

1. Which of the following characteristics apply to copper being used as a conductor?

   a. good mechanical strength
   b. ability to resist corrosion
   c. easily joined together
   d. all of the above

2. Which of the following is a disadvantage of using aluminum wire?

   a. conduction of electricity
   b. corrodes easily
   c. thicker insulation necessary
   d. all of the above

3. What are the four factors that should be considered in sizing circuit conductors?

4. How would you determine the voltage drop in a circuit?

5. What is a disconnect switch?

6. The most popular conductor used in the industry is _____.

7. On what application would you use the following types of electrical enclosures?

   a. general duty
   b. waterproof
   c. explosion proof

8. What is the difference between a fusible disconnect switch and a non-fusible disconnect switch?

9. What determines the type of insulation surrounding the conductor?

   a. application
   b. size
   c. number of conductors
   d. none of the above

10. How many hot legs would a three-pole breaker supply to an appliance?

11. When determining the size and installation of an electrical circuit, which of the following guides should be followed?

    a. *National Electrical Code*®
    b. local codes
    c. state codes
    d. all of the above

12. What is the purpose of a disconnect switch?

13. What is the difference between a fusible load center and a breaker load center? Which is primarily used in the industry today?

14. What is a circuit breaker?

15. How can a faulty circuit breaker be detected?

16. How are circuit breakers attached to the main lugs of an electrical panel?

17. What is a distribution center and what is its primary purpose?

18. A distribution center is used in which of the following applications?

    a. residential
    b. commercial
    c. industrial
    d. both b and c

19. Which of the following are true about the American Wire Gauge?

    a. smallest wire is No. 12
    b. largest wire is No. 4/0
    c. wires larger than 4/0 are sized in circular mils
    d. both b and c

20. True or False: Fusible load centers are usually installed in residences today.

21. What connects the breaker to the source of power in a breaker panel?

    a. breaker clips
    b. breaker connects directly to incoming power source
    c. main lug
    d. breaker screws

22. What would you do in a situation where there is no spare opening in a breaker panel for the installation of equipment?

23. True or False: In all areas of the United States, air-conditioning technicians can install the conductor from the distribution panel to the equipment.

24. What size wire would be required for a circuit requiring 100 amperes?

25. What would be the voltage drop in an electrical circuit if the current draw was 65 amps and the circuit was 750 feet long with No. 6 TW conductor?

26. What size wire would be required for the power supply of the following packaged unit?

    Compressor - 45 amps
    Indoor fan motor - 6 amps
    Condenser fan motor - 6 amps

27. What is the purpose of a main breaker?

28. What size wire would be required for an AK012 with a power supply of 208/240-3ϕ-60 Hz with the specifications given in Figure 8.5?

29. What size wire would be used for a fan motor if the figure for full load amps of the motor is 14 amperes?

30. What is the resistance of a No. 4 wire if the circuit is 500 feet long?

LAB MANUAL REFERENCE

For experiments and activities dealing with material covered in this chapter, refer to Chapter 8 in the Lab Manual.

Basic Electric Motors

OBJECTIVES

After completing this chapter, you should be able to

- Explain magnetism and the part it plays in the operation of electric motors.
- Explain torque and the purpose of different types of single-phase motors.
- Explain the operation of a basic electric motor.
- Understand how to operate, install, reverse the rotation, if possible, and diagnose problems in a shaded-pole motor.
- Understand the purpose of capacitors in the operation of a single-phase motor and be able to explain the difference between a starting and running capacitor.
- Correctly diagnose the condition of any capacitor and, using capacitor rules, be able to substitute a capacitor if a direct replacement is not available.
- Explain how to operate, install, troubleshoot, and repair (if possible) split-phase and capacitor-start motors.
- Explain how to operate, install, troubleshoot, and repair (if possible) permanent split-capacitor motors.
- Explain how to operate, install, troubleshoot, and repair (if possible) capacitor-start–capacitor-run motors.
- Understand how to operate, install, reverse, and troubleshoot three-phase motors.
- Identify the common, start, and run terminals of a single-phase compressor motor.

KEY TERMS

Capacitor
Capacitor-start motor
Capacitor-start–capacitor-run
   motor
Delta winding
Electromagnet
Flux
Hermetic compressor
Induced magnetism
Magnetic field
Magnetism
Microfarad

Permanent magnet
Permanent split-capacitor motor
Rotor
Running capacitor
Shaded-pole motor
Split-phase motor
Squirrel cage rotor
Star winding
Starting capacitor
Stator
Three-phase motor
Torque

INTRODUCTION

The electric motor changes electric energy into mechanical energy. Motors are used to drive compressors, fans, pumps, dampers, and any other device that needs energy to power its movement.

There are many different types of electric motors with different running and starting characteristics. Most single-phase motors are designed and used according to their running and starting torque. **Torque** is the strength that a motor produces by turning, either while starting or running. This chapter covers most types of motors available today and how they are used in the heating, cooling, and refrigeration industry. All electric motors should be properly grounded.

We begin our study with a discussion of magnetism, an effect that is needed to operate motors, relays, contactors, and other electric devices.

9.1 MAGNETISM

Magnetism is the physical phenomenon that includes the attraction of an object for iron and is exhibited by a permanent magnet or an electric current. Magnetism is produced in many different ways, but regardless of how it is produced, the effect is basically the same. The magnetic field of the earth, for example, is the same as the magnetism in a horseshoe magnet, the magnetism produced by a transformer, and the magnetism produced by an electromagnet. A good example of magnetism is the ability of a horseshoe magnet to pick up articles made of iron. The most common example of magnetism is the reaction of a compass to the Earth's magnetic field.

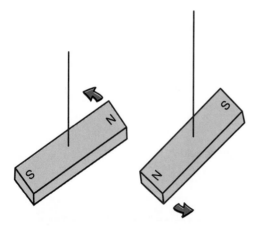

FIGURE 9.1 Repulsion of like poles of two bar magnets

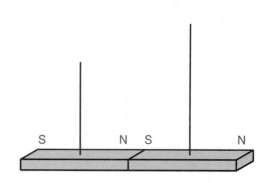

FIGURE 9.2 Attraction of unlike poles of two bar magnets

All magnets have two poles, a north pole and a south pole. If the north pole of a bar magnet is brought close to the north pole of another bar magnet, they will repel each other, as shown in Figure 9.1. If the south pole of one bar magnet is brought close to the north pole of another bar magnet, they will attract each other and come together, as shown in Figure 9.2. Therefore, like poles of magnets repel each other and unlike poles attract.

Magnetic Field

The magnetic lines of force of a magnet that flow between the north and south poles are called **flux**. These lines of force are highlighted in Figure 9.3. The area that the magnetic force operates in is called a **magnetic field**. Magnetic fields can flow through materials, depending on the strength of the magnetic field. A magnetic field is best conducted through soft iron. That is why certain parts of motors and other electric devices are made of soft iron.

Induced Magnetism

Induced magnetism is created when a piece of iron is placed in a magnetic field. The important fact to remember about a magnetic field is that the closer an object is to the magnet, the stronger the magnetic field is on that object. Therefore, if we insert an iron bar within two or three inches of a magnetic field, we induce a stronger field than if we placed the bar six inches from the field.

Two types of magnets are in use today: the permanent magnet and the electromagnet. The **permanent magnet** is a piece of magnetic material that has been magnetized and can hold its magnetic strength for a reasonable

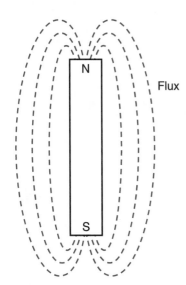

FIGURE 9.3 Magnetic field of a bar magnet

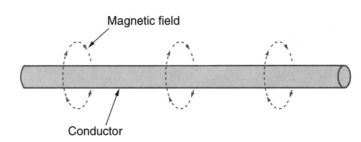

FIGURE 9.4 Magnetic field created around a current-carrying conductor

length of time. The permanent magnet must be made of a magnetic material, such as iron, nickel, cobalt, or chromium. Some nonmagnetic materials such as glass, rock, wood, paper, and air cannot be made magnetic but can be penetrated by a magnetic field.

The **electromagnet** is a magnet produced through electricity. When an electron flow is in a conductor, a magnetic field is created around the conductor, as highlighted in Figure 9.4. The larger the electron flow, the stronger the magnetic field. Therefore, if we take an iron core and wind a current-carrying conductor around it, the iron core will become a magnet, as shown in Figure 9.5. In Figure 9.5 the magnetic field is highlighted. The electron flow and the number of turns of the conductor around the core determine the strength of an electromagnet. Figure 9.6 shows an electromagnet that is used as a solenoid in a contactor.

Magnetism is important in the heating, cooling, and refrigeration industry because of its many uses in the operation of electric devices. Motors require magnetism to create a rotating motion. Relays and contactors use magnetism to open and close a set of contacts. All of the devices discussed in this chapter use magnetism in some way.

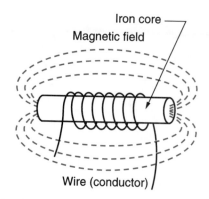

FIGURE 9.5 Magnetic field of an iron core when a current-carrying conductor is wound around the core

FIGURE 9.6 An electromagnet used as a solenoid in a contactor

9.2 BASIC ELECTRIC MOTOR

Electric motors are common devices in the heating, cooling, and refrigeration industry. Motors are used to create a rotating motion and drive components that need to be turned. Motors power compressors, pumps, fans, timers, and any other device that must be driven with a rotating motion.

In an electric motor, electric energy is changed to mechanical energy by magnetism, which causes the motor to turn. The method by which magnetism causes motors to rotate uses the principle that like poles of magnets repel and unlike poles attract. Suppose a simple magnet is placed on a pivot and used as a **rotor** (the rotating part of an electric motor) and a horseshoe magnet is used as a **stator** (the stationary part of a motor), as shown in Figure 9.7(a). Movement will be obtained by the repulsion and

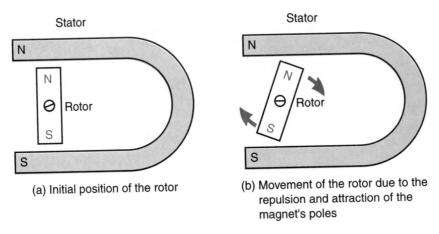

(a) Initial position of the rotor

(b) Movement of the rotor due to the repulsion and attraction of the magnet's poles

FIGURE 9.7 A simple electric motor

attraction of the poles of a magnet. The rotor would turn until the unlike poles are attracted to each other, as shown in Figure 9.7(b).

To make an electric motor move continuously, we must have a rotating magnetic field, which is produced by the reversal of the poles, or the polarity, in the rotor or stator. An alternating current of 60 hertz changes direction 120 times per second. Therefore, the current would change the polarity of the stator poles on each reversal of current. If the rotor has a permanent polarity, as shown in Figure 9.8, then the changes of polarity in the stator would cause the rotor to move. Therefore, if alternating current changes direction, causing a polarity change, 120 times a second, then the motor will turn in a continuous motion because the poles of the stator will be continuously repelling and attracting the permanent poles of the rotor. Figure 9.8 shows the motor in one complete cycle of current or ⅟₆₀ second. The movement of the motor is

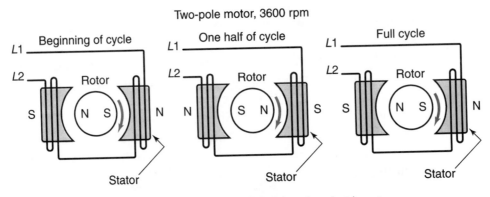

FIGURE 9.8 Complete cycle of operation of an electric motor

FIGURE 9.9 Windings of an electric motor

FIGURE 9.10 Squirrel cage rotor *(Courtesy of Magnetek, Inc., St. Louis, MO)*

caused by the magnetic field of the stator as it rotates through its alternations of current. Figure 9.9 shows the windings of an ordinary electric motor. The windings of the motor are a part of the stator.

In motors, the rotor is not a permanent magnet, as we stated in the previous explanation. The **squirrel cage rotor** shown in Figure 9.10 is the most commonly used rotor today. The squirrel cage rotor derives its name from its cagelike appearance. In the squirrel cage rotor, copper or aluminum bars are evenly spaced in the steel portion of the rotor and connected by an aluminum or copper end ring. The squirrel cage rotor produces an inductive magnetic field within itself when the stator is energized.

The most common motors operate much like a transformer, with the stator being the primary magnetic field and the rotor being a movable secondary magnetic field. The rotor will have magnetism induced into it from the stator, and its magnetic poles will be permanent. The magnetic poles of the stator are moving at the rate of the alternations of the current.

9.3 TYPES OF ELECTRIC MOTORS

The industry uses all kinds of AC motors to rotate the many different devices that require rotation in a complete system. Different motors are needed for different tasks because not all motors have the same running and starting characteristics. This fact, along with the increased cost of stronger motors, allows the industry to use the right motor for the right job. For example, many compressors require a motor with a high starting torque and a good running efficiency. Small propeller fans use motors with a low starting torque and average running efficiency.

Motor Strength

The starting methods, or strength, is generally used to classify motors into types. Motors are selected mainly because of the starting torque (power) required for the motor to perform its function. Five general types of motors are used in environmental systems: shaded-pole, split-phase, permanent split-capacitor, capacitor-start–capacitor-run and capacitor-start, and three-phase. There are others, such as the repulsion-start–induction-run and series motors. However, these are outdated or not commonly used in the industry. The starting torques of the five general types of induction motors, expressed as a percentage of their running torque, are as follows: shaded-pole, 100%; split-phase, 200%; permanent split-capacitor, 200%; capacitor-start–capacitor-run and capacitor-start, 300%; and three-phase, 600%.

Motor Speed

The following formula can be used to determine the speed of an electric motor with a load:

$$\text{Speed} = \frac{\text{Flow reversals/second} \times 120}{\text{Number of poles}}$$

One cycle of alternating current has two flow reversals. If 60 hertz alternating current is being used, there are 120 flow reversals per second. For example, if a four-pole motor is used in an application, its calculated rpm is

$$\text{Four-pole motor speed} = \frac{60 \times 120}{4} = 1800$$

The actual rpm of a four-pole motor is 1750 rpm. Motor speeds that are common to the industry are:

Two-pole motors	3450 rpm
Four-pole motors	1750 rpm
Six-pole motors	1050 rpm
Eight-pole motors	900 rpm

Open and Enclosed Motors

Motors are commonly either open or enclosed. The open motor, shown in Figure 9.11, has a housing and is used to rotate a device such as a fan or a pump that is itself not enclosed in any type of housing. The enclosed motor

FIGURE 9.11 Open motor *(Courtesy of Magnetek, Inc., St. Louis, MO)*

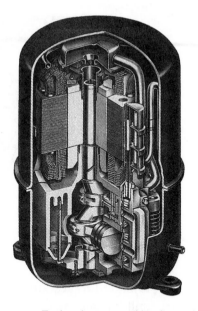

FIGURE 9.12 Enclosed motor used in a hermetic compressor *(Courtesy of Tecumseh Products Co.)*

is housed within some type of shell. The most common enclosure of a motor is a completely sealed hermetic compressor, as shown in Figure 9.12. Any starting apparatus used on an enclosed motor must be mounted outside the enclosure. The starting apparatus of an open motor is usually mounted within the motor itself.

In the following sections, we will discuss the five basic motor types in detail.

Motor Dimensions

The National Electrical Manufacturers' Association (NEMA) has established standard motor dimensions. The standards are useful when a technician is forced to locate a replacement motor for a particular application.

9.4 SHADED-POLE MOTORS

Most single-phase induction motors require a starting winding to create a starting torque that enables the motor to start. In most cases, the starting winding is located 90 electrical degrees from the main winding. A **shaded-pole motor** uses a shaded pole made of a closed turn of a heavy copper wire banded around a section of each stator pole. A shaded-pole motor is shown in Figure 9.13. Shaded-pole motors are used when very small starting and

FIGURE 9.13 Shaded-pole motor *(Courtesy of Magnetek, Inc., St. Louis, MO)*

running torques are required, such as in a furnace fan, a small condensing unit fan, and an open-type propeller fan. These motors are easily stalled, but in most cases, because of the small locked rotor amperes (i.e., the current draw of the motor when power is applied but the motor does not turn), they can stall and still not burn out the windings.

Operation

Figure 9.14 shows the stator of a shaded-pole motor. At one side of each pole, a small groove has been cut into the stator and banded by a solid copper wire or band, as shown in Figure 9.15.

When the shaded-pole motor is starting, a current is induced into the shaded pole from the main windings. The shaded poles produce a magnetic

FIGURE 9.14 Stator and rotor of shaded-pole motor

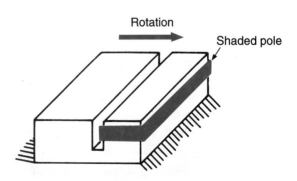

FIGURE 9.15 Shaded pole

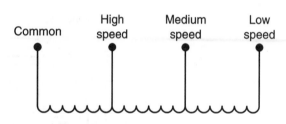

FIGURE 9.17 Schematic diagram of three-speed, shaded-pole motor

FIGURE 9.16 Schematic wiring diagram of the shaded pole-motor

field that is out of phase with the magnetic field of the main winding, and a rotating magnetic field is produced that is sufficient to give the desired starting torque. When the motor approaches full speed, the effect of the shaded pole is negligible. The rotation of the shaded-pole motor is from the unshaded edge of the pole toward the shaded edge of the pole.

Figure 9.16 shows the schematic diagram of a single-speed, shaded-pole motor. A single-speed, shaded-pole motor has only one winding, with the exception of the shaded pole, and is relatively simple. Figure 9.17 shows the schematic diagram of a three-speed, shaded-pole motor. The main winding is the speed winding for high-speed operation. For medium-speed operation, the main winding is put in series with the medium-speed winding, which increases the number of poles and produces fewer revolutions per minute. For low-speed operation, the main winding is put in series with the medium- and low-speed windings, which increases the number of poles in the stator and further reduces speed.

Reversing

Shaded-pole motors are difficult to reverse because to do so, you must disassemble them. The rotation of the shaded-pole motor is determined by the location of the shaded poles. Figure 9.18 shows a layout of a single-speed, shaded-pole motor. When the shaded poles are on the left side of the main poles, as in Figure 9.18, the rotation will be toward the shaded poles, or clockwise. On the other hand, when the shaded poles are on the right side of the main poles, as shown in Figure 9.19, the rotation will again be toward the shaded poles, but in this case, the rotation will be counterclockwise. Therefore, for reversing of the shaded-pole motor, the stator must be reversed to change the positions of the shaded poles, and this usually means disassembling the motor.

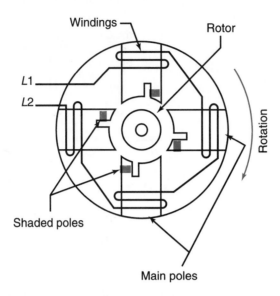

FIGURE 9.18 Layout of a shaded-pole motor, with clockwise rotation in the direction of the shaded poles

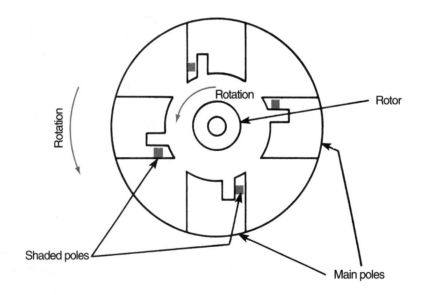

FIGURE 9.19 Layout of a shaded-pole motor, with counterclockwise rotation in the direction of the shaded poles

Troubleshooting

Shaded-pole motors are easy to identify because of the copper band around the shaded pole, previously shown in Figure 9.15. A single-speed, shaded-pole motor is easily diagnosed for trouble because of its simple winding patterns, previously shown schematically in Figure 9.16. Multispeed shaded-pole motors are more difficult to troubleshoot because of the additional speed windings, which were previously shown in Figure 9.17. The shaded-pole motor can be checked with an ohmmeter to determine the condition of the windings. Because a shaded-pole motor has stalled does not mean the windings are faulty. If this condition should occur, the motor probably needs lubrication. The shaded-pole motor is simple and easy to troubleshoot. It is used in many applications in the industry.

9.5 CAPACITORS

The **capacitor** consists of two aluminum plates with an insulator between them. The insulator prevents electrons from flowing from one plate to the other, but it permits the storage of electrons. Figure 9.20 shows the schematic symbol for a capacitor. Capacitors are used to boost the starting torque or running efficiency of single-phase motors.

Two Types Used in the Industry

Two types of capacitors are used primarily in the industry: the electrolytic or starting capacitor and oil-filled or running capacitor (Figure 9.21). **Starting capacitors** consist of two aluminum electrodes (plates) with a chemically treated paper, impregnated with a nonconductive electrolyte,

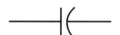

FIGURE 9.20 Symbol for a capacitor

(a) Starting capacitors

(b) Running capacitors

FIGURE 9.21 Common capacitors used in the industry (a) Starting capacitors (b) Running capacitors *(Courtesy of Aerovox, Inc.)*

between them. They can be purchased in ranges from 75 to 600 micro-farads (µF) and from 120 to 300 volts. A **microfarad** is the unit of mea-surement for the strength of a capacitor; all capacitors are rated according to their strength in microfarads. The electrolytic capacitor is used to assist a single-phase motor in starting.

The oil-filled capacitor consists of two aluminum electrodes with paper between them and an oil-filled capacitor case. It is available in microfarad ranges of about 2 to 60 and voltage ranges of 240 to 550. The oil-filled capacitor can be used for small or moderate torque starting, but it is more commonly used to increase a motor's running efficiency.

The major difference between the two types of capacitors is in their application. A starting capacitor is built in a relatively small case with a dielectric—a nonconductor of electric current. It is used for only a short period of time on each cycle of the motor. Therefore, a starting capacitor has no need to dissipate heat, although its capacity is larger than that of its counterpart, the running capacitor.

The **running capacitor** is designed to stay in the motor circuit for the entire cycle of operation. Therefore, it must have some means of dissipating the heat. The oil in the capacitor case is used for this purpose. The oil-filled capacitor is physically larger than the starting capacitor but smaller in capac-ity than the starting capacitor. Both capacitors are in wide use in the industry.

Troubleshooting

Short capacitor life and malfunctions can be caused by several different factors. High voltage can cause a capacitor to overheat. This can damage the plates and short the electrodes. Starting capacitors can be damaged by faulty starting apparatus that would keep the capacitor in the line circuit long enough to damage the capacitor. Excessive temperature can shorten the life of capacitors or cause permanent damage due to poor ventilation, starting cycles that are too long, or starting cycles that occur too frequently. The cause of the malfunction should be corrected as soon as possible. The capacitors themselves are frequently the cause of the problem.

 CAUTION Before handling or checking a motor capacitor, short from one terminal to another with a 20,000 ohm, four-watt resistor.

All capacitors used on single-phase motors are designed specifically to assist the motor in proper operation. However, in some cases to replace a capacitor with an exact replacement is impossible. If this situation should occur, use the following guidelines for replacing the capacitor:

1. The voltage of any capacitor used for replacement must be equal to or greater than that of the capacitor being replaced.
2. The strength of the starting capacitor replacement must be at least equal to but not more than 20% greater than that of the capacitor being replaced.
3. The strength of the running capacitor replacement may vary by plus or minus 10% of the strength of the capacitor being replaced.
4. If capacitors are installed in parallel, the sum of the capacitors is the total capacitance.
5. The total capacitance of capacitors in series may be found in the following formula:

$$C = \frac{C_1 \times C_2}{C_1 + C_2}$$

These rules are intended only as a guide. Remember, it is always preferable to use an exact replacement.

 CAUTION When making electrical connections to a running capacitor, make sure that power supplying the capacitor is connected to the marked terminal.

Many methods for testing capacitors are in common use in the industry today. A capacitor can be checked by using an ohmmeter. The ohmmeter should be placed on a high-ohm scale and both leads should be connected to the terminals of a discharged capacitor. If the needle of the meter shows a deflection to the right end of the scale and back to infinity, the capacitor is probably good. If the needle comes to rest on 0 ohms, the capacitor is shorted. If the needle of the meter does not move, it indicates an open capacitor.

In case of doubt, another method can be used to check the capacitor. By briefly applying voltage to a capacitor, reading the amperage, and then substituting the values into the following formula, we can obtain the exact capacitance:

$$\text{microfarads} = \frac{2650 \times \text{amperes}}{\text{volts}}$$

When performing this test, put a fuse in the circuit to prevent overloading due to a shorted capacitor, as shown in Figure 9.22. Starting capacitors

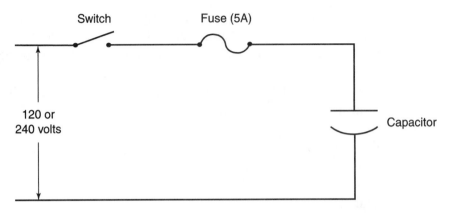

FIGURE 9.22 Schematic diagram of electric circuit to check the capacitance of a capacitor

FIGURE 9.23 Capacitor tester *(Courtesy of Sealed Unit Parts Co., Allentown, NJ)*

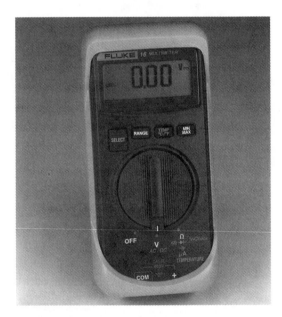

FIGURE 9.24 Digital volt-ohm meter capable of reading the capacitance of a capacitor

should be put in the circuit for approximately five seconds only. Many commercial capacitance testers are available on the market, one of which is shown in Figure 9.23. Many new digital volt-ohm meters test capacitors. Figure 9.24 shows a digital volt-ohm meter that can check the capacitance of a capacitor.

9.6 SPLIT-PHASE MOTORS

Two general classifications of split-phase motors are used in the industry. The resistance-start–induction-run motor and the capacitor-start–induction-run motor are types of split-phase motors in common use today. Each of these motors has different operating characteristics while being similar in construction. **Split-phase motors** use some method of splitting the phase of incoming power to produce a second phase of power, giving the motor enough displacement to start. The split-phase motor uses two windings to displace the phase and create the needed displacement between the run and start windings to produce rotation.

Resistance-Start–Induction-Run Motor

The resistance-start–induction-run motor shown in Figure 9.25 has both a starting winding to assist the motor in starting and a running winding to continue rotation after the motor has reached a certain speed. Most single-phase motors have some method of beginning the rotation; in a split-phase motor, rotation is started by splitting the phase to make a two-phase current. The single-phase current is split between the running and the starting winding, which puts one of the windings out of phase by 45 to 90 degrees. The starting windings are used to assist the split-phase motor in starting. They are also used until the motor has reached a speed that is about 75% of its full-capacity speed. The starting windings then drop out of the circuit by the use of a centrifugal switch. After that occurs, the motor operates at full speed on the main or running winding alone.

FIGURE 9.25 Split-phase motor *(Courtesy of Magnetek, Inc., St. Louis, MO)*

FIGURE 9.26 Cutaway view of an electric motor *(Courtesy of Magnetek, Inc., St. Louis, MO)*

A cutaway view of a split-phase motor is shown in Figure 9.26. Without the capacitors, it closely resembles a split-phase motor.

The split-phase motor can be operated on 120 volts–single-phase–60 hertz or 208/240 volts–single-phase–60 hertz. Some split-phase motors can operate on either voltage by making simple changes in their wiring if so desired. Thus, they are dual-voltage motors. Split-phase motors can be reversed by reversing the leads of the starting winding at the terminals in the motor.

Split-phase motors are used when a high starting torque is not required. They are used in such equipment as belt-driven evaporator fan motors, hot-water pumps, small hermetic compressors, grinders, washing machines, dryers, and exhaust fans.

Operation

The phases in a split-phase motor are split by the makeup of the starting windings. The starting winding is designed with smaller wire and more turns than is the running winding, which has a greater inductance. Therefore, the running winding is displaced from the starting winding because of its greater inductance. This displacement causes a resistance to current flow to build up in the running winding. The phase displacement means the current reaches the two windings at separate times, allowing one winding to lead, in this case the starting winding. However, some manufac-

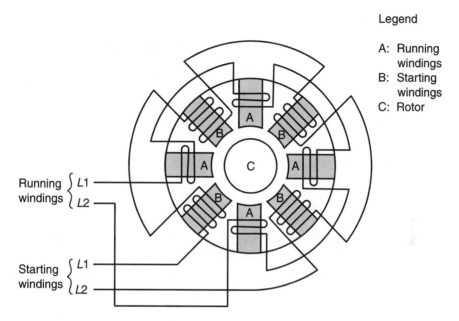

Legend

A: Running windings
B: Starting windings
C: Rotor

Running windings { L1
Running windings { L2

Starting windings { L1
Starting windings { L2

FIGURE 9.27 Layout of a split-phase motor

turers allow current to reach the running winding first by designing an increased resistance into the starting winding and a decreased induction into the running winding. Whatever method is used, the motor basically operates on the same principle: splitting the phase.

The operation of a split-phase motor, referring to Figure 9.27, is as follows:

1. Power is applied to the running and starting windings in parallel. The motor itself splits the phase by the counter electromotive force (emf) in the running winding, which acts as a resistance to hold back the current flow to the running winding. On the alternation of power, the starting winding creates a higher magnetic field than the running winding.

2. In half of a cycle the alternations are changed. The running winding has the stronger magnetic field, moving the rotor a certain distance depending on the number of poles in the motor. For the motor in Figure 9.27, the distance is one-fourth of a rotation.

3. As the alternations continue at the rate of 60 cycles per second, the motor continues to rotate with the magnetic field of the stator. Therefore, the rotor, with its magnetic field permanent, attempts to keep up with the rotating magnetic field of the stator.

4. The motor is equipped with a centrifugal switch that drops the starting winding out of the circuit when the motor has reached 75% of its full speed.

Troubleshooting

Split-phase motors are one of the most reliable types of motors used in the heating, cooling, and refrigeration industry. They are used on most types of single-phase equipment. The split-phase motor is easy to troubleshoot if the service technician has a good understanding of its operation. The three probable areas of trouble are the bearings, the windings, and the centrifugal switch.

The bearings of any motor often give trouble because of wear and improper maintenance. Identification of a motor with bad bearings is simple. The motor will have trouble turning and in some cases may be locked down completely.

The windings of a split-phase motor can be shorted, open, or grounded. This is easily diagnosed with an ohmmeter.

The centrifugal switch, shown in Figure 9.28, is the hardest section to diagnose for troubles because it stays in the circuit only a short time. The centrifugal switch has a tendency to stick in an open or closed position because of wear and often must be replaced. The centrifugal switch can

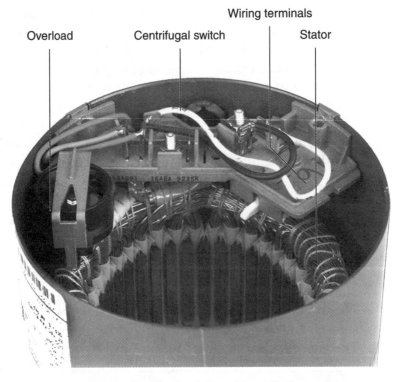

FIGURE 9.28 Centrifugal switch used in a split-phase motor

usually be heard when it drops in after the motor is cut off. Hence, it can be checked effectively in this manner. If the centrifugal switch does not drop out of the circuit, the motor will pull an excessively high ampere draw and cut off on overload. One sure method of checking the centrifugal switch is by disassembling the motor and making a visual inspection.

Capacitor-Start–Induction-Run Motor

The **capacitor-start motor**, shown in Figure 9.29, produces a high starting torque, which is needed for many applications in the industry. The open capacitor-start motor operates like a split-phase motor except that a capacitor is inserted in series with the centrifugal switch and the starting winding. The centrifugal switch breaks the flow of current to the starting capacitor and starting winding. The centrifugal switch opens when the motor has reached a speed that is 75% of its full speed. Figure 9.30 shows a schematic diagram of the motor.

Capacitor-start motors are used on pumps, small hermetic compressors, washing machines, and some types of heavy-duty fans.

Open Type

As we have said, the open capacitor-start motor is similar in design to the split-phase motor with the exception of the capacitor. Therefore, troubleshooting the open capacitor-start motor is similar to checking the split-phase motor except for checking the capacitor. There are four possible

FIGURE 9.29 Capacitor-start motor *(Courtesy of Magnetek, Inc., St. Louis, MO)*

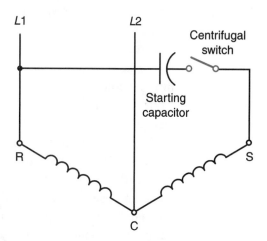

FIGURE 9.30 Schematic diagram of an open capacitor-start motor with a centrifugal switch. Relays may be used instead of centrifugal switch.

areas of trouble in the open capacitor-start motor: windings, bearings, centrifugal switch, and capacitor.

The windings can be easily checked with the use of an ohmmeter by checking for shorts, opens, and grounds.

The bearings of a motor usually fail because of lack of maintenance or wear. Motor bearings will usually become tight or lock down completely. This condition can be determined by trying to turn the motor. If the motor has a tight place in its rotation or will not turn at all, the bearings are faulty in the motor.

Due to the constant opening and closing of the centrifugal switch, it is often the culprit in motor problems. The centrifugal switch may stick in an open or closed position or its contacts may be defective. A centrifugal switch, in some cases, can be checked with an ohmmeter to determine its position, open or closed. In other cases, the motor will have to be disassembled to check the switch.

The capacitor is easy to check with an ohmmeter. The capacitor is often mounted in one end bell of the motor rather than on top of the stator.

Enclosed Type

When capacitor-start motors are used in small hermetic compressors, a centrifugal switch cannot be used because of the oil used to lubricate the compressor. Instead, an external relay is used to break the power going to the starting winding and the starting capacitor. In this case, the capacitor-start motor is an enclosed motor with a starting relay. By inserting a capacitor in the starting winding, a phase displacement is created between the running and starting windings, causing the motor to rotate.

The enclosed capacitor-start motor has an external relay to drop the starting capacitor out of the circuit. This capacitor should be checked to determine its condition.

The condition of the windings of an enclosed motor can easily be checked with an ohmmeter. The windings have a set of terminals on the outside of the casing that lead to the windings. Use an ohmmeter to check across these terminals to determine if the windings are shorted, open, or grounded.

The enclosed motor can also be locked down due to worn bearings or to internal failure of some component of the motor. This condition can be detected with an ammeter or by the humming sound of the motor on an attempted start.

9.7 PERMANENT SPLIT-CAPACITOR MOTORS

Permanent split-capacitor motors, also known as *PSC motors,* are simple in design and have a moderate starting torque and a good running efficiency, which makes them a popular motor in the industry. Figure 9.31 shows a PSC motor used as a fan motor. Figure 9.32 shows a hermetic compressor that uses a PSC motor to power the compressor.

The starting winding and the running capacitor of the PSC motor are connected in series, as shown in Figure 9.33(a). The schematic diagram of the hookup is shown in Figure 9.33(b). The running and starting windings are in parallel, but the capacitor causes a phase displacement.

Permanent split-capacitor motors are used on compressors, where the refrigerant equalizes on the "off" cycle, on direct-drive fan motors, and in other applications in the industry. It has a relatively low cost in comparison with other motors because it does not have a switch to drop the starting winding out of the circuit. The PSC motor can be used only when a moderate starting torque is required to begin rotation.

FIGURE 9.31 Permanent split-capacitor motor used as a fan motor *(Courtesy of Magnetek, Inc., St. Louis, MO)*

FIGURE 9.32 Hermetic compressor utilizing a PSC motor *(Courtesy of Tecumseh Products Co.)*

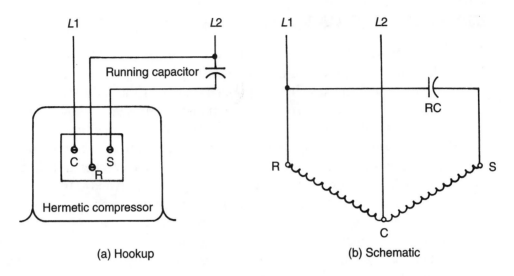

(a) Hookup   (b) Schematic

Legend
C: Common terminal
R: Running winding terminal
S: Starting winding terminal
RC: Running capacitor

FIGURE 9.33 Diagrams of permanent split-capacitor motor

Operation

The permanent split-capacitor motor has two windings: a running (main) winding and a starting (phase) winding. Both windings are wound with almost the same size and length of wire. A running capacitor is put in series with the starting winding. The capacitor causes the electron flow through the starting winding to shift it out of phase with the running winding. Therefore, a rotating magnetic field is set up, causing the rotor to turn.

Multispeed PSC motors contain additional running windings. The starting winding is in series with the running capacitor and in parallel with the running winding. Figure 9.34 shows a schematic diagram of a three-speed PSC motor. For high-speed operation, the starting and main windings are energized. Medium-speed operation is accomplished by energizing the starting winding with the main and medium-speed windings connected in series. For low-speed operation, the main, medium, and low windings are connected in series with each other, and all are connected in parallel to the starting winding.

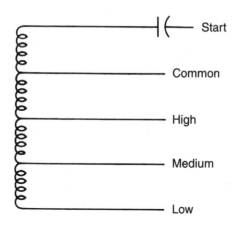

FIGURE 9.34 Schematic diagram of a three-speed PSC motor

Troubleshooting

The PSC motor usually gives trouble-free operation for long periods. The three most common failures in a PSC motor are in the bearings, windings, or capacitor.

The bearings of a PSC motor often become faulty because of wear or lack of proper maintenance. Bearings in any motor can be diagnosed with little trouble by rotating the motor by hand and noticing rough places in the movement or the shaft being locked in one position.

The windings of a motor become faulty because of overheating, overloading, or a faulty winding. A bad motor winding can be easily checked with an ohmmeter. The windings could be shorted, open, or grounded. The service technician should use care in diagnosing problems with the windings of PSC motors because they are often built with several speeds.

A bad capacitor can keep a PSC motor from starting or can pull a high ampere draw when running. Capacitors can be checked by one of the methods covered in Section 9.5. In most cases, faulty PSC motors will be replaced with new motors rather than repaired. The PSC motor is easy to troubleshoot with the right tools and knowledge.

Probably the most difficult aspect of PSC motors is their design. PSC motors are often built with several speeds. Service technicians must pay careful attention when replacing a faulty PSC motor because if the motor is connected incorrectly, permanent damage can occur. Most PSC motors are furnished with a wiring diagram to ensure correct installation. Motor manufacturers, however, make only a limited number of motors to replace the many different motors in the field. Thus, a service technician may have to adapt the replacement motor to a specific application.

9.8 CAPACITOR-START–CAPACITOR-RUN MOTORS

The **capacitor-start–capacitor-run motor**, or *CSR motor,* produces a high starting torque and increases the running efficiency. It is actually a capacitor-start motor with a running capacitor added permanently to the starting winding. The starting winding is energized all the time the motor is running. The capacitor-start–capacitor-run motor takes the good running characteristics of a permanent split-capacitor motor (see Section 9.7) and adds the capacitor-start feature. This produces one of the best all-around motors used in the industry.

Capacitor-start–capacitor-run motors are used almost exclusively on hermetic or semi-hermetic compressors. Rarely will this type of motor be used as an open-type motor because of the cost of the components necessary to produce it. Most open-type motors do not use a starting relay but use the centrifugal switch instead. Open types of motors are usually built as permanent split-capacitor motors or capacitor-start motors. Occasionally, a CSR motor will be used in an open-type motor when an extremely high starting torque is required.

Operation

The CSR motor begins operation on a phase displacement between the starting and running windings, which allows rotation to begin. The running capacitor lends a small amount of assistance to the starting of the motor, but its main function is to increase the running efficiency of the motor. Figure 9.35 shows a schematic diagram of this motor with its starting components.

Troubleshooting

The capacitor-start–capacitor-run motor is sometimes difficult to troubleshoot because of the number of components that must be added to a regular motor to produce it. The windings, bearings, potential relays, starting capacitor, and running capacitor must all be checked.

The windings of a CSR motor can be easily checked with an ohmmeter to determine if the windings are shorted, open, or grounded. In most cases, the windings will be enclosed in a hermetic casing and the terminals will be on the outside of the casing. However, the type of motor makes little difference in checking the winding as long as the technician uses the correct terminals.

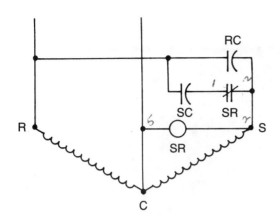

FIGURE 9.35 Schematic diagram of a capacitor-start–capacitor-run motor

Legend

C: Common terminal
R: Running winding terminal
S: Starting winding terminal
RC: Running capacitor
SC: Starting capacitor
SR: Starting relay (potential)

The bearings of a CSR motor can be worn so badly that the motor will not turn or will turn only with a great deal of difficulty. The bearings of hermetically sealed motors are enclosed and therefore harder to check, but the condition of the bearings can be determined by a whining sound, or by the motor pulling a larger-than-normal ampere draw. Care should be taken not to condemn the bearings of a motor because of a high ampere draw unless you are sure this is the problem.

The starting relay can be checked by diagnosing the condition of the contacts and the coil. The contacts can be checked with an ohmmeter or by visual inspection. On an ohmmeter the contacts should show zero resistance. The visual inspection is easy once the relay is disassembled. Then the condition of the contacts can be determined: sticking, pitting, or misalignment. The coil is checked like the windings of a motor.

The starting and running capacitors are easily checked with an ohmmeter to determine their condition.

Troubleshooting a CSR motor is done by checking all components of the motor. These motors must be correctly checked to prevent other components from being destroyed. For example, a capacitor will be destroyed if the contacts or coil of a starting relay are bad.

9.9 THREE-PHASE MOTORS

Three-phase motors are rugged, reliable, and more dependable than other types of motors. The most common type and the type often used in heating, cooling, and refrigeration is the squirrel cage induction type, shown in Figure 9.36. This motor will be the only three-phase motor discussed in this chapter.

FIGURE 9.36 Three-phase induction motor *(Courtesy of Reliance Electric Co., a Rockwell Automation Business)*

Three-phase motors are considerably stronger than single-phase motors because of the three phases that are fed to the motor. Three-phase current actually gives three hot legs to the device, rather than the two hot legs supplied by single-phase power. Therefore, instead of having a two-phase displacement, a three-phase displacement is available without using starting components. Three-phase motors are common to the industry; thus the technician should understand their operation.

Operation

Three-phase motors operate on the same principles as the single-phase with the exception of the three-phase displacement. A rotating magnetic field is produced in the stator. This interacts and causes a magnetic field in the rotor. However, the three-phase motors require no starting apparatus, because none of the phases are together. In the sine wave of the three-phase motor, none of the phases peaks at the same time. Each phase is approximately 120 electrical degrees out of phase with the others. For this reason, there is no need to use any device to cause a phase displacement, as is needed in the starting of single-phase motors.

Three-phase motors can be purchased in any voltage range desired. For example, a dual-voltage, three-phase motor can be operated on two difference voltages with minor modifications in the wiring.

Three-phase motors have two basic types of windings. They are the **star winding** or wye (Y) winding, as shown in Figure 9.37, and the **delta winding**, as shown in Figure 9.38. There is no operational difference between the two types, but it does allow designers more latitude in three-phase motor design.

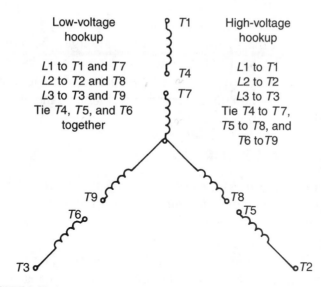

Low-voltage
hookup

*L*1 to *T*1 and *T*7
*L*2 to *T*2 and *T*8
*L*3 to *T*3 and *T*9
Tie *T*4, *T*5, and *T*6
together

High-voltage
hookup

*L*1 to *T*1
*L*2 to *T*2
*L*3 to *T*3
Tie *T*4 to *T*7,
*T*5 to *T*8, and
*T*6 to *T*9

FIGURE 9.37 Schematic diagram of the star winding of a three-phase motor

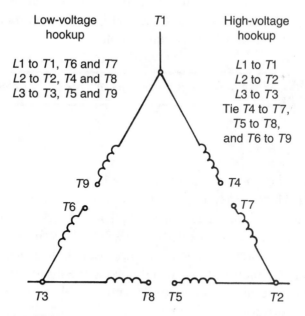

Low-voltage
hookup

*L*1 to *T*1, *T*6 and *T*7
*L*2 to *T*2, *T*4 and *T*8
*L*3 to *T*3, *T*5 and *T*9

High-voltage
hookup

*L*1 to *T*1
*L*2 to *T*2
*L*3 to *T*3
Tie *T*4 to *T*7,
*T*5 to *T*8,
and *T*6 to *T*9

FIGURE 9.38 Schematic diagram of the delta winding of a three-phase motor

Troubleshooting

A three-phase motor can be checked by reading the resistance of the winding with an ohmmeter. If a resistance reading of 0 ohm occurs, the motor is shorted. A reading of infinite resistance indicates an open winding. A reading of some measurable resistance is usually from 1 ohm to 50 ohms, depending on the size of the motor. The larger the motor, the smaller the resistance. The smaller the motor, the larger the resistance of the winding. Care should be taken because of the chance of a spot burnout in the winding. Experience should give service technicians the ability to diagnose any type of electric motor.

9.10 HERMETIC COMPRESSOR MOTORS

Hermetic compressors are becoming increasingly popular because of their low cost. Hermetic motors are of the induction type. They are designed for single- and three-phase current. There are four basic types of single-phase motors used in hermetic compressors. The split-phase is used on small equipment (fractional horsepower). The capacitor-start is also used on small equipment. The permanent split-capacitor is used on most window units and small residential units. The capacitor-start–capacitor-run is used on any application that requires a good starting and running torque. Many hermetic compressors are built with three-phase motors; usually these are used on the larger equipment.

CAUTION Make sure all compressors are properly grounded.

Operation

Hermetic compressor motors are totally enclosed in a shell with refrigerant and oil. Hence, they require special considerations. Nothing can be used inside the shell that is capable of causing a spark or that has to move on the crankshaft, such as a centrifugal switch. Therefore, no starting apparatus can be incorporated inside the compressor shell. Starting relays and capacitors must be mounted and wired outside the motor. It must be remembered that hermetic motors operate the same as other motors with the exception of the enclosure.

CAUTION Oil and refrigerant can spray out of a hermetic compressor when an electrical terminal of the compressor is vented.

CAUTION The protective covering of the electrical terminals of a hermetic compressor should always be in place in the event of terminal venting.

Terminal Identification

All single-phase motors have a common, a start, and a run terminal. These terminals are sometimes wired directly into an open-type motor and are difficult to find. The common is the junction point of the start and run terminals. The start and run terminals are connected to one end of the windings while the common is connected to the other end. The schematic diagram of a single-phase compressor motor is shown in Figure 9.39 with the terminals identified. Of course, each of the windings of a three-phase hermetic motor is the same because no starting apparatus is required.

In single-phase motors, it is important for the service technician to determine the common, start, and run terminals. This task can be performed simply and easily by using an ohmmeter to obtain the resistance of each winding with respect to common. Figure 9.40 shows the resistance values of a single-phase motor after the resistance has been measured at each

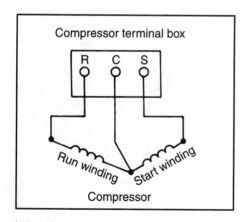

Legend

R: Run terminal
C: Common terminal
S: Start terminal

FIGURE 9.39 Schematic diagram of a single-phase compressor with the terminals identified

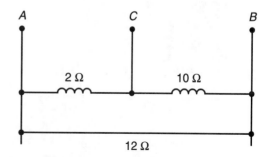

FIGURE 9.40 Terminals of a single-phase compressor with ohmic values given

terminal on the compressor. To find the run, start, and common terminals, the following procedure should be followed:

1. Find the largest reading between any two terminals. The remaining terminal is common (in Figure 9.40 the reading between *A* and *B* is largest; *C* is common).
2. The larger reading between common and the other two terminals identifies start (*C* to *A* is 2 ohms and *C* to *B* is 10 ohms; therefore, common to *B* is larger and *B* is start).
3. The remaining terminal is run (*A* is run).

This procedure is important, especially in installing the external electric devices, although it is not necessary if a good, readable diagram is available. In a three-phase motor, the resistances between all three terminals are the same.

Troubleshooting

Troubleshooting a hermetic compressor motor is often difficult because of its physical makeup and because it is totally enclosed in a shell and cannot be visually inspected. Small hermetic compressors usually have some type of external overload, as shown in Figure 9.41, whereas large hermetic compressors usually have internal overloads, as shown in Figure 9.42. The winding layouts of single-phase hermetic compressors are similar regardless of motor size. The only difference is the size of the windings, which will

FIGURE 9.41 Small hermetic compressor with external overload *(Courtesy of Tecumseh Products Co.)*

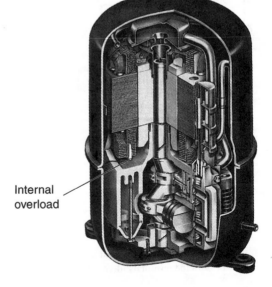

Internal overload

FIGURE 9.42 Large hermetic compressor with internal overload (cutaway) *(Courtesy of Tecumseh Products Co.)*

vary the resistance readings of the motor windings. Three-phase hermetic compressor motors are generally produced in sizes above 3 horsepower. Through experience, the service technician will be able to determine the approximate resistance of the motor windings in a hermetic compressor.

Electrical troubleshooting of hermetic compressor motors is done by taking a resistance reading of the windings with a good ohmmeter. Determining the condition of the windings is easy if the problem with the motor is open windings, shorted windings, or grounded windings. Figure 9.43 shows a schematic representation of these three conditions.

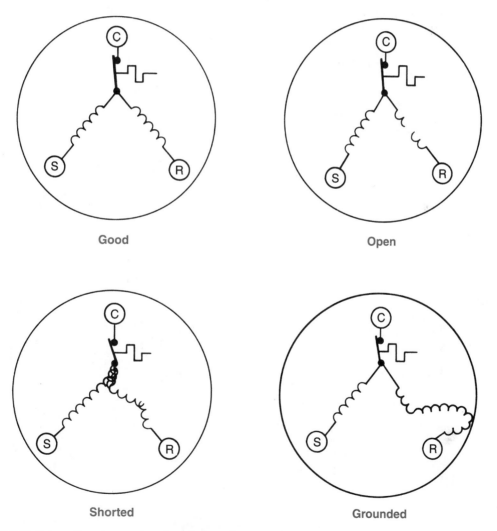

FIGURE 9.43 Schematic representation of good, shorted, open, and grounded compressor windings with internal overload

CAUTION If arcing sounds (sizzling, sputtering, or popping) are heard inside a compressor, immediately move away; this sound indicates a possible compressor terminal venting situation.

Most single- or three-phase hermetic compressor motors have three terminals on the outside of the casing that connect the motor to the external power wiring, as shown in Figure 9.44. Some large hermetic compressors have more than three terminals, such as dual-voltage, part winding motors or two-speed motors, as shown in Figure 9.45. The resistance readings of single-phase motor windings are not the same because the compressor has

Spade-type hermetic push-on terminals

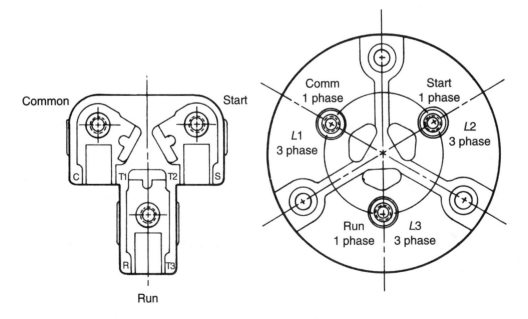

Screw terminals

FIGURE 9.44 Several terminal arrangements on hermetic compressors

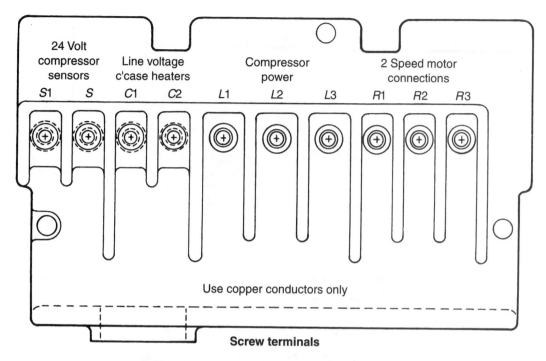

FIGURE 9.45 Terminals on a large hermetic compressor

a start winding and a run winding connected by a common wire, as shown in Figure 9.46. The physical makeup of a single-phase motor will allow the service technician to match the resistance readings to determine the condition of the windings. The sum of the resistance readings of the start to common terminals and the run to common terminals should equal the resistance reading obtained between the run and start terminals, as discussed in the terminal section. If the readings do not match, a spot burnout of the winding is likely. Three-phase motors will have the same resistance in each winding; if not, the motor is bad because of the spot burnout. The service technician must be careful, however, before condemning a hermetic motor whose winding resistance readings vary, because the problem may actually be bad connections, a faulty meter, or a misreading of the meter. A good service technician should use every possible diagnostic tool to ensure that no good hermetic compressor is condemned.

 When removing a compressor, make sure that electrical **CAUTION** power supplies have been disconnected and the refrigerant recovered.

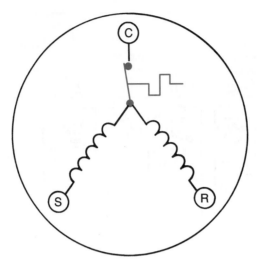

Resistance readings

C to R: 2 ohms
C to S: 8 ohms
R to S: 10 ohms

FIGURE 9.46 Schematic of windings of a single-phase compressor with internal overload (ohm readings for winding shown)

Diagnosing an open, shorted, or grounded hermetic compressor motor is easy because the resistance readings obtained are definite and exact. An open winding in the compressor motor means there is no continuity or no complete circuit; it gives an infinite resistance reading, as shown in Figure 9.47. A shorted winding in a compressor motor means the winding has burned together; it gives a zero ohm reading, as shown in Figure 9.48. A grounded winding in a compressor motor means that part of the winding is contacting the compressor body; it gives a resistance reading between the shell and the terminals of a compressor, as shown in Figure 9.49. Good contact on the compressor shell must always be maintained if the motor is grounded; therefore, any paint must be removed from a small section of the compressor. The open and shorted windings should be read on a low ohm scale (R × 1), but the grounded winding should be read on the R × 10,000 scale or higher. A grounded compressor can be dangerous because the technician or customer can receive an electrical shock if he or she touches the casing of a slightly grounded compressor. A resistance reading as high as 500,000 ohms indicates a grounded compressor that should be changed. Grounded compressors, if allowed to operate, will often operate at a higher-than-normal temperature; the warmer the windings, the lower the resistance of the ground in most cases.

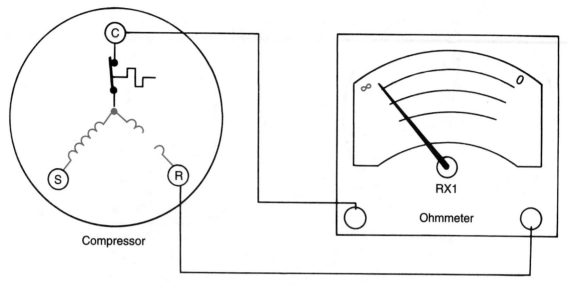

FIGURE 9.47 Open compressor winding being checked with an ohmmeter

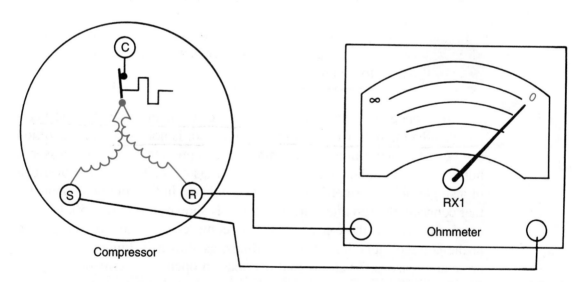

FIGURE 9.48 Shorted compressor winding being checked with an ohmmeter

CAUTION To ensure safety and prevent damage to the motor, restart only after determining the cause of stoppage.

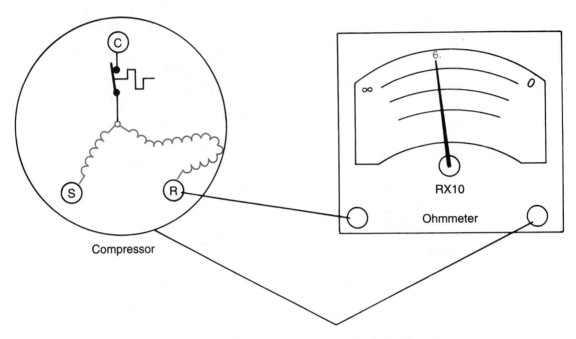

FIGURE 9.49 Grounded compressor windings being checked with an ohmmeter

> **CAUTION** Before resetting a circuit breaker or fuse, check for a short circuit to ground.

Before condemning the compressor, service technicians should make certain that the internal overload of the compressor is not open. This condition can be easily determined by touching the compressor; if the compressor is hot, it is a good indication that the overload is open. The internal overload of a single-phase hermetic compressor is located in the common conductor that connects the run and start windings, as shown in Figure 9.50. Internal overloads used in three-phase hermetic compressor motors are connected at the common junction of the windings, as shown in Figure 9.51. There are many reasons for an internal overload to open in a hermetic compressor: for example, low refrigerant charge, locked down compressor, faulty starting components, and high discharge pressure.

> **CAUTION** When troubleshooting electric motors or hermetic compressor motors that are extremely hot, make sure they have ample time to cool before condemning them.

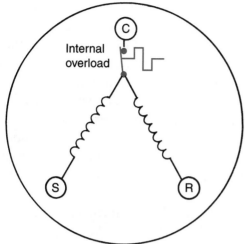

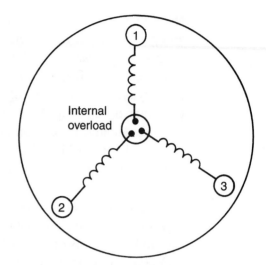

FIGURE 9.50 Schematic of a single-phase compressor with internal overload

FIGURE 9.51 Schematic of a three-phase compressor with internal overload

Mechanical failures in hermetic compressors often seem like electrical problems, especially when the compressor is locked down or when the internal overload opens because of some mechanical failure. The technician must make certain the problem is truly electrical before an accurate diagnosis can be made.

9.11 SERVICE CALL PROTOCOL

All heating, refrigeration, and air-conditioning systems at one time or another will require servicing. There are many types of service procedures that are performed each day for customers by industry technicians. Service procedures often performed by technicians in the industry are pre-season startups, preventive maintenance calls, or inoperative system calls. No matter what the reason for the call, technicians must always keep in mind that they are performing a service for the customer and that without the customer and service calls, there is no need for their company or their job. When customers choose a company to service their heating, refrigeration, and air-conditioning equipment, a great deal of trust and confidence is placed in the company and its employees. The technician must make certain that his or her actions and attitude reinforce the decision that the customer has made. A satisfied customer will do more advertising for the company and technician by word of mouth than many other methods of high-cost advertising. When the customer calls and asks for a specific technician, the technician knows that he or she has impressed the customer and pleased his or her employer.

Most technicians have the technical skills required to diagnose and repair most heating, refrigeration, and air-conditioning system problems. You can be very efficient in repairing the system, but you can still fail to satisfy the customer because of your actions or attitudes. The technician is constantly being evaluated by customers, employers, and coworkers. The customer wants to be treated in a fair and courteous manner while the system problem is being resolved. The employer is concerned about the profit margin and wants the technician to work faster and still maintain a high level of accuracy. Coworkers should be friends but in many cases conflicts occur over how fast one works, callbacks, and pay scales. The technician has a difficult task in keeping his or her customers, employer, and coworkers happy.

One of the most important aspects of the technician's job is the treatment of the customer. The technician must first treat the customer in a courteous manner. Often this will be difficult because the customer will be angry because of the problems with his or her heating, refrigeration, or air-conditioning system. For example, the customer may have spent a cold, uncomfortable night without heat, or it may be 105°F in the house without air conditioning, or food is spoiling because of an inoperative freezer, or this is a callback, or the cost of the repair is higher than expected. The technician will often have to listen to unhappy customers and try to be understanding even before correcting the problem. The technician must be patient and listen to the customer's complaint while trying to obtain as much information as possible about the problem. The technician should also avoid disrupting the activities of the customer whenever possible.

The technician's appearance is of utmost importance because often the technician is judged by this alone. The customer will be much more comfortable with a technician who is clean and neat in appearance. Oftentimes, because of the location and service procedure, it will be difficult to maintain a clean and neat appearance the entire day. It is the responsibility of the technician to try to remain clean and neat regardless of the working conditions (under houses, in attics, or in muddy, sloppy conditions). Changing a compressor, cleaning a furnace, or cleaning a dirty condenser can make it difficult to remain clean and neat. Many companies require employees to wear uniforms for this purpose. It is not practical to change clothes every time you work in dirty areas or work with dirty components, but you must do everything possible to maintain a neat and clean appearance at all times.

The technician must treat the customer's property with respect. The technician will be required to work in close proximity to walls and floor coverings and must do so without marring or damaging these surfaces. Every effort must be made to prevent any type of damage to the customer's structure or furnishings; make sure hands are clean before working on a thermostat mounted on an interior wall, remove mud from shoes or remove shoes before entering the structure, dust clothing before entering the structure, and do not use furniture as a resting place for tools or components. It is the technician's responsibility to prevent damage to the customer's property. Damage to the customer's structure or furnishings will have to be corrected by the company. Most companies have liability insurance to cover such costs. The technician should also make certain to check the service procedures performed to ensure that no danger exists from gas leaks, loose wires, or fire in the equipment that was serviced.

Upon completion of the service procedure, the technician should fully explain the problem and the service performed to correct it. If there are several options for correcting the problem, all options should be discussed and the customer allowed to choose the option he or she prefers. The technician should be fair about the amount of time spent and parts used for which the customer will be billed. For example, it is unfair to bill the customer for 10 pounds of refrigerant when only 5 were required; it would also be unfair to bill the customer for more time than was required. It is also dishonest to replace parts that are still in good condition and charge the customer for them. The technician must always make the correct ethical decisions regarding the customer to ensure continued trust.

The technician should never forget that the company will be judged by the technician's actions, appearance, and attitude. No greater compliment can be paid a service technician than for the customer to call back and request a particular technician for a service call.

9.12 SERVICE CALLS

Service Call 1

Application: Residential conditioned air system

Type of Equipment: Gas-fired forced-air furnace

Complaint: No heat

Service Procedure:

1. The technician reviews the work order from the dispatcher for available information. The work order information reveals that the furnace is an upflow gas-fired furnace located in a closet within the living space of the residence.

2. The technician informs the customer of his or her presence (if the customer is available) and obtains specifics about what the system is doing.
3. Upon entering the residence, the technician makes certain that no dirt or foreign material are carried into the structure. The technician also takes care not to mar interior walls.
4. The customer informs the technician that the main burner is igniting and operating for several minutes and cutting off. The blower is not operating.
5. The technician checks the limit and fan controls and determines that they are operating correctly. The technician checks the voltage available to the fan motor and determines that the voltage supply is correct.
6. The technician turns the power supply off and locks or labels it.
7. The fan motor is only moderately warm, indicating that the motor has been off for some time.
8. The technician checks the motor windings with an ohmmeter and reads infinite resistance, determining that the motor windings are open.
9. The technician changes the fan motor and checks the operation of the furnace.
10. The technician informs the customer of the problem and that it has been corrected.
11. A subsequent courtesy call by the technician to the customer to make certain the equipment is operating properly builds confidence and good will between the company and the customer.

Service Call 2

Application: Commercial refrigeration (walk-in cooler)

Type of Equipment: Air-cooled condensing unit with evaporator

Complaint: Warm walk-in cooler

Service Procedure:

1. The technician reviews the work order from the dispatcher for available information. The work order information reveals that the temperature of the walk-in cooler is above a safe temperature for the stored product.
2. The technician informs the business manager of his or her presence and obtains specifics about what the refrigeration unit system is doing.
3. Upon entering the business, the technician makes certain that the normal business activities are not interrupted and that his or her appearance is clean and neat.

4. The manager informs the technician that the refrigeration unit is not maintaining the correct temperature.
5. The technician makes a visual inspection of the indoor unit and finds the evaporator fan motor operating.
6. The technician proceeds to the outdoor condensing unit and observes that the compressor runs only a short period of time and cuts off. The condenser fan motor is not running.
7. With the compressor running, the technician knows that power is available to the condensing unit.
8. The technician turns the power off to the condensing unit and locks or labels it.
9. The technician finds that the motor turns freely, but is extremely hot. The technician makes the determination that the motor bearings are good.
10. The technician determines that the condenser fan motor is a PSC motor and checks the running capacitor with an ohmmeter or capacitor tester. The technician finds that the capacitor is open.
11. The technician removes the faulty running capacitor and replaces it with the correct replacement capacitor.
12. The technician restores power to the condensing unit, observes the operation of the fan motor, then checks the amp draw of the motor and compares the reading with the nameplate running ampacity of the motor to ensure that the motor is operating properly.
13. The technician interrupts power to the unit and oils the condenser fan motor.
14. The technician informs the manager of the problem and that it has been corrected.
15. A subsequent courtesy call by the technician to the manager to make certain the equipment is operating properly builds confidence and good will between the service company and the customer.

Service Call 3

Application: Residential conditioned air system

Type of Equipment: Packaged heat pump

Complaint: No cooling

Service Procedure:

1. The technician reviews the work order from the dispatcher for available information. The work order information reveals that the unit is a packaged heat pump. The customer will leave the key under a flower pot on the right side of the front door.

2. Upon entering the residence, the technician makes certain that no dirt or foreign material is carried into the structure. The technician also takes care not to mar interior walls.

3. The technician sets the thermostat to cool and observes the operation of the equipment. The technician immediately notices that the indoor fan motor is not operating.

4. The technician goes outside to check the unit, specifically the indoor fan motor, based on the indication that the motor is not running when the thermostat is set to cool. The technician finds that the compressor and outdoor fan motor are operating, but the indoor fan motor is not operating.

5. The technician turns the power supply off and locks or labels it.

6. The technician removes the unit cover, allowing access to the indoor fan motor.

7. The technician finds that the indoor fan turns with difficulty and the motor is hot to the touch.

8. The technician has determined by the fan being difficult to turn that the bearings are bad and that the indoor fan motor should be replaced. (Occasionally, oiling the motor will free the bearings, but this is usually only a temporary correction and, in most cases, the best solution is to replace the motor.)

9. The technician replaces the fan motor with the correct motor.

10. The technician turns the power on and checks the operation and amp draw of the new fan motor. The indoor fan motor is operating properly, so the technician turns the power off and replaces the access cover.

11. The technician turns the power on to the unit.

12. The technician returns the thermostat setting to normal and leaves a note for the customer indicating the service performed.

13. The technician replaces the key in the location designated by the customer.

Service Call 4

Application: Commercial conditioned air system

Type of Equipment: Commercial and industrial fan coil unit

Complaint: Blower motor does not operate

Service Procedure:

1. The technician reviews the work order from the dispatcher for available information. The work order information reveals that the motor operating the blower in the fan coil unit is not operating.

2. The technician informs the maintenance supervisor of his or her presence and obtains any specific information about the fan coil unit.

3. The technician visually observes the blower motor and controls and finds that the motor is controlled by a magnetic starter, which has opened due to a possible overload. The technician resets the overload and the blower motor rotates at a very low rate of speed, and then the magnetic starter opens again due to an overload.

4. The technician turns the power off and locks or labels it.

5. The technician checks the motor and finds that it turns freely, but is extremely hot to the touch.

6. The technician has determined that the bearings of the motor are good and the overheating is due to other factors.

7. The technician checks the power at the magnetic starter and finds that only single phase is available and that the motor requires 240 volts–three phase–60 hertz.

8. The technician checks the voltage at the distribution center and checks the fuses with an ohmmeter. One of the three fuses has an infinite reading and is bad.

9. The technician replaces the bad fuse with one of equal specification.

10. The technician restores power to the magnetic starter and observes that the fan operates.

11. The technician checks the amp draw of the fan motor and determines that the motor is operating properly.

12. The maintenance supervisor is informed that the problem was a blown fuse which created a single-phasing condition and overloaded the motor, and that the problem has been corrected.

Service Call 5

Application: Residential conditioned air system

Type of Equipment: Air-cooled condensing unit

Complaint: No cooling

Service Procedure:

1. The technician reviews the work order from the dispatcher for available information. The work order reveals that the system is not cooling.

2. The technician informs the customer of his or her presence and obtains specifics about what the conditioned air system is doing.

3. Upon entering the residence, the technician makes certain that no dirt or foreign material is carried into the structure. The technician also takes care not to mar interior walls.

4. The thermostat is set to the cool position and the indoor fan motor is operating, but the system is not cooling.

5. The technician proceeds to the air-cooled condensing unit and finds that the condenser fan motor is operating, which normally indicates that the contactor is closed.

6. The cover is removed from the condensing unit and the technician determines that the compressor is not operating and is very hot to the touch.

7. A voltage check is made at the compressor terminals and the correct voltage is available.

8. The technician makes a resistance check of the compressor motor and reads 25 ohms between the start and run terminals of the compressor, but reads infinite resistance between common and start and common and run. These resistance readings indicate that the internal overload of the compressor is open (it is not allowing voltage to reach the windings).

9. All starting components and capacitors are checked by the technician. If all are good, the technician will have to allow the internal overload ample time to close. After the internal overload has reset, the technician will make a resistance check of the windings. The resistance reading of the run to start terminals of the compressor should equal the readings of the sum of the common to start and common to run terminals.

10. The windings are good, so the technician tries to start the compressor. If the compressor fails to start and pulls a high current draw or the internal overload opens again, the compressor is probably mechanically seized or locked down.

11. The technician informs the customer that the compressor is bad and will have to be replaced. The technician gives the homeowner an estimate of the replacement cost and time when the replacement can be made.

Service Call 6

Application: Residential conditioned air system

Type of Equipment: Heat pump

Complaint: No heat

Service Procedure:

1. The technician reviews the work order from the dispatcher for available information. The work order indicates that the breaker for the outdoor section of the heat pump is overloading, while the indoor unit is operating. The customer will leave the key under the mat at the back door.

2. Upon entering the residence, the technician makes certain that no dirt or foreign material is carried into the structure. The technician also takes care not to mar interior walls.

3. The technician sets the thermostat to heat and observes the operation of the equipment. The technician notices that the indoor fan motor is operating, but the unit is not heating.

4. The technician goes to the outdoor unit and observes that the unit is not operating. The voltage is checked at the line connections of the disconnect switch of the outdoor unit and no voltage is available. The disconnect is turned off. The technician locates the electrical panel and resets the breaker that feeds the outdoor unit.

5. The disconnect is closed and nothing happens. A voltage check is again made which indicates that the breaker has broken the circuit again.

6. The technician determines that a short circuit exists somewhere in the outdoor unit. A resistance check must be made of each load that receives voltage from the disconnect. The resistance checks indicate that the resistance reading from the common to run terminals of the compressor is 0 ohms. This indicates that the compressor motor windings are shorted and the compressor must be replaced.

7. The customer is informed that the compressor is bad and will have to be replaced. The technician gives the homeowner an estimate of the replacement cost and a time when the replacement can be made.

SUMMARY

The single most important operating principle of an electric motor is the rotating magnetic field produced by alternating current. An alternating current is applied to the stator of the motor to produce a magnetic field. The magnetic field interacts with the rotor to produce a magnetic field in the rotor. When these two magnetic fields act together, they produce a rotating movement in the motor.

There are basically five types of motors used in the industry: shaded-pole, split-phase, permanent split-capacitor, capacitor-start–capacitor-run and capacitor-start, and three-phase.

The shaded-pole motor is a low-starting torque motor used on some propeller types of fans. These motors are easy to identify because of the copper band around the shaded pole. They are easily diagnosed for trouble because of their simple winding patterns.

Split-phase motors are relatively low-torque motors. They are simple and inexpensive devices. A split-phase motor can be used on a small hermetic compressor with the starting winding being dropped out by a starting relay. It can be used with any open types of motors that do not require a high starting torque.

The permanent split-capacitor motor has a low starting torque and a running capacitor in the starting winding. The running capacitor remains in the circuit at all times to produce good running efficiency. The permanent split-capacitor motor is used on most residential air conditioners of five horsepower or under and on direct-drive fan motors. It is relatively inexpensive because it does not have a switch to drop the starting winding.

The capacitor-start and the capacitor-start–capacitor-run motors are similar in design. They have a high starting torque. The addition of a running capacitor to a capacitor-start motor produces the capacitor-start–capacitor-run motor, which has good running efficiency.

Three-phase motors are commonly used on large pieces of equipment. They operate much like single-phase motors except that they have three basic phase displacements without the use of any starting apparatus. They have better starting and running characteristics than single-phase motors.

Hermetic compressor motors are becoming popular because of their low cost. They are used in many cooling units, especially for the smaller systems. Hermetic motors must have all their starting apparatus wired and mounted externally because it cannot be contained in the compressor shell. A hermetic compressor motor may be any one of the five basic motor types discussed in this chapter. The enclosure distinguishes the hermetic compressor from the other motor types.

Most single-phase motors require some method of producing a second phase of electricity in the motor to make it start. The design of the windings in a split-phase motor allows the use of a centrifugal switch to drop the starting winding out of the circuit after the motor has attained 75% of its full speed. Other types of single-phase motors use capacitors to create the second phase. The permanent split-capacitor motor incorporates a running capacitor to aid in the starting of the motor. The capacitor-start–capacitor-run motor incorporates the running capacitor along with a starting capacitor, using a potential relay to drop out the starting capacitor. The capacitor-start motor operates as a split-phase motor except that a starting capacitor is added to the start winding.

The service technician will often need to diagnose why a hermetic compressor will not run. Frequently, the fault lies in the hermetic motor. Hermetic compressor motors can easily be diagnosed with a good ohmmeter as open, shorted, or grounded. Spot burnouts are harder to diagnose, but you can still identify them easily by using the resistance readings of the motor.

CAUTION All electric motors should be properly grounded.

CAUTION Before starting an electric motor that has blown a fuse or tripped a circuit breaker, make sure that the motor does not have a short to ground before resetting or replacing the safety device.

REVIEW QUESTIONS

1. What is magnetism?

2. Torque is _____.
   a. strength that a motor produces by turning
   b. rotating motion
   c. a motor under load
   d. all of the above

3. A magnetic field is _____.
   a. the area around an electric motor
   b. the area around an electric wire
   c. the area in which a magnetic force operates
   d. all of the above

4. True or False: A permanent magnet is a piece of material that has been magnetized and can hold its magnetic strength for a reasonable length of time.

5. How is an electromagnet produced?

6. Which of the following produces the best electromagnet?
   a. cobalt
   b. soft iron
   c. paper
   d. wood

7. Unlike poles of a magnet _____ each other and like poles _____ each other.

8. What part does polarity play in the operation of an electric motor?

9. What part of a motor produces an inductive magnetic field within itself to facilitate the rotating motion?
   a. squirrel cage rotor
   b. motor bearings
   c. capacitor
   d. all of the above

10. What part does the frequency of alternating current play in the operation of an electric motor?

11. What would be the speed of a two-pole motor if there were 7200 flow reversals per minute?
    a. 1100 rpm
    b. 1800 rpm
    c. 2800 rpm
    d. 3600 rpm

12. What are the five types of single-phase motors used in the industry?

13. Which of the following correctly lists the motor's starting torque from lowest to highest?
    a. capacitor-start, split-phase, shaded-pole, three-phase
    b. shaded-pole, three-phase, permanent split-capacitor, capacitor-start
    c. three-phase, split-phase, shaded-pole, capacitor-start
    d. shaded-pole, split-phase, capacitor-start, three-phase

14. Which of the following is a common use of a shaded-pole motor?
    a. furnace fan motor
    b. propeller fans
    c. small pumps
    d. compressors

15. How does a shaded-pole motor operate?

16. How can a shaded-pole motor be reversed?

17. What determines the rotation of a shaded-pole motor?
    a. location of windings
    b. location of shaded pole
    c. location of rotor
    d. none of the above

18. Draw a diagram of a three-speed, shaded-pole motor.

19. What enables a split-phase motor to develop enough torque to begin rotation?

20. What removes the starting winding from the electrical circuit of an open-type split-phase motor once it reaches 75% of its operating speed?
    a. disconnect switch
    b. phase-out switch
    c. centrifugal switch
    d. starting switch

21. What are the three probable areas of trouble in a split-phase motor?

22. What is the unit of measurement for the strength of a capacitor?
    a. microamp
    b. microtorque
    c. microfarad
    d. microwatt

23. What is the purpose of a capacitor?
    a. boost starting torque
    b. increase running efficiency
    c. increase motor speed
    d. both a and b

24. What is the difference between a running and a starting capacitor?

25. List the five capacitor replacement rules.

26. Explain the operation of a permanent split-capacitor motor.

27. How are a PSC motor and a capacitor-start–capacitor-run motor similar?
    a. both use a starting capacitor
    b. both use a running capacitor
    c. both use starting relays
    d. none of the above

28. What are the advantages and disadvantages of using the following types of motors?
    a. shaded-pole motor
    b. PSC motor
    c. split-phase motor
    d. capacitor-start–capacitor-run motor

29. What are the similarities between an open-type split-phase motor and a capacitor-start motor?
    a. both have a capacitor
    b. both have a relay
    c. both have a centrifugal switch
    d. none of the above

30. Which of the following is an advantage in using a three-phase motor?
    a. higher starting torque
    b. stronger
    c. more dependable
    d. all of the above

31. Draw a wiring diagram of a capacitor-start–capacitor-run motor.

32. True or False: All starting apparatuses are mounted externally to the hermetic compressor shell.

33. What is the process in troubleshooting any electric motor?

34. Which of the following is the capacitance of an 88 μF and a 108 μF starting capacitor connected in series?
    a. 196 μF
    b. 96 μF
    c. 48 μF
    d. 40 μF

35. Which of the following is the capacitance of two 20 μF running capacitors connected in parallel?
    a. 10 μF
    b. 20 μF
    c. 30 μF
    d. 40 μF

36. If a capacitor produces 15 A on a 240-volt supply, which of the following is its microfarad rating?
    a. 166 μF
    b. 15 μF
    c. 200 μF
    d. 3450 μF

37. Which of the following capacitors could be used to replace a 35 μF, 370 V running capacitor?
    a. 35 μF, 330 V
    b. 35 μF, 390 V
    c. 30 μF, 440 V
    d. 40 μF, 370 V

38. Which of the following capacitors could be used to replace a 188 μF, 250 V starting capacitor?

    a. 188 μF, 120 V
    b. 259 μF, 120 V
    c. 200 μF, 250 V
    d. 300 μF, 250 V

39. Which of the following capacitors or combination of capacitors could be used to replace a 45 μF, 370 V running capacitor?

    a. 40 μF, 440 V
    b. 30 μF, 250 V
    c. 30 μF, 370 V
    d. 15 μF, 440 V

40. Which of the following capacitors or combination of capacitors could be used to replace an 88 μF, 250 V starting capacitor?

    a. 180 μF, 250 V
    b. 180 μF, 330 V
    c. 150 μF, 120 V
    d. 150 μF, 250 V

41. Find the common, start, and run terminals of the following hermetic compressors.

    a.
    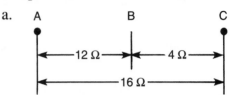

    b.
    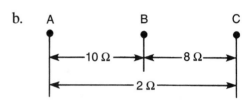

    c.
    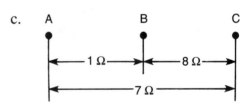

42. Briefly explain the procedure for troubleshooting hermetic compressor motors.

43. What are the electrical failure categories for hermetic compressor motors?

44. What precautions should be taken when checking hermetic compressor motors?

45. What would be the highest allowable resistance reading for a grounded compressor motor?

PRACTICE SERVICE CALLS

Determine the problem and recommend a solution for the following service calls. (Be specific; do not list components as good or bad.)

Practice Service Call 1

Application: Residential conditioned air system

Type of Equipment: Packaged heat pump

Complaint: Intermittent heat

Symptoms:

1. All components of equipment operating properly except the indoor fan motor.
2. Indoor fan motor is turning slower than normal.
3. Indoor fan motor is drawing 12 amps (Nameplate amps 3.2 A).
4. Indoor fan motor cuts off approximately every 10 minutes for a considerable period of time.
5. Ohmmeter reading of run winding is 2 ohms.
6. Ohmmeter reading of start winding is 12 ohms.
7. Ohmmeter reading of capacitor is infinite.

Practice Service Call 2

Application: Commercial refrigeration (walk-in freezer)

Type of Equipment: Air-cooled condensing unit with evaporator (unit voltage 240 V-1ø-60 Hz)

Complaint: No refrigeration

Symptoms:

1. Compressor out due to opening of high-pressure switch. (Technician resets high-pressure switch.)
2. Compressor operates after resetting the high-pressure switch, but the condenser fan motor does not operate.
3. The condenser fan motor is cold to the touch.
4. Voltage to the condenser fan motor is 240 volts.
5. Ohm reading of the start winding of the condenser fan motor is 0 ohms.

Practice Service Call 3

Application: Commercial and industrial conditioned air system

Type of equipment: Commercial and industrial fan coil unit (unit voltage 240 V-3ø-60 Hz)

Complaint: Fan does not operate

Symptoms:

1. Motor supplied with 240 V-3ø-60 Hz.
2. Resistance reading of motor is between T_1 and T_4 and T_2 and T_5 = infinite ohms, T_3 and T_6 = 2 ohms and T_7 and T_9 = infinite ohms. (Motor is dual voltage, wye winding.)
3. Motor is cold to the touch.

Practice Service Call 4

Application: Domestic refrigeration

Type of Equipment: Frost-free refrigerator

Complaint: No refrigeration

Symptoms:

1. Compressor does not run.
2. Compressor starting components are good.
3. Resistance readings of compressor terminals are: C to S = 12 ohms, C to R = 0 ohms, and S to R = 0 ohms.

Practice Service Call 5

Application: Residential conditioned air system

Type of Equipment: Gas furnace with air-cooled condensing unit

Complaint: No cooling

Symptoms:

1. Indoor fan operating normally.
2. Condenser fan motor operating normally.
3. Compressor is extremely hot to the touch.
4. Resistance readings of the compressor terminals are: C to R = 2 ohms, C to S = 13 ohms, and R to S = 15 ohms.
5. Resistance reading between the run terminal and compressor housing = 200 ohms.

Practice Service Call 6

Application: Residential conditioned air system

Type of Equipment: Air-cooled packaged unit

Complaint: No cooling

Symptoms:

1. Indoor fan motor operating normally.
2. Condenser fan motor operating normally.
3. Compressor is extremely hot to the touch.
4. Resistance readings of the compressor terminals are: R to S = 15 ohms and C to R & S = infinity.
5. All capacitors and starting components are good.
6. Compressor cools and upon supplying power has a current draw of 85 amps. (Nameplate full load amps = 24 A)

LAB MANUAL REFERENCE

For experiments and activities dealing with material covered in this chapter, refer to Chapter 9 in the Lab Manual.

10

Components for Electric Motors

OBJECTIVES

After completing this chapter, you should be able to

- Identify and explain the operation of motor starting relays and other starting components that are used on single-phase hermetic compressor motors.
- Select the correct potential relay for an application with information available on the potential relay to be replaced.
- Troubleshoot and install motor starting relays on hermetic compressor motors.
- Lubricate and identify the types of bearings used in electric motors.
- Identify the type of motor drives used on industry applications.
- Calculate the variables in a V-belt drive application to obtain the desired equipment rpm.
- Recognize and adjust a V-belt application to the proper tension and alignment.

KEY TERMS

Back electromotive force
Ball bearings
Bearing
Current or magnetic relay
Direct drive
Full-load amperage
Hot-wire relay

Locked rotor amperage
Potential relay
Sleeve bearings
Solid-state relay
Starting relay
V-belt

INTRODUCTION

In Chapter 9 we discussed electric motors and some of their starting components. In this chapter we discuss the starting components used on single-phase hermetic motors. Single-phase hermetic motors and other special motors require some type of external starting component because they are enclosed in a sealed case. Four types of starting relays are used on this type of motor: current, potential, hot-wire, and solid-state. These devices are used on most single-phase hermetic compressor motors with the exception of permanent split-capacitor motors.

Electric motors must have bearings to permit smooth and easy rotation. The ball bearing or the sleeve bearing is used in most motors. Motors also must have some means of transferring their rotating motion to the device being powered by the motor. A direct-drive hookup transfers the rotating motion directly from the motor to the device. The belt-drive hookup transfers the rotating motion to the device by a belt connection.

Magnetic starters and push buttons are used to stop and start electric motors. A magnetic starter opens and closes sets of contacts to stop and start loads. The magnetic starter also incorporates overload protection for the device it controls. Push-button switches are used to control magnetic starters in most cases.

10.1 STARTING RELAYS FOR SINGLE-PHASE MOTORS

Single-phase motors, with the exception of the permanent split-capacitor motor, must have some means of dropping the starting winding (or the starting capacitor in the case of a capacitor-start–capacitor-run motor) out of the circuit. In an open-type motor, this is accomplished simply and easily by a centrifugal switch mounted in the motor. The switch opens the starting circuit once the motor reaches 75% of its full speed. In enclosed motors some type of **starting relay** must be used.

Basically, four types of starting relays are used in the industry. The first three types—current, hot-wire, and solid-state relays—are generally used on small hermetic motors. The fourth type, the potential relay, can be adapted to any size motor. Starting relays are essential to the operation of a hermetic compressor motor with the exception of the permanent split-capacitor motor.

Each of the four types of starting relays uses a different method to drop the starting circuit in or out. A potential relay operates on the principle of

back electromotive force. The back electromotive force is the amount of voltage produced in the starting winding of a motor. The current or magnetic type of relay operates on the current or amperage that the motor uses to start. Hot-wire relays use current flow to produce heat across a thermal element, which operates the starting circuit. The solid-state relay uses a positive temperature coefficient (PTC) material that effectively removes the start winding or component from the circuit. Each of these relays must be correctly sized and matched to the application, except the solid-state relays that can be used over certain horsepower ranges. Each relay is designed to remove the starting circuit when the motor reaches approximately 67% to 75% of full speed.

10.2 CURRENT OR AMPERAGE RELAYS

On all electric motors, the starting amperage is greater than the running amperage because the rotor is at a standstill on startup. The starting amperage of an electric motor is usually stated as the **locked rotor amperage** (LRA). The locked rotor amperage is the maximum amp draw of the motor when the motor is in a locked rotor condition. The running amperage of an electric motor is usually stated as the **full-load amperage** (FLA). Therefore, the ampere draw of the motor is high at the time of the initial startup, but as the motor gains speed, the ampere draw decreases. The **current or magnetic relay** uses the electrical characteristics of the motor to remove the starting circuit electrically once the motor has established a good running speed, approximately 75% of full speed.

Operation

The current relay is built much like a solenoid, with copper wire wrapped around a steel hollow core holding a steel plunger, as shown in Figure 10.1. Figure 10.2 shows a cutaway view of a current relay. The contacts of the current relay are normally open, as shown schematically in Figure 10.3. The contacts of most current relays are protected by covers that cannot be removed, so it is almost impossible to visually inspect the contacts.

Figure 10.4 shows the position of the contacts when the motor is in the starting phase. As the speed of the motor increases, the amperage decreases. When the motor has reached 67% to 75% of its full speed, the amperage will be low enough to cause the magnetic field strength of the relay coil to decrease enough to drop the relay contacts out of the starting circuit.

FIGURE 10.1 Current relays

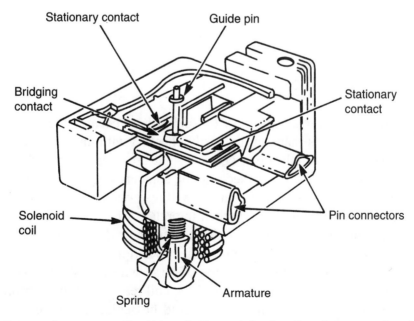

FIGURE 10.2 Cutaway of current-type relay *(Adapted with permission from Texas Instruments, Inc., Attleboro, MA)*

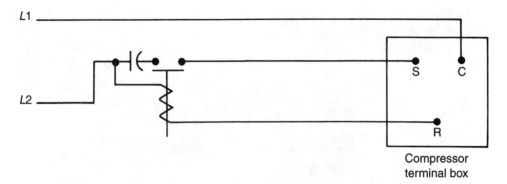

FIGURE 10.3 Schematic diagram of a current relay when it is de-energized

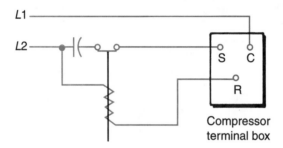

FIGURE 10.4 Schematic diagram of a current relay when it is energized

Troubleshooting

Most current relays are easy to troubleshoot because they have a coil and a set of contacts than can be easily checked with an ohmmeter. The current relay has normally open contacts that are easily checked by turning the relay upside down and checking the contacts with an ohmmeter. If the contacts are good, the relay contacts will read open when the relay is right side up. They will read closed when the relay is inverted. It is imperative when checking a current relay with an ohmmeter that the ohm readings of the contacts be obtained from the correct position of the relay.

The coil of the relay is made of large wire and should have a very low resistance, around 0 to 1 ohm. If the coil reads any higher, more than likely the coil is bad and the relay should be replaced. The current relay and

FIGURE 10.5 Commercial relay and motor tester

motor starting components should be completely checked to ensure that no other components are bad. Figure 10.5 shows a commercial relay tester that can be used to check current relays.

10.3 POTENTIAL RELAYS

The **potential relay** is gaining popularity because it can be adapted easily to most any compressor. The back electromotive force produced by the starting winding of a motor is the controlling factor of a potential relay.

Operation

When a single-phase motor is operating, a voltage is produced across the starting windings above and beyond the voltage being applied to the motor. The starting windings actually act as a generator to produce the back electromotive force of a motor. The back electromotive force of a motor corresponds to the motor speed. The potential relay is designed to open, dropping the starting circuit, when the motor reaches a certain back electromotive force that is predetermined by the manufacturer of the motor.

Figure 10.6 shows two potential relays that look much like ordinary general-purpose relays. The wiring diagram for a potential relay is shown in Figure 10.7. The potential relay has an advantage over other relays because its contacts are normally closed when the unit starts, so there is no arcing. Arcing is an electric spark that is produced across two sets of

FIGURE 10.6 Potential relays

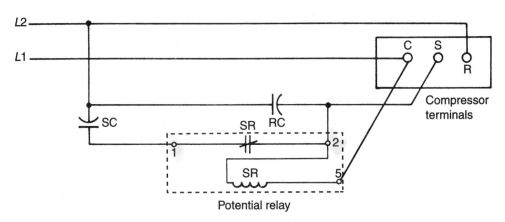

Potential relay

FIGURE 10.7 Schematic diagram of a potential relay

contacts. As the motor speed increases, so does the back electromotive force. When the speed approaches 75% to 80% of full speed, the back electromotive force is large enough to drop the starting circuit by energizing the potential relay coil. The potential relay is rated by, and operates on, three voltage ratings: pickup voltage—the minimum voltage required to energize the coil and open the NC contacts; dropout voltage—the mini-

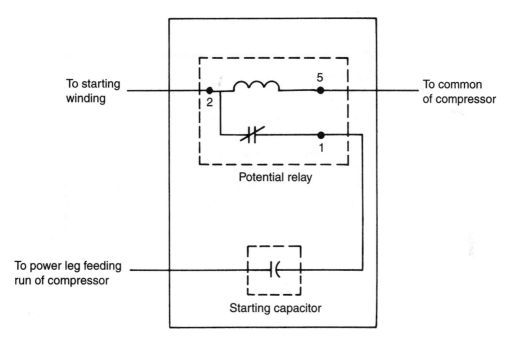

To starting winding

5

2

1

To common of compressor

Potential relay

To power leg feeding run of compressor

Starting capacitor

FIGURE 10.8 Schematic diagram of a hard-starting kit

mum voltage the coil requires once it is energized to keep the contacts from closing; and continuous coil voltage—the maximum voltage at which the coil can continuously operate.

On many permanent split-capacitor hermetic motors that are experiencing difficulty starting, a hard-starting kit can be installed to eliminate the problem, making the PSC motor a capacitor-start–capacitor-run motor. A hard-starting kit consists of a potential relay and a starting capacitor. Figure 10.8 shows the schematic diagram of a hard-starting kit, and Figure 10.9 shows a hard-starting kit as it would come from a manufacturer.

Troubleshooting

The potential relay is easy to troubleshoot because only two parts of the relay must be checked. The coil of the potential relay can be checked with an ohmmeter across 2 and 5 (Figure 10.7), which could check out as good, open, or shorted. The contacts of the potential relay across 1 and 2 should be checked, and they should be closed. Motor current will increase if the potential relay does not drop the starting circuit at the proper time. This condition is dangerous to the motor and the starting apparatus. The

FIGURE 10.9 Hard-starting kit

potential relay should always be checked completely to prevent this condition from occurring. A service technician should always check the operation of a unit after the dropout of a starting relay. The correct size relay should always be used for replacement of a faulty relay. If the starting relay is working properly, there will be no ampere draw through the starting circuit after the motor has reached full speed. Figure 10.10 shows a commercial potential relay tester.

FIGURE 10.10 Commercial potential relay tester

The use of potential relays on single-phase hermetic motors has increased over the past 10 years. Potential relays vary greatly in number of terminals, coil group, pickup voltage, and mounting position. An explanation of the number system of the two major manufacturers of potential relays should be helpful in the field.

The two major brands of potential relays are General Electric and RBM. Explanations of their numbering codes are given in Figures 10.11 and 10.12, respectively. Pay careful attention to the coil group and the pickup voltage when you cross-reference a potential relay.

Explanation of G.E. Potential Relay Code

3ARR3 − = Potential Relay Type   EXAMPLE: 3ARR3-A5C3

1st digit after − = Letter indicating number of term. + bracket

    A = 5 screw terminal "L" Bracket.   D = 3 screw terminal Flat Bracket.
    B = 5 screw terminal Flat Bracket.   U = 5 quick connect term.
    C = 3 screw terminal "L" Bracket.   "L" Bracket.

2nd digit = Number designating Coil Group

    2 = 168 Continuous volts.   6 = 420 Continuous volts.
    3 = 332 Continuous volts.   7 = 130 Continuous volts.
    4 = 502 Continuous volts.   8 = 214 Continuous volts.
    5 = 253 Continuous volts.   10 = 375 Continuous volts.

3rd digit = Letter indicating calibration (Hot Pick Up)

A = 260 − 280	H = 365 − 395	R = 180 − 190
B = 280 − 300	J = 120 − 130	S = 190 − 200
C = 300 − 320	K = 130 − 140	T = 200 − 220
D = 320 − 340	L = 140 − 150	U = 220 − 240
E = 340 − 360	M = 150 − 160	V = 240 − 260
F = 350 − 370	N = 160 − 170	W = 210 − 230
G = 360 − 380	P = 170 − 180	

    Room Temperature calibration is 5 to 7% lower than these values.

4th digit = Number indicating Mounting Position (See note, end of section)

    1 = Face Down
    2 = Face Up
    3 = Face Out − Numbers Horizontal
    4 = Face Out − Rotated 90° clockwise from number 3 position.
    5 = Face Out − Numbers upside down − Horizontal.
    6 = Face Out − Rotated 90° counterclockwise from Number 3 position.

FIGURE 10.11 Explanation of G.E. potential relay code *(Courtesy of Tecumseh Products Co.)*

Explanation of RBM Potential Relay Code

128− = Potential Relay Type   EXAMPLE: 128-122-1335CA

1st & 2nd digit after− = Numbers indicating Type of Bracket.
  11 = Flat Bracket remote (Tecumseh).
  12 = "L" Bracket (Tecumseh).
  16 = "L" Bracket for "FB" model compressors.
  20 = "L" Bracket for Tecumseh Twins — 1½ HP and Larger.
  21 = "L" Bracket for capacitor box mounting.
  29 = Flat Bracket (Marion) was "14" (under cover).

3rd digit = Contact structure.
  2 = SPNC − less than 1½ HP.
  6 = SPNC − 1½ HP and Larger.

4th & 5th digits = Number of terminals, type and location
  11 = 3 screw terminal.
  12 = 4 screw terminal (seldom used).
  13 = 5 screw terminal.
  23 = 5 quick connect terminals.

6th digit = Number indicating Coil Group
  1 = 130 Continuous voltage   5 = 395 Continuous voltage
  2 = 170 Continuous voltage   6 = 420 Continuous voltage
  3 = 256 Continuous voltage   7 = 495 Continuous voltage
  4 = 336 Continuous voltage

7th digit = Number indicating Mounting Position (See note)
  1 = Face Down
  2 = Face Up
  3 = Face Out − Horizontal − Numbers upside down.
  4 = Face Out − 90° clockwise from No. 3 position.
  5 = Face Out − Horizontal − Numbers right side up.
  6 = Face Out − 90° counterclockwise from Number 3 position.

8th digit = Letter indicating calibration (Hot Pick Up)
  A = 260 − 280 volts.   G = 360 − 380 volts.   P = 170 − 180 volts
  B = 280 − 300 volts.   H = 365 − 395 volts.   R = 180 − 190 volts.
  C = 300 − 320 volts.   J = 120 − 130 volts.   S = 190 − 200 volts.
  D = 320 − 340 volts.   K = 130 − 140 volts.   T = 200 − 220 volts.
  E = 340 − 360 volts.   L = 140 − 150 volts.   U = 220 − 240 volts.
  F = 350 − 370 volts.   M = 150 − 160 volts.   V = 240 − 260 volts.
                                                  W = 210 − 230 volts.

Room Temperature calibration is 5 to 7% lower than these values.

9th digit = Letter to indicate customer's part number to be stamped on relay.

NOTE:—Mounting of Relay

As noted above, the 4th digit in the code number of G.E. relays and the 7th digit for RBM relays indicates the position in which the relay is to be mounted. It is of utmost importance that the relay be mounted in the required position. Mounting in any other position can change the relay's operating characteristics enough so that the compressor will not start properly. This can result in burning out the compressor motor.

FIGURE 10.12 Explanation of RBM potential relay code *(Courtesy of Tecumseh Products Co.)*

FIGURE 10.13 Hot-wire relay

10.4 HOT-WIRE RELAYS

The **hot-wire relay,** shown in Figure 10.13, operates on the principle that electric energy can be converted to heat. The relay uses two bimetal strips. One strip operates the starting circuit; a second strip acts as an overload for the running winding.

Operation

Figure 10.14 shows the schematic of a hot-wire relay. From L to A is the actual hot wire, which would heat up the bimetal elements B and C. C goes to the starting winding and B to the running winding. If the hot wire reaches a temperature high enough to cause element C to warp, the starting circuit is dropped out. This is the correct procedure. If the hot wire reaches an even higher temperature, this causes element B to warp and open the circuit to the running winding. Then the compressor would cut off on overload by the relay. Some manufacturers allow the hot wire to stretch as it gets hot to open and close the contacts. Both types (warping and stretching) are similar in operation. Figure 10.14 shows a hot-wire relay hooked up in a system. Hot-wire relays are sized or rated by the motor's voltage and horsepower.

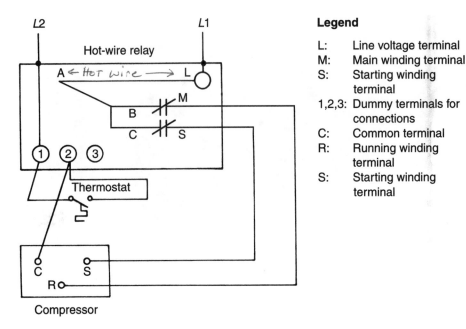

FIGURE 10.14 Schematic diagram of a hot-wire relay connected to a compressor

Troubleshooting

The hot-wire relay is probably the hardest starting relay to check. The relay is hard to check because it is hard to completely detect the heat being supplied to the relay. Even the slightest temperature difference will cause the relay to show an overload.

The relay contacts are checked with an ohmmeter. The diagnosis of the hot wire and thermal elements of the relay is done while the unit is operating or starting. The first thing to check for is to see if the motor is running. If the motor is running but cutting out on overload, the bearings inside the motor might be tight.

The starting winding should drop in and out quickly and is, therefore, hard to detect. The amperage drawn by the starting winding should be checked to determine if it is too high. Usually, the starting winding will draw a higher amperage than the running winding. As for other starting relays, in troubleshooting the hot-wire relay, the relay and all other motor components should be checked.

10.5 SOLID-STATE STARTING RELAYS AND DEVICES

The advancement of solid-state controls has produced another type of current-sensitive relay, a PTC starting switch. Certain ceramic materials

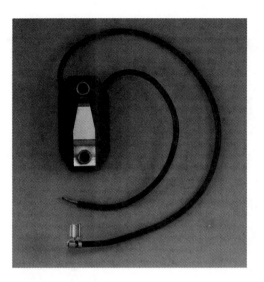

FIGURE 10.15 Solid-state starting relay *(Courtesy of Sealed Unit Parts Co., Inc., Allenwood, NJ)*

FIGURE 10.16 Solid-state starting relay installed on a compressor *(Courtesy of Tecumseh Products Co.)*

(PTC) increase their resistance as they heat up from current passing through them. This is the basis of the **solid-state relays** that are replacing many of the other varieties on fractional horsepower motors. A PTC solid-state starting relay is shown in Figure 10.15. The solid-state starting relay used on small split-phase and capacitor-start motor compressors is simple in its operation. The device is used to drop out the starting windings and starting capacitor of a small hermetic compressor motor. Figure 10.16 shows a solid-state relay installed on a fractional-horsepower hermetic compressor. The flexibility of the solide-state relay has greatly expanded its usage because one relay will cover a range of horsepowers. Manufacturers are producing a PTC relay with a built-in starting capacitor, as shown in Figure 10.17, and a built-in starting capacitor and overload, as shown in Figure 10.18.

Permanent split-capacitor (PSC) motors often need assistance in starting. A potential relay and starting capacitor can be used to change a normal PSC motor into a capacitor-start–capacitor-run motor with a high starting torque. On some occasions a positive temperature coefficient start device is used to increase the starting torque of a PSC motor. The PTC start device is easier to install and less expensive than the conventional potential relay and starting capacitor. Figure 10.19 shows the PTC start device. This device is wired in parallel with the run capacitor. The PTC start device performs much like a small starting capacitor; it momentarily increases the

FIGURE 10.17 Solid-state relay with built-in capacitor

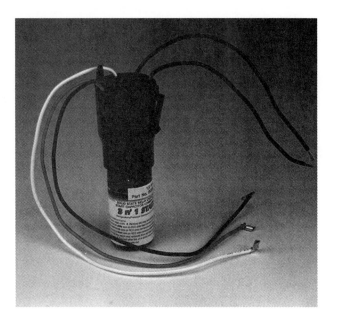

FIGURE 10.18 Solid-state relay with built-in capacitor and overload

FIGURE 10.19 Positive temperature coefficient starting device *(Courtesy of Bill Johnson)*

current in the motor starting winding. As the material heats up, its resistance increases quickly to the point where it becomes a nonconductor and the motor returns to PSC operation. This device requires a three-minute cool-down period between starts. To check this device, the service technician need only determine if it is dropping out of the circuit or if it is being energized on startup. It should be replaced if it is not working properly.

An electronic motor starting relay has been introduced that can be used as a universal replacement for many motor starting relays in the industry. This electronic relay operates on a time function principle rather than sensing voltage or current. The time function design allows this relay to be used as a replacement for many potential and current starting relays. This relay is shown in Figure 10.20. The wiring diagram is shown in Figure 10.21 with the relay replacing a potential relay and in Figure 10.22 replacing a current-type relay.

Operation

The solid-state starting relay is placed in series with the start winding of the fractional horsepower hermetic compressor, as shown in Figure 10.23. The solid-state starting relay normally has a very low resistance. As the compressor motor starts, the current flows to the start winding. The resistance of the solid-state starting relay rapidly rises very high. Therefore, the current flow to the start winding is reduced greatly, which takes the starting winding out of the circuit. Most solid-state starting relays can also be used with a capacitor placed in series with the relay and start winding, as shown

FIGURE 10.20 Universal motor starting relay *(Courtesy of Sealed Unit Parts Co., Inc., Allenwood, NJ)*

WIRING AS A POTENTIAL RELAY

REPLACES ALL POTENTIAL MOTOR STARTING
RELAYS FOR COMPRESSORS RATED AT
120 TO 270 VAC SINGLE PHASE
UP TO 5 H.P. ANY PICK-UP VOLTAGE

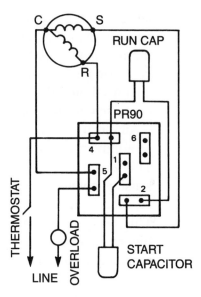

DIAGRAM NOTE:
A wire from the RUNNING
TERMINAL of the compressor
motor must be connected to #4
on the relay.

FIGURE 10.21 Wiring of universal relay replacing potential relay *(Courtesy of Sealed Unit Parts Co., Inc., Allenwood, NJ)*

PRO-90 WIRING AS A
CURRENT STARTING RELAY

FIGURE 10.22 Wiring of universal relay replacing a current relay *(Courtesy of Sealed Unit Parts Co., Inc., Allenwood, NJ)*

in Figure 10.24. One advantage of the solid-state starting relay is that one model covers a variety of horsepower ranges. For example, one relay could be used from ½ horsepower to ⅕ horsepower, and another from ¼ horsepower to ⅓ horsepower. Some manufacturers produce this type of relay to cover compressor motors above ⅓ horsepower. The major advantage of this type of relay is that the service technician can stock three basic relays and cover many applications and sizes.

These starting devices are also available with a built-in starting capacitor (refer again to Figure 10.17). When the relay capacitor is used, the existing overload must remain on the compressor for protection. A 3-in-1 relay is

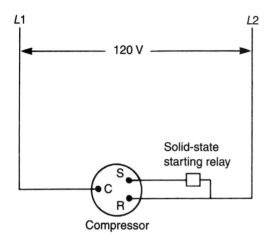

FIGURE 10.23 Schematic diagram of compressor and solid-state starting relay (split-phase motor)

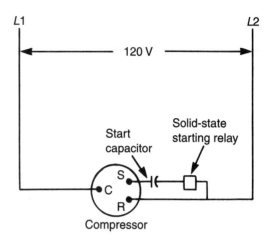

FIGURE 10.24 Schematic diagram of compressor and solid-state starting relay (capacitor-start motor)

also available (refer again to Figure 10.18) that contains the solid-state relay, a capacitor, and an overload.

The solid-state PTC-type starting relay is usually limited to domestic refrigeration applications, but it can be used on other applications if the manufacturer's installation instructions are followed. The solid-state relay usually requires 3 to 10 minutes to cool down between operating cycles, so it is not feasible for short-cycling applications.

Troubleshooting

The solid-state starting relay is easy to troubleshoot because of its simplicity. On many occasions, the relay will show some visible signs that would indicate the condition of the relay. In some cases, the relay will be charred, severely burned, and cracked; in other instances, the relay will only be slightly burned. The relay can also be checked with an ohmmeter: if the resistance is very low, the relay is good; a high resistance across the relay indicates a bad relay. The technician must make certain the relay is cool when performing the resistance check. The service technician could also check to make certain the relay is dropping the starting winding out of the circuit by using a clamp-on ammeter if a wire is connecting the relay to the compressor terminal. If the relay pushes directly onto the compressor terminals, this check is practically impossible; hence, an amp reading of the compressor motor must be taken, and a judgment must be made whether the start winding remained in the circuit.

10.6 MOTOR BEARINGS

All rotating electric equipment has some type of bearing to allow for smooth and easy rotation. A **bearing** is that part of a rotating electric device that allows free movement. The two types of bearings used in the heating, cooling, and refrigeration industry are ball bearings and sleeve bearings, shown in Figure 10.25. The ball bearing is the most efficient because it produces less friction. The sleeve bearing is cheaper, however, and is more commonly used.

Ball Bearings

Ball bearings are designed with an inner and an outer ring, which enclose the balls by use of a separator. The inner ring is the bore in which the shaft is pressed.

Ball bearings are lubricated in three ways. The permanently lubricated bearings are sealed at the factory and only rarely require additional lubrication. The packed lubrication of ball bearings is done by hand. The bearing must be disassembled and hand packed with grease every two to five years. Many ball bearings are equipped with a grease fitting and should be lubricated every two years. Lubrication is of prime importance to a motor. And note that overlubrication is as damaging as underlubrication. Figure 10.26 shows a ball bearing pressed into an end bell of a motor.

FIGURE 10.25 Sleeve and ball bearings used in the industry

FIGURE 10.26 Mounted ball bearing

Ball bearings are used on most heavy loads. However, ball bearings cannot be used in a hermetic compressor because of the danger of sparks. Some of the many advantages of using ball bearings are mountings in any position, antifriction design, versatility, and the ability to carry a large load. Ball bearings are often permanently lubricated from the factory or require infrequent lubrication. Thus, they require less maintenance than a sleeve bearing. The life of a ball bearing is usually long and trouble free, but misuse can rapidly destroy a ball bearing.

Sleeve Bearings

Sleeve bearings are brass or bronze cylinders in which a shaft rotates. The bearings have more friction than the ball bearing and thus are used for lighter duty. Lubrication of sleeve bearings is accomplished by oil wick lubrication, yarn-packed lubrication, or oil ring lubrication.

The oil wick lubrication has a wick that extends into an oil reservoir. The wick picks up oil from the reservoir and transfers it to the shaft, as shown in Figure 10.27. The reservoir should be filled with oil twice a year.

The yarn-packed lubrication is merely yarn packed around the shaft to lubricate it, as shown in Figure 10.28. This bearing should be lubricated every few months.

The oil ring lubrication is used on large motors. In this device a ring rotates through an oil reservoir, picking up oil and transferring it to the shaft, as shown in Figure 10.29. The oil level of the reservoir should be checked monthly.

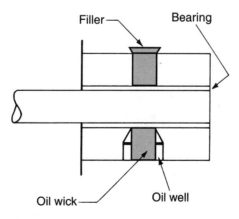

FIGURE 10.27 Oil wick lubrication of a sleeve bearing

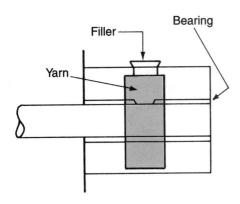

FIGURE 10.28 Yarn-packed lubrication of a sleeve bearing

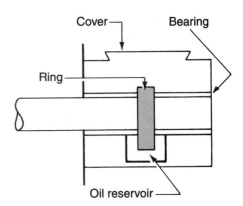

FIGURE 10.29 Oil ring lubrication of a sleeve bearing

Hermetic compressors use sleeve bearings because with sleeve bearings there is no danger of sparks. Care must be taken in mounting motors equipped with sleeve bearings because of the lubrication problems when the bearings are mounted in a vertical position. Sleeve bearings are commonly used in many applications in the industry.

10.7 MOTOR DRIVES

A motor drive is the connection between an electric motor and a component that requires rotation. Electric motors are used to drive most devices that require rotation. There are two basic driving devices: direct drive and belt and pulley drive.

Direct Drive

Direct-drive methods require that a device turn with the same revolutions per minute as the motor. Fan motors and pumps are often direct drive; hermetic compressors are always direct drive. Direct-drive hookups require a close fit between the motor and the device. In a hermetic compressor, the crankshaft is made in one piece, with the motor on one end and the compressor portion on the opposite end. Direct-drive applications usually require a coupling, except with hermetic compressors and fan motors.

The two types of direct-drive couplings used in the industry are the flexible-hose coupling, as used on oil burners, and flange couplings, as used on some open types of compressors and hot-water pumps. Figure 10.30 shows both commonly used types of direct-drive couplings.

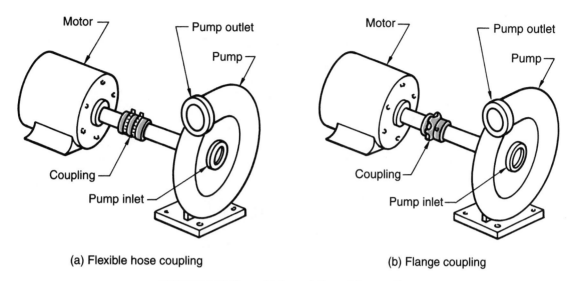

(a) Flexible hose coupling   (b) Flange coupling

FIGURE 10.30 Two widely used direct-drive couplings

 CAUTION When removing a device from the shaft of an electric motor, make sure that the motor shaft or the device being removed is not damaged.

V-Belt

Although many types of belts are used to drive devices, only the **V-belt** will be discussed here. It is used almost exclusively in the industry. By using two pulleys and connecting them with a V-belt, rotation can be transferred from the motor to the device, as shown in Figure 10.31. V-belts are used in many applications in the industry, such as driving open types of compressors, fan motors, and pumps.

 CAUTION The technician should never wear loose clothing or neckties when working around rotating equipment. Long hair should also be tied back.

There are basically three sizes of V-belts used at present. Size FHP V-belts are ⅜ inch wide and used on fractional-horsepower motors. Size A-section is ½ inch wide and used for most jobs requiring one- to five-horsepower motors. Size B-section is ²¹⁄₃₂ inch wide and used when the motor pulley is

FIGURE 10.31 V-belt connection between fan and motor (*Courtesy of Grainger*)

larger than five inches in diameter. More than one V-belt is often used to pull large devices. When belt changes are needed, a matched set should be used in a dual-belt application. The equipment's revolutions per minute can be changed by changing the pulley size or sizes. The following formula can be used:

$$\text{diameter of equipment pulley} = \frac{\text{motor rpm} \times \text{motor pulley size}}{\text{rpm of equipment}}$$

This formula can be transposed to solve for any of the stated values by simple algebra. For example, if the motor rpm were 1750 with a five-inch pulley, what size equipment pulley would be required to produce a speed of 1000 rpm?

$$\text{diameter of equipment pulley} = \frac{\text{motor rpm} \times \text{motor pulley size}}{\text{rpm of equipment}}$$

$$\text{diameter of equipment pulley} = \frac{1750 \times 5}{1000}$$

$$\text{diameter of equipment pulley} = 8.75 \text{ inches}$$

V-belts should always be properly aligned and the tension adjusted. Figure 10.32(a) shows an incorrectly aligned V-belt. The proper amount of

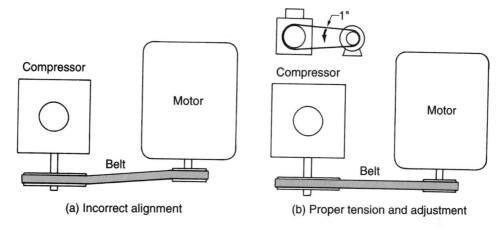

(a) Incorrect alignment   (b) Proper tension and adjustment

FIGURE 10.32 V-belt alignment

tension for most belts is shown in Figure 10.32(b). Frequently a V-belt will slip, causing erratic equipment speed or breakage. In these cases, the belt should be replaced and the tension adjusted correctly.

 Always use caution when working around an electric motor that is transferring rotating motion with V-belts.

10.8 SERVICE CALLS

Service Call 1

Application: Commercial refrigeration (walk-in cooler)

Type of Equipment: Air-cooled condensing unit with evaporator

Complaint: No refrigeration

Service Procedures:

1. The technician reviews the work order from the dispatcher for available information. The work order information reveals that the walk-in cooler of the meat market is hot with rapid deterioration of stored product. The condensing unit is located outside behind the structure.
2. The technician informs the manager of his or her presence and obtains any additional information about the problem.

3. The manager has noticed that the compressor is starting and running only a short time before cutting off. The condenser fan motor is running normally.

4. The equipment is 240 V-1ø-60 Hz. With the condenser fan motor picking up power from the terminals of the compressor, the technician assumes that voltage is being applied to the compressor. The technician measures the current draw of the compressor while running. The current draw is seven times that of the normal running amps of the compressor.

5. The compressor overload breaks power to the compressor after only a few seconds of operation.

6. The technician turns the power off and locks or labels it when working with live circuits.

7. The technician checks the compressor motor and determines that the motor is in good condition.

8. The starting capacitor is hot to the touch. The technician checks to determine if the capacitor is being removed from the circuit by checking the amp reading of the capacitor; if the capacitor is reading a current draw then the capacitor is remaining in the circuit. The potential relay is checked to determine the condition of the relay contacts and coil. The technician reads the resistance between terminals 1 and 2 and 2 and 5. The reading between 1 and 2 reads 0 ohms. Terminals 1 and 2 are the normally closed contacts of the relay and should read 0 ohms. The reading between 2 and 5 reads infinite ohms. Terminals 2 and 5 are the relay coil terminals and should have a very high but measurable resistance. The potential relay coil is open, allowing the starting capacitor to remain in the circuit and causing the motor to overload.

9. The part number of the faulty relay is an RBM 128-122-1335CA. The relay must be replaced with a potential relay with the continuous coil voltage of 256 and a coil calibration of 300 to 320 volts. The technician finds the correct relay from the available supply on the service truck.

10. The technician replaces the relay and checks to make certain that the new relay is removing the capacitor from the circuit by observing the current draw of the capacitor. There should be a current draw through the capacitor while the motor is starting, but when the motor reaches a certain speed, the capacitor current draw should be 0 amps.

11. The technician checks the current draw of the compressor to make certain the compressor is operating properly.

12. The technician informs the manager of the problem and that it has been corrected.

Service Call 2

Application: Domestic refrigeration

Type of Equipment: Domestic chest-type freezer

Complaint: Freezer defrosting

Service Procedure:

1. The technician reviews the work order from the dispatcher for available information. The work order information reveals that the freezer is located in the utility room under the carport and will be unlocked. The customer has commented that the compressor tries to start but does not run. The freezer has been unplugged.
2. The technician plugs the freezer in and notices that the compressor tries to start but does not and the overload takes it out of the circuit. When checking a chest-type freezer the technician knows that only three or four components will need to be checked when the compressor is attempting to start: the compressor, relay, overload, and capacitor, if used. The overload is seldom the problem, so the technician will check that last.
3. The technician unplugs the freezer.
4. The technician reads the resistance of the compressor and determines that the compressor is electrically good. The relay is a current type relay and must be checked next. The relay has a normally open set of contacts and a magnetic coil that closes the contacts when the compressor tries to start.
5. The technician checks the resistance of the coil and reads infinity. The technician determines that the current-type relay coil is open and not energizing the starting winding, thus preventing the compressor from starting. The freezer motor is a split-phase motor and does not require a capacitor.
6. The technician obtains the correct relay from the supplier and replaces the faulty relay.
7. The technician starts the freezer and checks the current draw of the compressor to make certain the freezer is operating properly.
8. The technician leaves a note informing the homeowner of the service performed.

Service Call 3

Application: Residential conditioned air system

Type of Equipment: Packaged heat pump

Complaint: No heat

Service Procedure:

1. The technician reviews the work order from the dispatcher for available information. The work order reveals that the unit is a packaged heat pump. Someone will be home.
2. Upon entering the residence, the technician makes certain that no dirt or foreign materials are carried into the structure. The technician also takes care not to mar interior walls. The customer informs the technician that the auxiliary heat light has been on when the unit is operating.
3. This heat pump has a capacitor-start–capacitor-run compressor motor. With the indoor fan motor operating properly, the technician knows that voltage is available to the unit.
4. The technician removes the compressor access panel. The compressor is hot to the touch, indicating that it had been attempting to start and failed, resulting in an open internal overload.
5. The technician must determine why the compressor is not starting. The compressor and motor are found to be in good electrical condition. The starting components of the compressor must be checked to determine if they are good.
6. The technician uses an ohmmeter to check both starting and running capacitors and finds them good. The potential relay contacts must be checked. An ohm reading of infinity is read across terminals 1 and 2, indicating that the normally closed contacts are open and the start capacitor is not being introduced to the motor starting circuits.
7. The technician turns the power supply off and locks it or labels it.
8. The technician replaces the faulty potential relay with the proper relay.
9. Power is restored to the unit and the technician checks the operation of the compressor, making certain that the capacitor is being removed from the circuit.
10. The technician reads the current of the motor and makes certain the compressor is operating properly.
11. The technician replaces the compressor access cover.
12. The technician informs the customer of the corrected problems.

Service Call 4

Application: Commercial refrigeration (refrigerator)

Type of Equipment: Air-cooled condensing unit mounted on top with evaporator

Complaint: No refrigeration

Service Procedure:

1. The technician reviews the work order from the dispatcher for available information. The work order information reveals that the temperature of the refrigerator is above a safe temperature for the product stored.
2. The technician informs the owner of the sandwich shop of his or her presence and obtains specifics about the operation of the refrigerator.
3. The technician makes every attempt not to disturb the activity of kitchen personnel.
4. The compressor is hot to the touch and not operating, but the condenser fan motor is operating, indicating that power is available to the compressor.
5. The technician turns the power supply off and locks or labels it.
6. The technician determines that the compressor and motor are in good condition.
7. The technician observes that the compressor attempts to start. The technician measures the current draw of the compressor; the current draw is extremely high.
8. The technician checks the current-type relay. The resistance reading for the coil is slightly above 0 ohms, indicating that the coil is good. The resistance of the relay contacts is infinite when the relay is inverted, indicating that the relay is bad.
9. The technician replaces the current-type relay.
10. The technician restores power and checks the operation of the compressor.
11. The technician informs the owner of the sandwich shop of the problem and that it has been corrected.

Service Call 5

Application: Ventilation system

Type of Equipment: Exhaust fan

Complaint: Exhaust fan not running

Service procedure:

1. The technician reviews the work order from the dispatcher for available information. The work order information reveals that the exhaust fan is not operating, causing a stuffy condition in the room.

2. The technician informs the maintenance supervisor of his or her presence and obtains additional information about the problem. The supervisor informs the technician that the motor is operating, but the fan is not.

3. The technician turns the power supply off and locks or labels it.

4. The technician visually inspects the exhaust fan and locates a broken belt.

5. The technician obtains the correct replacement and installs it, adjusting the tension.

6. The technician makes certain that the exhaust fan is operating properly.

7. The technician informs maintenance supervisor of the problem and that it has been corrected.

SUMMARY

Many types of electric motors require special electrical devices to keep the starting winding or starting component in the circuit until the motor has reached approximately 75% of full speed. Starting relays are used in most cases to drop the start windings and/or starting component from the electrical circuit once the motor has reached operating speed. It is essential that the starting components and/or starting windings be removed from the circuit at the correct time to prevent damage to the motor or starting components.

There are four types of relays in common use in the industry today. The current-type relay operates on the principle that amperage flowing through a wire will produce a magnetic field. At its highest point, this magnetic field will close a set of contacts, thus putting the necessary starting components in the circuit and removing them at the correct time. The potential relay has a normally closed set of contacts that are opened and closed by the back electromotive force of the motor. This back emf corresponds to the motor speed, thus dropping out the starting components at the right time. Electronic relays used as starting relays are becoming increasingly popular in the industry. They have several advantages over the regular type of starting relay in that they can be used over a variety of sizes while incorporating other motor components if needed. The hot-wire relay is seldom used in the industry because better methods are available for dropping the starting components out of the circuit.

All rotating heating, cooling, and refrigeration equipment must have bearings to operate efficiently and smoothly. Ball bearings and sleeve bearings are commonly used today for this purpose. Correct lubrication is essential to ensure long life and efficient service from the bearings.

In most cases in the industry, electric motors drive some type of equipment. Direct-drive applications are common, as are V-belt applications. Direct-drive applications require very accurate fit with little vibration. The device being driven must turn at the same revolutions per minute (rpm) as the motor. V-belt connections have a certain amount of tolerance, but they must be correctly adjusted. They can be used to alter the revolutions per minute of the equipment by changing the pulleys.

REVIEW QUESTIONS

1. What is the purpose of a starting relay?
   a. to start an electric motor
   b. to remove the starting winding or component from the circuit
   c. to protect the motor from starting overloads
   d. to prevent the motor from starting under heavy loads

2. Explain the operating principle of a potential relay.

3. Explain the operating principle of a current relay.

4. What is the difference between a current and potential relay?

5. Explain the operating principle of a hot-wire relay.

6. The contacts of the current relay are _____.

7. The controlling factor for a potential relay is _____.
   a. back electromotive force
   b. line voltage
   c. voltage drop
   d. none of the above

8. True or false: The contacts of the potential relay are normally open.

9. What is the approximate speed at which the starting windings or components should be removed from the motor electrical circuit?
   a. 50%
   b. 75%
   c. 90%
   d. 100%

10. As a single-phase motor's speed is increased from a stationary position, the current draw _____.
    a. increases
    b. decreases

11. As a single-phase motor's speed is increased from a stationary position, the back electromotive force_____.
    a. increases
    b. decreases

12. Which of the following statements reflects the correct terminal identification of a potential relay?
    a. 1 & 2 coil, 2 & 4 contacts
    b. 2 & 4 coil, 3 & 6 contacts
    c. 1 & 5 coil, 2 & 5 contacts
    d. 2 & 5 coil, 1 & 2 contacts

13. A G.E. potential relay with the number 3ARR3-D4F6 would have a continuous coil voltage of _____.

    a. 214
    b. 332
    c. 420
    d. 502

14. An RBM potential relay with the number 128-122-2324K would have a continuous coil voltage of

    _____.

    a. 130
    b. 170
    c. 256
    d. 336

15. A G. E. potential relay with a number 3ARR3-C3A1 could be replaced with which of the following RBM potential relays?

    a. 128-212-1167AB
    b. 128-112-1161AB
    c. 128-151-1111CD
    d. 128-212-1147AB

16. An RBM potential relay with a number 128-122-1161BC could be replaced with which of the following G. E. potential relays?

    a. 3ARR3-C4B2
    b. 3ARR3-C9D2
    c. 3ARR3-C6B2
    d. 3ARR3-C2B2

17. What are the two types of bearings used in the industry?

18. Which of the following is *not* an advantage of the ball bearing?

    a. more efficient than a sleeve bearing

    b. requires less maintenance than a sleeve bearing
    c. cheaper than a sleeve bearing
    d. longer life than a sleeve bearing

19. What is the main reason that sleeve bearings are used in the industry?

    a. cost
    b. maintenance
    c. life
    d. efficiency

20. What is a direct-drive application?

21. How is the direct-drive application used in the industry?

22. Why are V-belts popular in the industry?

23. What is the correct tension on a V-belt?

24. True or False: If the starting relay is working properly, there will be no ampere draw through the starting circuit after the motor reaches full speed.

25. What is the purpose of a hard-start kit in conjunction with a hermetic compressor?

26. What electrical components make up a hard-start kit?

27. Name the three ways in which ball bearings may be lubricated.

28. True or False: Overlubrication of a motor and its bearings is as damaging as underlubrication.

29. Name three ways in which sleeve bearings may be lubricated.

30. What are the common sizes of V-belts and what are their applications?

31. What is the compressor speed of an open-type compressor if the motor rpm is 1750, the motor pulley diameter is 4 inches, and the pulley diameter of the compressor is 16 inches?

    a. 438
    b. 468
    c. 498
    d. 528

32. What size motor pulley would be required if the motor turns 1750 rpm, the fan pulley is 8 inches, and the desired rpm of the fan is 650 rpm?

    a. 3
    b. 4
    c. 5
    d. 6

33. What is one situation in which a ball bearing cannot be used?

34. Explain the operation of a solid-state PTC starting relay.

35. What advantages does the solid-state starting relay have over the conventional current-type relay?

PRACTICE SERVICE CALLS

Determine the problem and recommend a solution for the following service calls. (Be specific; do not list components as good or bad.)

Practice Service Call 1

Application: Commercial refrigeration

Type of Equipment: Frozen food display with air-cooled condensing unit (240 V-1ø-60 Hz)

Complaint: No refrigeration

Symptoms:
1. Condenser fan motor operating normally.
2. Evaporator fan motor operating properly.
3. Internal overload cycling compressor on and off.
4. All starting components in good condition.
5. Compressor motor in good condition.

Practice Service Call 2

Application: Domestic refrigeration

Type of Equipment: Frost-free domestic refrigerator (Compressor has split-phase motor)

Complaint: No refrigeration

Symptoms:
1. Compressor not operating.
2. Correct voltage available to compressor.
3. Contacts of current-type relay are good.
4. Compressor cycled on and off by compressor overload.
5. Compressor draws locked rotor current when start is attempted.

Practice Service Call 3

Application: Residential conditioned air system

Type of Equipment: Air conditioner using air handling unit with air-cooled condensing unit (240 V-1ø-60 Hz and CSR motor with internal overload)

Complaint: No cooling

Symptoms:
1. Indoor electrical components operating normally.
2. Condenser fan motor operating properly.
3. 240 V-1ø-60 Hz available to compressor.
4. Compressor starts and runs for a short period of time.
5. Compressor and motor are in good condition.

Practice Service Call 4

Application: Commercial refrigeration

Type of Equipment: Glass door refrigerator (240 V-1ø-60 Hz and CSR motor with external overload)

Complaint: No refrigeration

Symptoms:
1. Evaporator fan motor operating normally.
2. Proper voltage available to compressor.
3. Compressor tries to start, but does not.
4. Compressor and motor are in good condition.
5. Start and run capacitor good.
6. Condenser fan motor operating properly.

Practice Service Call 5

Application: Residential conditioned air system

Type of Equipment: Oil-fired furnace with a belt-drive blower motor

Complaint: No heat

Symptoms

1. Blower motor not operating.
2. Correct voltage available to blower motor.
3. Blower motor drawing locked rotor current.
4. Blower can't be turned by hand.

Practice Service Call 6

Application: Commercial refrigeration

Type of Equipment: Ice machine (120 V-1ø-60 Hz with CS motor)

Complaint: No ice production

Symptom:

1. Correct voltage available to compressor.
2. Compressor attempts to start but is cut off by external overload.
3. Current-type starting relay is good.
4. Compressor and motor are in good condition.

LAB MANUAL REFERENCE

For experiments and activities dealing with material covered in this chapter, refer to Chapter 10 in the Lab Manual.

11

Contactors, Relays, and Overloads

OBJECTIVES

After completing this chapter, you should be able to

- Explain the parts and operation of contactors and relays.
- Explain the application of contactors and relays in control systems.
- Correctly install a contactor or relay in a control system.
- Draw a simple schematic wiring diagram using contactors and/or relays to control loads in a control system.
- Understand the types and application of overloads.
- Troubleshoot contactors and relays.
- Identify the common types of overload used to protect loads.
- Explain the operation of the common overloads.
- Determine the best type of overload for a specific application.
- Draw schematic wiring diagrams using the proper overload to protect loads.
- Troubleshoot common types of overloads.
- Explain the operation of a magnetic starter.
- Size the overload devices to be used in a magnetic starter for motor protection.
- Wire a magnetic starter using switches, thermostats, and push-button stations.
- Troubleshoot magnetic starter and push-button stations.

KEY TERMS

Coil	Current overload
Contactor	Fuse
Contacts	Inductive load

Internal compressor overload

Line break overload

Magnetic overload

Magnetic starter

Mechanical linkage

Overload

Pilot duty overload

Push-button station

Relay

Resistive (noninductive) load

INTRODUCTION

Control systems used on modern heating, cooling, and refrigeration systems use many different control components to obtain automatic control. The purpose of a control system is to automatically control the temperature of some medium. The function of a control system is to stop and start electric loads that control the temperature of the medium. In the case of an air-conditioning or heating system, the primary purpose is to control the temperature within a certain area. In a refrigeration system, the purpose is to control air temperature or water temperature.

In an air-conditioning or refrigeration system, the compressor is the largest load and usually requires a contactor or magnetic starter to energize it. Loads that require more control and are too large for line voltage thermostats or manual switches use relays for the proper control. Relays and contactors work similarly. The main difference between them is their current-carrying capacity. The contactor can handle a large ampacity. Relays are usually limited in the ampacity they can carry.

In all control systems there must be a means of protecting the loads. The overload is used to protect loads such as compressors, heaters, fan motors, and pumps. The basic overload device is the common fuse. However, the fuse is inadequate to protect effectively all the important loads in a control system. Thus, more effective devices for overload protection are used. As we will see, overloads come in many designs and sizes.

All electric control components serve a definite purpose in the total control system. Fortunately, many manufacturers now use simple control systems with fewer components in their residential air-conditioning control systems. However, the commercial and industrial control systems are fairly complex, with more components and better control than the residential equipment. Therefore, it is essential that heating, cooling, and refrigeration technicians become familiar with control components and understand them so they can diagnose faulty components and perform effective control system troubleshooting.

In this chapter we look at the components of a control system that control the loads in the system. In succeeding chapters we will discuss other control system components and the methods for troubleshooting these systems.

11.1 CONTACTORS

A **contactor** (Figure 11.1) is used to control an electric load in a control system. Contactors make or break a set of contacts that controls the voltage applied to some load in cooling systems. They isolate the voltage controlling its magnetic coil from the voltage applied to the load, making it possible to control a higher voltage or power-consuming load, with a lower voltage or power-consuming control circuit (pilot duty). A contactor consists of a coil that opens and closes a set of contacts due to the magnetic attraction created by the coil when it is energized. Magnetic starters are also used to start and stop large loads in cooling systems. The major difference between magnetic starters and contactors is that the magnetic starter houses its own overload. Magnetic starters will be discussed later in this chapter.

Applications

The largest electric load in any cooling system that requires control is the compressor. In smaller equipment, several other loads might be connected in parallel with the compressor. Larger systems usually maintain a switching device for each component. The contactor used in a small residential

FIGURE 11.1 Contactor

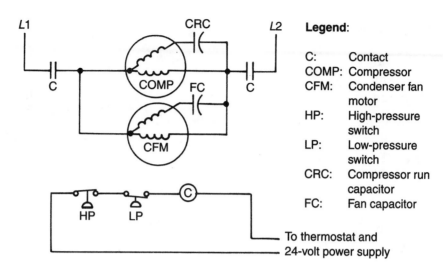

FIGURE 11.2 Schematic diagram of a small residential condensing unit with a contactor controlling the compressor and condenser fan motor

air-conditioning unit probably controls the compressor and condenser fan motor. Figure 11.2 shows a wiring diagram of a small residential unit. Large air-conditioning units usually have several contactors. A large condensing unit, for example, might use two contactors for the compressor and three for the condenser fan motors. Large electric resistance heating systems also have several contactors. For example, they might use a contactor for each section of heaters and for some method of controlling the fan or air supply.

Operation

Different manufacturers design contactors in different ways. But all contactors accomplish the same purpose: opening and closing a set of contacts. The armature of a contactor is the portion that moves. The movement of the armature can be accomplished in basically two ways, with a sliding armature or a swinging armature. The sliding armature is shown in Figure 11.3, and the swinging armature is shown in Figure 11.4. The sliding armature mounts between two slots in the frame of the contactor and moves up and down in these slots. The swinging armature is mounted on a pivot or hinge and moves up and down in a swinging motion.

The armature of a contactor is connected by a mechanical linkage to a set of contacts that causes a completed circuit when the armature is pulled into the magnetic field produced by the coil. This operation is true for both the sliding armature and the swinging armature. The magnetic field that closes

FIGURE 11.3 Contactor with a sliding armature

FIGURE 11.4 Contactor with a swinging armature

a contactor is created by a coil wound around a laminated iron core, as shown in Figure 11.5. When the coil is energized, a magnetic field is created around the laminated core. The core then becomes an electromagnet of sufficient strength to attract the armature closing the contacts. Both types of contactors use the same principle of operation. Some contactors have springs mounted between the armature and the stationary contacts to ensure the contactor opens when the coil is de-energized.

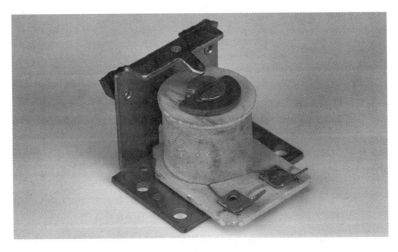

FIGURE 11.5 Solenoid coil of a contactor

Coils

Coil characteristics depend on the type of wire and the manner in which it is wound. The potential coil is energized by a certain voltage being applied to it. Coils of this type are designed to be operated on 24 volts, 120 volts, 208/240 volts, and occasionally 480 volts. The coil is identified by the voltage marked on it (refer again to Figure 11.5). The potential coil is used on many special relays in the industry. The connection of a coil is usually made directly on the terminals of the coil, but in some cases, the connections are jumped to a section of the contactor frame.

Contacts

The **contacts** of a contactor make a complete circuit when the contactor is energized, allowing voltage to flow to the controlled load. Contactors are rated by the ampere draw they can carry. There are two types of loads that a contactor can control: an inductive load, such as a motor, which has a higher ampere draw on startup than while running; and a resistive load, which has a constant ampere draw, such as a resistance heater. Some contactors are rated for both inductive loads and resistive loads, so care should be taken when selecting a replacement contactor. The ratings of contactors are usually marked on the contactor frame.

Contacts are made of silver and cadmium, which resists sticking. The contacts are connected to a strong backing by mechanical or chemical bonding. The chemical composition of contacts is such that they operate at cool temperatures of up to 125% of their current-carrying capacity. Contactors are usually manufactured with two or three poles and in many cases four. The fourth pole is used to interlock some load device into the system or can be left unused. A two-pole contactor is required for single-phase systems. A contactor with at least three poles is required for three-phase systems.

Troubleshooting

The diagnosis of a faulty contactor encompasses three sections of the contactor: the coil, the contacts, and the mechanical linkage. A defect in any contactor part can cause the total contactor to be faulty.

Coil. The contactor coil must be in good condition to create a strong enough electromagnetic force to pull in the contactor. The coil of a contactor rarely becomes so weak that it does not close the contacts, unless there is excessive friction to the **mechanical linkage.** The coil can be diagnosed as good, open, or shorted. The open and shorted conditions indicate a bad coil and can be checked with an ohmmeter. If the coil is shorted, the resistance reading will be zero. If the coil is open, the resistance will be infinite. A measurable resistance usually indicates a good coil.

A coil can also be checked by applying voltage to it and observing the contactor to see if it closes. The voltage reading of a coil should be taken before checking the coil to see if the contactor should be closed. Care should be taken when the diagnosis leads to a shorted contactor coil. If voltage is applied to it, the coil will cause a direct short and other damage could result.

Contacts.   The contactor contacts must be in good condition to ensure that the proper voltage reaches the load. In most cases, a visual inspection is sufficient to diagnose bad contacts. Figure 11.6 shows a good set and a bad set of contacts for comparison.

A voltage reading taken across the contacts of the same pole will show the voltage drop across the contacts. Figure 11.7 shows the proper procedure for testing a set of contacts in this manner. The voltage indicated on the meter is the voltage drop across the contacts (the voltage lost to the equipment). The 20 volts shown on the meter are considered to be excessive. Hence, the contactor should be replaced or repaired. Any voltage above 5% of the rated voltage for the equipment is considered to be excessive. The contactor must be closed with voltage applied to make this check.

FIGURE 11.6 A good set and a dirty, pitted set of contacts *(Courtesy of Square D Company)*

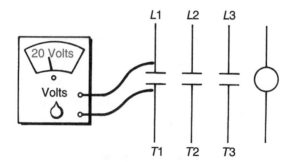

FIGURE 11.7 Procedure for testing a set of contacts with a voltmeter

Mechanical Linkage. Probably the easiest faults to diagnose with a contactor are problems with the mechanical linkage. In most cases, any trouble with the mechanical linkage can be detected by visual inspection. Or problems can be detected by breaking the power supply and manually moving the armature of the contactor to see if the movement is free and without excessive friction. The mechanical linkage of a contactor will usually fail because of wear, corrosion, or moisture. In many cases when a contactor coil burns out, it will heat the coil and cause the varnish of the coil to gum up the contactor. For a contactor to operate properly, it must seat the contacts accurately and have free moving parts.

Repairing. Contactors can be repaired by using replacement parts from the manufacturer or a wholesaler if time permits. However, it is often difficult to find all the necessary parts, such as contacts and coils, because it is almost impossible for a wholesaler to stock all the components needed for all contactors. Most manufacturers do sell a kit that will completely replace the contact portion of the contactor. But since parts are difficult to obtain, it is usually advisable to purchase a new contactor instead of repairing the faulty one. However, be careful to choose the correct contactor for the particular application and size.

11.2 RELAYS

Relays are used to open and close a circuit to allow the automatic control of a device or circuit. Relays are similar to contactors with the exception of the pole configuration and the amount of current that each device can effectively handle. Relays can be used to control most any device in the system within a certain ampacity limit.

Operation

Relays are built with the same components as a contactor. These include a coil, contacts, and some type of mechanical linkage to open and close the contacts when the relay coil is energized. When voltage (or current in some cases) is applied to a relay, it will close because of the magnetic field created in the coil and iron core. This magnetic field causes the armature of the relay to be attracted to the electromagnet created by the coil and its core. Figure 11.8 shows several different commonly used relays. The coil of a relay can be energized by voltage or current draw. The general type of relay controlling a device is closed by voltage. The current relay is used to control some starting device when used with a small hermetic motor.

Relays can be purchased with most any type of pole configuration. Normally open or normally closed contacts are both used in control circuits. The normally open contact opens when the relay is de-energized and closes when the relay is energized. The normally closed contact closes when the relay is de-energized and opens when the relay is energized. The normal position of the relay denotes the position of any controlling device in the de-energized position. The most common types of pole configuration for relays are single-pole–single-throw, single-pole–double-throw, double-pole–single-throw, and double-pole–double-throw. Any of these configurations can be normally open or normally closed.

FIGURE 11.8 Several different types of relays used in the industry

Applications

Relays can be used to control indoor fan motors, condenser fan motors, damper motors, starting capacitors, and control lockouts. They are used for any device that requires an automatic means of opening and closing a circuit.

The indoor fan relay is a good example of the use of a relay. On the cooling cycle, the indoor fan must be energized. This is accomplished by the use of an indoor fan relay, as shown in Figure 11.9. The indoor fan relay will energize when the system switch and the fan switch or the cooling thermostat are closed, starting the fan motor. The indoor fan relay can also be used to control a two-speed fan motor, using high speed on cooling and low speed on heating, with a thermal fan switch controlling the fan motor on heating, as shown in Figure 11.10. Relays can also be used to control large contactors that cannot be effectively energized with 24 volts in a control system. The control relay in Figure 11.11 is a good example of this application.

The potential or voltage-type relay is energized when voltage is applied to the relay coil. This relay is used to control some load device by opening and closing its contacts. The voltage-type relay can be used for many purposes, such as indoor fan relays, condenser fan relays, control relays, and lockout relays.

Three types of relays are used to assist in starting motors, as we saw in Chapter 10. The potential relay uses voltage to energize its coil and drop the starting apparatus out of the circuit. The current relay uses current flow

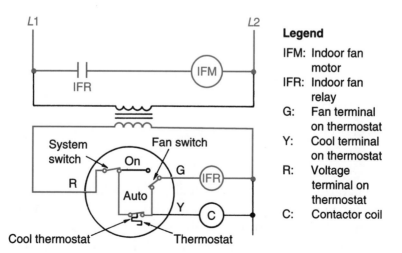

FIGURE 11.9 Schematic diagram of an indoor fan relay circuit

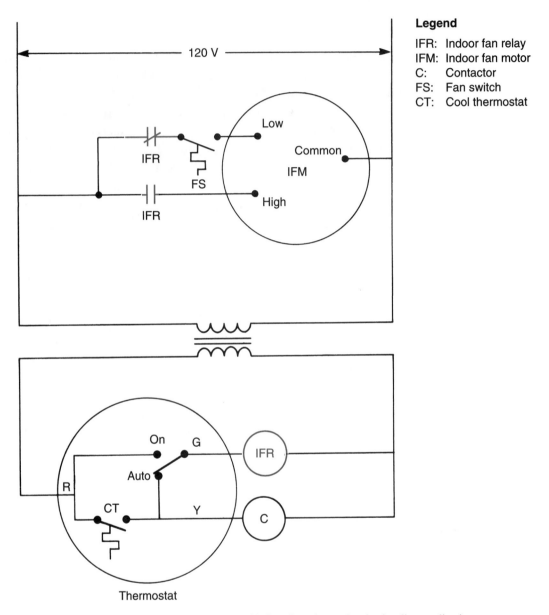

Legend
IFR: Indoor fan relay
IFM: Indoor fan motor
C: Contactor
FS: Fan switch
CT: Cool thermostat

FIGURE 11.10 Schematic diagram of indoor fan relay on heating/cooling application

to energize the circuit that contains the starting apparatus and then drops the circuit out when the current has dropped. The thermal relay uses heat to open and close starting circuits when it is used with a motor. Starting relays for motors are used throughout the industry. However, these relays should not be confused with the general type of relay used to control loads.

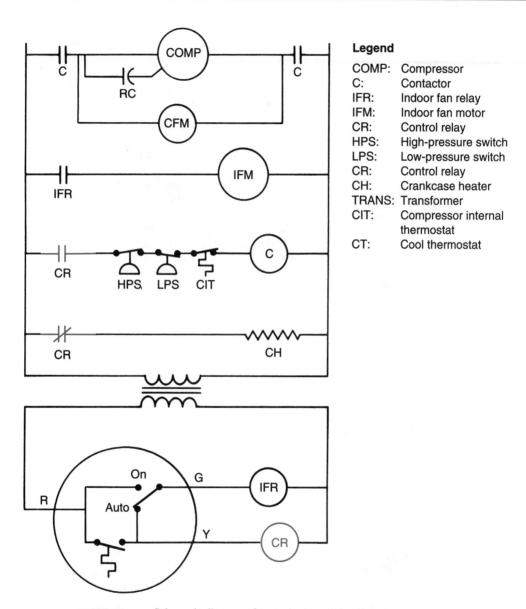

FIGURE 11.11 Schematic diagram of control relay used to energize contactor

Legend

COMP: Compressor
C: Contactor
IFR: Indoor fan relay
IFM: Indoor fan motor
CR: Control relay
HPS: High-pressure switch
LPS: Low-pressure switch
CR: Control relay
CH: Crankcase heater
TRANS: Transformer
CIT: Compressor internal thermostat
CT: Cool thermostat

Construction

The contacts used in relays are made like contacts in a contactor. The contact is made of a silver and cadmium alloy attached to some kind of strong backing that can withstand the pressure exerted by the armature.

A relay is usually mounted in a plastic enclosure. Hence, visual inspection is not as easy for relays as it is for contactors. The relay contacts

FIGURE 11.12 Contacts of a relay

usually cannot be seen unless the relay is disassembled or the cover is removed. The contacts of a relay are shown in Figure 11.12.

The relay armature can be swinging or sliding as shown in Figure 11.13. These devices operate in relays in the same way that they do in contactors.

The relay coil is built to produce enough magnetism to effectively close the contacts of the relay. The size of the relay coil is smaller than the contactor coil. Figure 11.14 shows a comparison of a coil used in a relay and a coil used in a contactor.

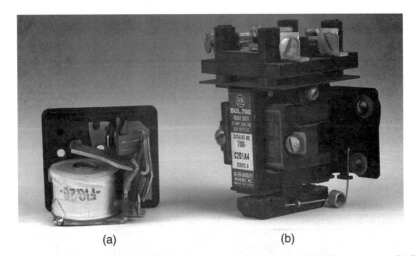

(a)   (b)

FIGURE 11.13 Two types of relay mechanical linkages used in the industry: (a) Swinging armature. (b) Sliding armature.

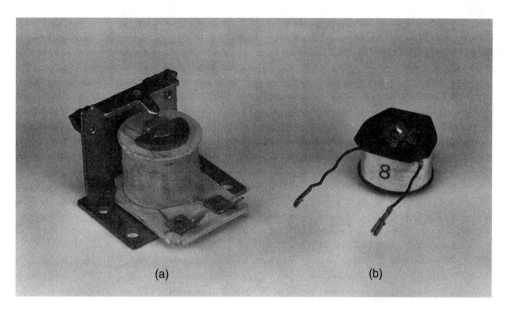

FIGURE 11.14 Comparison of solenoid coils of a (a) contactor and (b) relay

Troubleshooting

The diagnosis of a faulty relay is done in much the same way as the diagnosis of a faulty contactor. Diagnosis of a coil is the same whether it is for a relay or for a contactor. The relay contacts are usually hidden and cannot be visually inspected without disassembly. They must be checked with an ohmmeter. The contacts of a relay are not as heavy as the contacts of a contactor and therefore can take less punishment. Normally open and normally closed relay contacts will be completely melted if connected across line voltage with no load in series. This must be taken into consideration when troubleshooting the contacts of relays.

The mechanical linkage of a relay gives less trouble than that of a contactor because of the lighter weight of the armature. Any mechanical linkage problem in a relay will usually be caused by sticking contacts.

11.3 OVERLOADS

An **overload** is an electric device that protects a load from a high ampere draw by breaking a set of contacts. The simplest form of overload protection is the fuse. Fuses can be used to protect wires and noninductive loads, but they provide inadequate protection for **inductive loads**. A load that is purely resistive in nature with no coils to cause induction is called a **resistive** or **noninductive load**. The most common resistive load used in the industry is an electric heater.

Fuses

Fuses consist of two ends or conductors with a piece of wire that will melt and break the circuit if the current passing through it exceeds the amperage rating of the fuse. Fuses are available in many different styles and designs (again see Chapter 5). Fuses are used most commonly to protect wires, circuit components, and noninductive loads. Electric resistance heaters are the most common resistive loads protected by fuses. Figure 11.15 shows a schematic diagram of a set of resistance heaters protected by fuses. The system in Figure 11.15 controls three electric heaters by energizing contactors to start the heaters. The fuses in the circuit are used as safety devices for the heaters.

Circuit breakers are used for the same purpose as fuses but allow a high starting load. However, many control circuits use fuses for protection.

NOTE: Fuses are used as overloads for H1, H2, and H3

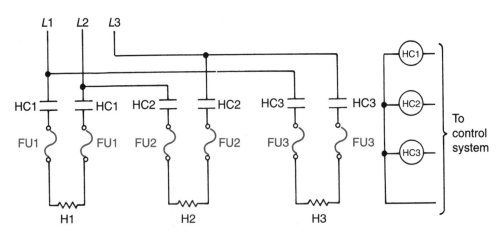

Legend

HC1: Heater contactor 1
HC2: Heater contactor 2
HC3: Heater contactor 3

FU1: Heater 1 fuse
FU2: Heater 2 fuse
FU3: Heater 3 fuse

H1: Heater 1
H2: Heater 2
H3: Heater 3

FIGURE 11.15 Schematic diagram showing fuses used as overloads for protection of heaters

Line Break and Pilot Duty Overloads

Overload devices used to protect inductive loads are more effective devices than fuses but also are more complex. Inductive loads require more amperage to start than to run. The amperage of a motor at the moment power is applied is largest because the rotor of the motor is in a stationary position, locked rotor condition. Figure 11.16 shows a graph of the ampere draw of a motor from start to full speed.

Overloads can be divided into two basic groups: line break and pilot duty. The **line break overload** breaks the power to a motor. A **pilot duty overload** breaks an auxiliary set of contacts connected in the control circuit. Overloads can be manual or automatic reset; the manual reset must be reset by the service technician or customer, but the automatic reset will reset automatically.

Line Break Overload. A line break overload is shown in Figure 11.17. One of the most common types of line voltage overloads is a metal disc mounted between two contacts. It is called a bimetal line break overload. Figure 11.18(a) shows a schematic diagram of a bimetal overload in closed position. If the current draw or temperature of the motor is sufficient to cause the disc to overheat and expand, the contact opens, as shown in Figure 11.18(b). This breaks the flow of power to the load. In some overloads, a heater or wire installed below the disc is sized to give off heat, as

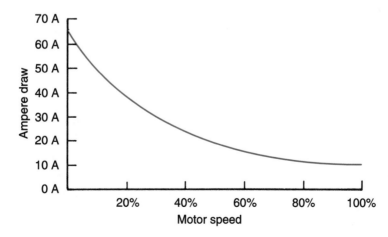

FIGURE 11.16 Ampere draw for motor speeds from locked rotor to full speed

FIGURE 11.17 Line break overload

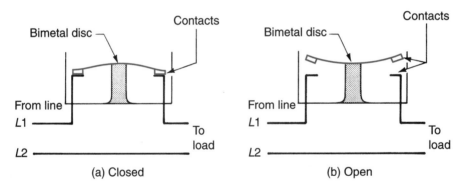

(a) Closed

(b) Open

FIGURE 11.18 Schematic diagram of bimetal line break overload in the closed and open position

shown in Figure 11.19. This gives a more accurate range of protection. Figure 11.20 shows a fractional horsepower compressor with an in-line overload. Figure 11.21 shows a three-phase overload for a compressor.

Other types of bimetal overloads are the two-wire and three-wire klixon overloads. The three-wire uses the current draw of both windings to open or close the overload and break the common to both windings, as shown schematically in Figure 11.22.

FIGURE 11.19 Bimetal overload with a heater installed below the disc

FIGURE 11.20 In-line overload on fractional horsepower compressor *(Courtesy of Tecumseh Products Co.)*

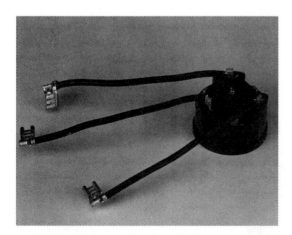

FIGURE 11.21 Three-phase overload for compressor

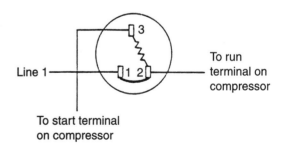

FIGURE 11.22 Schematic diagram of a three-wire bimetal overload

FIGURE 11.23 Internal compressor overload

The most popular line break overload for use in small central residential systems is an **internal compressor overload**, shown in Figure 11.23. The internal compressor overload is a small device inserted into the motor windings, as shown in Figure 11.24. This overload can sense the current draw of the motor, as well as the winding temperature, more effectively than external overloads. Figure 11.25(a) shows a schematic diagram of a compressor with a bimetal internal overload. Figure 11.25(b) shows a compressor with a three-wire bimetal external overload. Some small three-

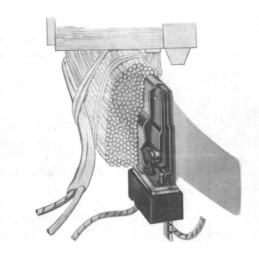

FIGURE 11.24 An internal overload fitted into a hermetic compressor motor winding *(Courtesy of Tecumseh Products Co.)*

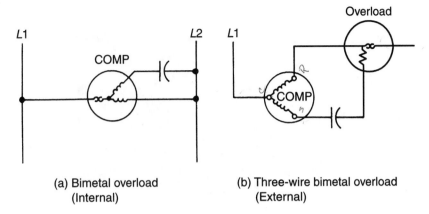

FIGURE 11.25 Schematic diagrams of bimetal overloads (a) Bimetal overload (Internal) (b) Three-wire bimetal overload (External)

phase hermetic motors also use an internal type of overload. Internal overloads should not be confused with internal thermostats. They are similar in appearance. The purpose of an internal thermostat is to break the control line if the windings overheat.

Pilot Duty Overload. The pilot duty overload breaks the control circuit when an overload occurs, which would cause a contactor to be deenergized, as shown in Figure 11.26. This type of overload is common on larger systems and still exists on smaller systems currently in the field.

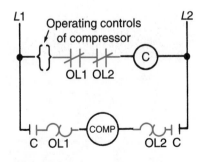

Legend

OL1:   Overload 1
OL2:   Overload 2
C: Contactor
COMP: Compressor

FIGURE 11.26 Schematic diagram showing pilot duty overload in the circuit

FIGURE 11.27 Current-type pilot duty overload

Two basic pilot duty overloads are being used in the industry today: the current overload and the magnetic overload. The current overload is shown in Figure 11.27. This **current overload** works similarly to the line break overloads except that a pilot duty set of contacts is opened rather than the line. In most cases, the bimetal disc of the overload would have to be so heavy that it could not control line voltage effectively. Therefore, in larger overloads pilot duty contacts are used.

The magnetic, or Heinemann, overload, as shown in Figure 11.28, is another type of pilot duty overload used in the industry. The **magnetic**

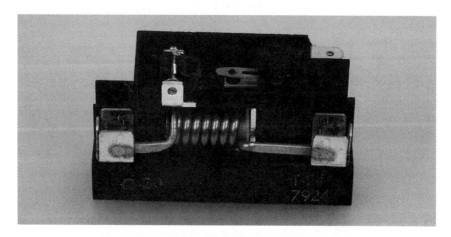

FIGURE 11.28 Magnetic overload *(Courtesy of Bill Johnson)*

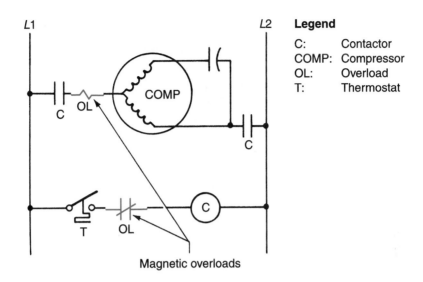

Legend

C: Contactor
COMP: Compressor
OL: Overload
T: Thermostat

FIGURE 11.29 Schematic diagram showing magnetic overloads protecting a compressor

overload consists of a movable metal core in a tube filled with silicone or oil. Surrounding the metal tube is a coil of wire. When the current increases, so does the magnetic field of the coil. The overload operates by the magnetic field created by the coil. The device is designed to create a magnetic field that is strong enough to pull the core up, opening the pilot contacts on overload.

The magnetic overload has a time-delay feature. There is a small hole drilled in the core. Once the field begins pulling the core in, the oil or silicone must go from one end of the tube to the other through the small hole. Thus, there is a short interval, due to the oil flow, between the time the motor starts up and the time the overload would break the circuit. Figure 11.29 shows a schematic diagram of magnetic overloads in a compressor circuit.

Many electronic solid-state overloads are used in the industry, and more are being developed. The design of the electronic overload varies with the application, but in most cases a temperature sensor is placed in the motor winding and connected to an external module mounted in the control panel of the unit. The sensors mounted in the motor windings will change resistance with a change in motor winding temperature. This change in resistance, when amplified by the electronic module overload, will open a set of pilot duty contacts. Most electronic overloads use two or three sensors mounted in the motor windings. This method of overload protection gives

FIGURE 11.30 Electronic overload with sensor

rapid response time to an overload condition. An electronic overload with a sensor is shown in Figure 11.30. A more complete explanation of the electronic overload is given in Chapter 18.

Troubleshooting

Most overloads are easy to troubleshoot because they have only one set of contacts along with a heater to heat the bimetal disc. An ohmmeter can be used to test the contacts in both pilot duty and line break overloads. In some cases, there will be a break in the heater or coil, depending on the type of overload you are checking. This element of an overload can also be checked with an ohmmeter.

The internal overload of a compressor is more difficult to diagnose because it cannot be isolated from the system. Most internal overloads break the common of the motor. If the motor is checked and there is an open starting and running winding, chances are that the common or the internal overload is open. If the compressor housing is cool, the motor might be damaged rather than the overload. Be sure when making this test that the compressor is cool. If the compressor is warm or hot, the overload could be open and working properly.

11.4 MAGNETIC STARTERS

A **magnetic starter**, shown in Figure 11.31, is composed of four sets of contacts, a magnetic coil to close the contacts when the coil is energized, and a set of overloads. Figure 11.32 shows a magnetic starter with the parts labeled. The coil is the heart of the system. All functions of the total control, and stopping and starting the equipment, are accomplished by the coil being energized or de-energized. The contacts open and close, depending on the action of the coil. The magnetic starter also has a means of overload

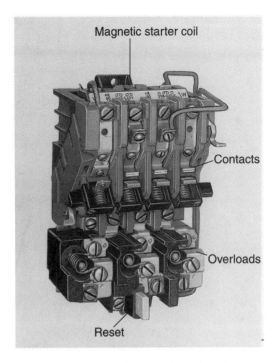

FIGURE 11.32 Parts of a magnetic starter *(Courtesy of Furnas Electric Co.)*

FIGURE 11.31 Magnetic starter *(Courtesy of Furnas Electric Co.)*

protection, which de-energizes the coil in the event that an overload occurs. Figure 11.33 shows a simple wiring diagram of a starter controlled by a thermostat.

In the heating, cooling, and refrigeration industry, many of the motors used are protected by internal overloads in the motor or by some external means, such as an overload relay, magnetic overload, or thermal elements. A contactor is a device that opens and closes automatically and allows voltage to flow to the equipment (refer again to Chapter 5). Most air-conditioning equipment uses a contactor with a separate means of overload protection. In many cases, especially with three-phase equipment, manufacturers use a magnetic starter, which incorporates a means of overload protection along with the ability to stop and start current flow to the equipment.

Types of Magnetic Starters and Their Operation

The overloads used in a magnetic starter are of three general classes: bimetal relay, thermal relay, and molten-alloy relay. The bimetal relay is composed of two metals welded together with different expansion qualities. One end is fastened securely and the opposite end can move. When

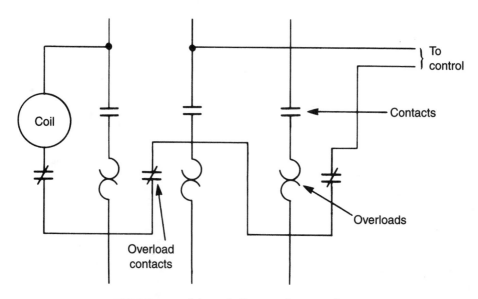

FIGURE 11.33 Schematic diagram of a magnetic starter

this element is heated above a certain temperature due to the current flow to the equipment, it will warp and thus break the circuit, causing the starter to open.

The thermal relay operates on the heat going through a wire to determine the current flow. If the heat is great enough to indicate an overload, the contacts of the relay shown in Figure 11.34 will open.

FIGURE 11.34 Thermal overload

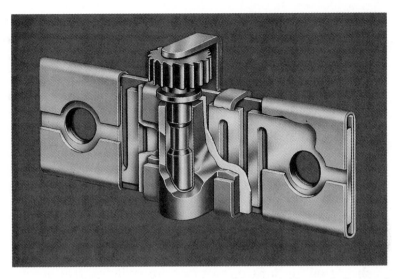

FIGURE 11.35 Molten-alloy overload element *(Courtesy of Square D Company)*

The molten-alloy overload element, shown in Figure 11.35, is nothing more than a link of material that is a good electrical conductor. A ratchet wheel is soldered to the conductor with a special alloy. When the current flow produces a higher-than-normal temperature, the solder melts, allowing the ratchet wheel to open the contacts.

No matter what type of overload relay is used, it should be sized correctly to work properly. Manufacturers of magnetic starters distribute much information on how to size starters and overloads (sometimes referred to as *heaters*).

Troubleshooting

Magnetic starters are used on most three-phase equipment because they contain overloads and can effectively control the operation of loads in a heating, cooling, or refrigeration system. Some common faults of magnetic starters occur in the contacts, solenoid coil heaters or overloads, and the mechanical linkage. No matter what part of the magnetic starter is faulty, it must be treated as a unit at the beginning of the diagnosis.

The service technician must first be certain that the magnetic starter is at fault. If the magnetic starter is at fault, it must be replaced, or the faulty section must be repaired. The contacts of a starter can be checked by visual inspection or by checking the voltage drop across each set of contacts. If the contacts are badly pitted, or if there is a voltage drop across the contacts, then they must be replaced.

The coil of a contactor can easily be faulty; it can be shorted, open, or grounded. The linkage that connects the movable contacts to the solenoid plunger can be worn or broken and will not close the contacts even if the coil is energized.

The overloads are difficult to troubleshoot because of the different types. Overloads are designed to open a control circuit when an overload exists. The overload circuit of a magnetic starter is connected within the control enclosure. The service technician must be familiar with the types of over-loads and their operation. The size of the overload should also be checked. The technician can easily check a magnetic starter step by step after it has been determined that the starter is not closing and should be.

11.5 PUSH-BUTTON STATIONS

Push-button stations or switches are switches that are controlled manually by the pressing of a button. Figure 11.36 shows a start-stop push-button station. Figure 11.37 shows two wiring diagrams using push-button switches as controls. They can have two switches (for stop and start) or many switches (for such functions as on, off, start, stop, jog, reverse, and forward). They are designed in many forms and with many different functions. There are relatively few instances in the heating, cooling, and refrigeration industry where push-button stations are used. They are used mainly for motor control, circuit control, and magnetic starters.

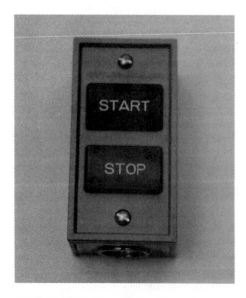

FIGURE 11.36 Stop-start push-button station

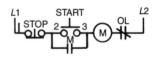

(a) Stop-start station

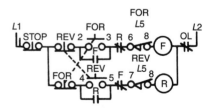

(b) Stop-forward-reverse station

FIGURE 11.37 Wiring diagram of some common push-button control systems

Push-button stations are easy to troubleshoot. The service technician must know, or be able to find out, the normal position of the switch. In some cases, push-button switches may be quite complex, but this will not occur very often. An ohmmeter can be used to diagnose the condition of the switch.

11.6 SERVICE CALLS

Service Call 1

Application: Residential conditioned air system

Type of Equipment: Packaged air-conditioning unit

Complaint: No cooling

Service Procedure:

1. The technician reviews the work order from the dispatcher for available information. The work order indicates that, according to the customer, no part of the unit is operating.
2. The technician informs the customer of his or her presence and obtains specifics about what the conditioned air system is doing. The customer is unable to provide any additional information. Upon entering the house, the technician makes certain that no dirt or foreign material is carried into the structure. The technician also takes care not to mar interior walls.
3. The technician begins the procedure by measuring the available voltage to the packaged unit. The technician measures 0 volts at the T1 and T2 terminals and between the neutral and T1 terminals of the disconnect. The reading between the neutral and T2 terminals of the disconnect is 120 volts. The technician has determined that voltage is not available to the unit.
4. The fuses of the fusible disconnect must be checked. The technician measures 240 volts across the fuse in line 1, and 0 volts across the fuse in line 2. The technician determines that the fuse in line 1 is blown.
5. The technician turns the power supply off and locks or labels it.
6. The blown fuse is changed with the correct size and type of ampacity.
7. The technician turns the power supply on.
8. The technician observes the operation of the unit to make certain the system is operating properly.
9. The technician informs the homeowner of the problem and that it has been corrected.

Service Call 2

Application: Residential conditioned air system

Type of Equipment: Gas-fired furnace with air-cooled condensing unit

Complaint: No cooling

Service Procedure:

1. The technician reviews the work order from the dispatcher for available information. The work order information reveals that the gas furnace is in a crawl space with the condensing unit located at the north end of the house. The housekeeper will be available to allow access to the residence.

2. The technician informs the housekeeper of his or her presence and obtain specifics about what the problems are with the conditioned air system. The housekeeper informs the technician that the indoor fan is not operating while the outdoor unit is still running.

3. The technician examines the setting of the thermostat, verifying the correct setting. Upon entering the house, the technician makes certain that no dirt or foreign material is carried into the structure. The technician also takes care not to mar interior walls.

4. An inspection of the condensing unit reveals that the compressor motor and condenser fan motor are both operating.

5. The technician proceeds to the crawl space to check the operation of the indoor blower motor.

6. The technician turns the power off and locks or labels it.

7. The technician removes the indoor blower access cover.

8. The technician visually inspects the blower assembly to make certain that the motor and fan turn easily, eliminating the possibility of a bad bearing in the motor. The technician also verifies that the motor is not overheating, indicating that the fan motor is not out on an internal overload.

9. The technician restores power to the furnace.

10. The technician measures the voltage available to the indoor fan motor. The voltage reading is 0, indicating that voltage is not available to the fan motor.

11. Using the wiring diagram, the technician locates the component(s) that are in series with the fan motor. In most cases, with conventional cooling systems, an indoor fan relay controls the motor during the cooling mode of operation.

12. The technician locates the indoor fan relay and finds that 24 volts are being supplied to the indoor fan relay coil. The contacts of the relay must be checked to determine their condition. The technician checks the voltage across the contacts and reads 120 volts, indicating that the contacts are open. Several methods are commonly used to check the condition of a set of contacts, including other voltage methods and resistance methods.

13. The technician replaces the indoor fan relay with a new relay with the correct coil voltage and contact configuration.

14. The technician oils the fan motor and checks the operation of the system.

15. The technician replaces the indoor blower access panel.

16. The technician informs the housekeeper of the problem and that it has been corrected.

Service Call 3

Application: Commercial refrigeration

Type of Equipment: Walk-in cooler with outdoor air-cooled condensing unit

Complaint: No refrigeration

Service Procedure:

1. The technician reviews the work order from the dispatcher for available information. The work order indicates that the evaporator fan is operating normally but the outdoor condensing unit is not operating.

2. The technician informs the store manager of his or her presence and obtains any additional information on the operation of the equipment.

3. The technician measures the voltage available to the condensing unit at the disconnect. The technician reads 240 volts, indicating that power is available.

4. Using a wiring diagram, the technician locates the component that controls the compressor and condenser fan motor (usually a contactor).

5. The technician checks the voltage being applied to the contactor coil and determines that 240-volt power is available, indicating that the contactor should be closed.

6. The technician, by visual inspection or a voltage reading, determines if the contacts are opened or closed. The contacts are open, so the contactor coil is bad.

7. The technician turns power supply off and lock or labels it.
8. The contactor must be replaced with one of the correct coil voltage and contact ampacity.
9. The technician checks the equipment for proper operation.
10. The technician informs the store manager of the problem and the corrective action taken.

Service Call 4

Application: Domestic refrigeration

Type of Equipment: Domestic chest-type freezer

Complaint: Food thawing

Service Procedure:
1. The technician reviews the work order from the dispatcher for available information. The work order information reveals that the food is thawing in the freezer and that the customer thinks power is available. The freezer is in the utility room off the kitchen and the key is under the mat at the back door.
2. The technician proceeds to the back door to gain access to the residence. Upon entering the house, the technician makes certain that no dirt or foreign material is carried into the structure. The technician also takes care not to mar interior walls.
3. The technician proceeds to the utility room and checks to make certain that power is available to the outlet to which the freezer is plugged in.
4. The thermostat and compressor are good.
5. The technician observes the compressor compartment and finds that the compressor is cold, indicating that no power is being applied to the compressor.
6. The technician checks the power at the compressor terminals and determines that the compressor is not receiving power.
7. The technician then checks from the power inlet at the compressor junction box to the compressor terminals. The overload reads voltage to one side, indicating that the overload is not allowing power to pass to the compressor. Therefore, the overload is faulty.
8. The technician replaces the overload with a correct replacement.
9. The technician leaves the customer a note describing what the problem was and stating that the freezer has been repaired.

Service Call 5

Application: Commercial and industrial conditioned air system

Type of Equipment: Commercial and industrial fan coil unit

Complaint: Fan will not run

Service Procedure:

1. The technician reviews the work order from the dispatcher for available information. The work order indicates that the technician is to check with the chief custodian upon entering the building.
2. The chief custodian informs the technician that the main fan for the HVAC system is not operating and needs to be repaired as soon as possible. The chief custodian has no other information.
3. The technician proceeds to the mechanical room that houses the main blower for the structure.
4. The technician examines the fan and, in the process, depresses the reset on the magnetic starter. The blower motor operates.
5. The technician checks the operation of the blower motor, ensuring proper operation.
6. The technician informs the chief custodian of the problem and demonstrates how building personnel could depress reset button before calling a service company. The technician also warns the chief custodian that continual resets are not normal and should be checked.

SUMMARY

Control systems used on modern heating, cooling, and refrigeration systems use many different control components to obtain automatic control. In this chapter we described four of these components: contactors, relays, overloads, and magnetic starters.

Contactors play an important part in the correct operation of equipment in the industry. Their purpose is to make and break a power circuit to a load, thus allowing the proper control of the equipment due to the contactor being energized or de-energized. The contacts, coil, and mechanical linkage are the main parts of the contactor. Each part must work properly for the contactor to work correctly. The coil can be faulty by being shorted or open. If it is good, a measurable resistance can be obtained on an ohmmeter. The contacts can be faulty due to wear, corrosion, or excessive friction. Whatever the diagnosis of a contactor, it should be repaired or replaced whenever it is faulty or gives erratic operation. When replacement is necessary, it is essential to select the correct contactor for the job.

The relay is used to control many loads in control systems. It is the most widely used device in the industry. The application and size of relays should be identified before any replacement is attempted. The relay has the same components as a contactor: coil, contacts, and mechanical linkage. The coil is easily diagnosed with an ohmmeter as good, open, or shorted. The contacts must be in good shape for proper operation and can be checked with an ohmmeter or voltmeter. The contacts of a relay are usually enclosed, which prevents a visual inspection. A faulty mechanical linkage in a relay usually results in sticking contacts.

Overloads play an important part in the industry because they protect expensive loads. The fuse is the simplest type of overload protection used. It is effective on noninductive loads, in protecting wires, and in protecting circuit components.

Overloads are divided into two basic types: line break and pilot duty. The line break overload breaks the line voltage to the components. It is used on small hermetic compressors and motors and is connected directly in the line voltage supply to the equipment or load. The pilot duty overload breaks a pilot duty set of contacts in the control circuit. Pilot duty overloads are arranged so that the line voltage feeds directly through them and on to the load. This line voltage will create an operating condition in the element in the overload—thermal, current, or magnetic—and open a set of pilot duty contacts. The internal overload is mounted directly in the windings of the motor and usually has no connections to the outside of the motor. The internal overload is connected in series with the common of the motor. All overloads must be properly sized to do an adequate job of protection.

A magnetic starter is much like a contactor; it has contacts that are opened or closed depending upon the action of the solenoid coil that energizes or de-energizes the magnetic starter controlling a load. The magnetic starter usually has three or four sets of contacts. It provides overload protection to the load it controls. The overload in the magnetic starter must be sized to the load it is controlling. The overload contacts remain in the starter and the overload elements, called heaters, must be correctly sized for the load. The heaters are installed separately in the starter. Push buttons and stations are used with magnetic starters in many applications but usually not in air-conditioning equipment. The most common type of push button is the simple start-stop station.

REVIEW QUESTIONS

1. Which of the following is the largest electrical load in an air-conditioning or refrigeration system?
   a. evaporator fan motor
   b. condenser fan motor
   c. compressor
   d. relay coil

2. Which of the following is the major difference between a relay and a contactor?
   a. ampacity rating
   b. number of contacts
   c. mechanical linkage
   d. none of the above

3. What is the major difference between a magnetic starter and a contactor?

4. What is the purpose of a contactor or relay?

5. Explain the operation of a contactor and a relay.

6. What are the two types of armatures used in contactors and relays?

7. What are the three major parts of a contactor or relay?

8. What is the proper procedure for checking the coil of a contactor or relay?

9. Contacts are usually made of _____.

10. What are the major reasons for replacing the contacts of a contactor?

11. A contactor or relay coil could be electrically diagnosed in which of the following conditions?
    a. open
    b. shorted
    c. good
    d. all of the above

12. True or False: A voltage reading taken across closed contacts of the same pole will show the voltage drop across the contacts.

13. The easiest faults to diagnose with a contactor are usually problems with the _____.

14. Why is it difficult to repair a contactor?

15. What do the terms "normally open" and "normally closed" mean in reference to relays and contactors?

16. True or False: The contacts in a relay are easily checked by a visual inspection.

17. True or False: The size of the relay coil is smaller than the contactor coil.

18. What is the purpose of an overload?

19. An electric heater is a (an) _____ load.
    a. resistive
    b. inductive

20. An electric motor is a (an) _____ load.
    a. resistive
    b. inductive

21. Which of the following is the simplest overload used in the industry?

    a. pilot duty
    b. line break
    c. fuse
    d. magnetic

22. The line break overload usually breaks _____.

    a. the control circuit
    b. the power voltage to the load
    c. the fuse element
    d. none of the above

23. The pilot duty overload breaks ____.

    a. the control circuit
    b. the power voltage to the load
    c. the fuse element
    d. none of the above

24. How does a line break overload operate?

25. True or False: The most popular line break overload used in small equipment is the two-wire klixon overload.

26. Which of the following types of overloads gives the fastest response time to an overload condition?

    a. magnetic
    b. current
    c. line break
    d. electronic

27. What is the advantage of an internal compressor overload?

28. What is the proper procedure for checking an internal compressor overload?

29. Name two basic types of pilot duty overloads and describe how they operate.

30. What is a magnetic starter?

31. True or False: Magnetic starters are primarily used on single-phase motors.

32. What are the three types of overloads used in magnetic starters in the industry? How does each type work?

33. True or false: Push-button stations are widely used in the industry.

34. What is the characteristic of the current draw of an electric motor from locked rotor (start) until the motor reaches full speed?

35. Draw a wiring diagram using a contactor to control a compressor; the contactor will be controlled by a single-pole–single-throw switch.

36. Draw a schematic diagram with a relay controlling a fan motor; the relay will be 24 volts and controlled by a single-pole–single-throw toggle switch.

PRACTICE SERVICE CALLS

Determine the problem and recommend a solution for the following service calls. (Be specific; do not list components as good or bad.)

Practice Service Call 1

Application: Residential conditioned air system

Type of Equipment: Packaged air conditioning

Complaint: No cooling

Symptoms:
1. Indoor fan motor operating properly.
2. Thermostat calling for cooling.
3. Compressor and condenser fan motor not operating.
4. Voltage to contactor coil is 24 volts.
5. Voltage reading across contacts $L1$ and T1 of contactor is 240 volts.
6. Compressor and condenser fan motor in good condition but not operating.

Practice Service Call 2

Application: Commercial refrigeration

Type of Equipment: Small chest-type ice cream display case

Complaint: Ice cream thawing

Symptoms:
1. Correct voltage available to ice cream box.
2. Thermostat in good condition.
3. Compressor and starting components in good condition.
4. Compressor is not operating.

Practice Service Call 3

Application: Commercial refrigeration

Type of Equipment: Walk-in cooler with outdoor air-cooled condensing unit

Complaint: No refrigeration

Symptoms:
1. Evaporator fan motor operating properly.
2. Correct voltage available to condensing unit.
3. Condenser fan motor and compressor in good condition but not operating.
4. Correct voltage available to contactor coil but contactor not closing.

Practice Service Call 4

Application: Residential conditioned air system

Type of Equipment: Gas furnace with air-cooled condensing unit

Complaint: No cooling

Symptoms:
1. Indoor blower operating properly.
2. Correct voltage available to contactor coil and contactor is closing.
3. No line voltage available to condensing unit.

Practice Service Call 5

Application: Commercial and industrial conditioned air system

Type of Equipment: Water chiller and fan coil units

Complaint: No chilled water available to fan coil units

Symptoms:
1. Chilled water pump not operating.
2. Correct voltage available to magnetic starter (208 V-3ø-60 Hz).
3. Magnetic starter closes but cuts out on overload.
4. Voltage reading across T1 and T2 or starter is 0 volts.
5. Voltage reading across overload element of magnetic starter is 240 volts.

LAB MANUAL REFERENCE

For experiments and activities dealing with material covered in this chapter, refer to Chapter 11 in the Lab Manual.

12

Thermostats, Pressure Switches, and Other Electric Control Devices

OBJECTIVES

After completing this chapter, you should be able to

- Explain the purpose of a transformer in a control circuit.
- Size a transformer for a control circuit.
- Troubleshoot and replace a transformer in a residential air-conditioning control circuit.
- Explain the basic function of a line and low-voltage thermostat in a control system.
- Identify the common types of thermostats used in the industry.
- Draw schematic diagrams using line and low-voltage thermostats as operating and safety controls.
- Install line and low-voltage thermostats on heating, cooling, and refrigeration equipment.
- Correctly set the heating anticipators and cooling anticipators, if adjustable, on a residential low-voltage control system.
- Explain the modes of operation and be able to correctly set or program a clock thermostat.
- Explain the function and operation of pressure switches.
- Install and correctly set the pressure switches in control systems used as operating and safety controls.
- Troubleshoot pressure switches.
- Understand, install, and troubleshoot the following controls in control systems used in the industry: 1) humidistats, 2) oil safety switches, 3) time-delay relays, 4) time clocks, and 5) solenoid valves.

KEY TERMS

Anticipators	Range
Clock thermostat	Snap action
Differential	Solenoid valve
High-pressure switch	Staging thermostat
Humidistat	Thermostat
Line voltage thermostat	Thermostat controlling element
Low-pressure switch	Time clock
Low-voltage thermostat	Time-delay relay
Oil safety switches	Transformer
Pressure switches	

INTRODUCTION

In the preceding chapter we discussed some commonly used devices that control loads in heating, cooling, and refrigeration systems. In this chapter we will discuss devices that control some small loads and contactors and relays that control larger loads.

Thermostats are extremely important to the industry because they are opened and closed by a change in temperature. In all phases of the industry, we must control temperature, and in most cases we do this by the use of a thermostat. Pressure switches are sometimes used to control temperature by the pressure-temperature relationship, but in most cases pressure switches are used as safety devices. Transformers are used to reduce line voltage to the low voltage that is used in many control systems.

Thermostats play an important part in most control systems. The thermostat is commonly used as the primary control to control the temperature within a given area. In some cases, it is also used as a safety device, as in motor protection. Thermostats that are used as primary controls can be heating thermostats or cooling thermostats or some combination of the two. Thermostats can also be used for a staging effect, that is, for operating equipment at different times, depending on the demands put on the system. Staging can provide a main stage for normal operation and a secondary stage for energizing a specific part of the control system at a specific time as the load dictates.

Pressure switches can be used for different purposes. Pressure switches are often used as safety devices to cut the equipment off if the pressure is dangerously high or low. In some cases in refrigeration systems, pressure switches are used as operating controls. Whatever the purpose of the pressure switch, it always reacts to a specific pressure in a certain way.

Transformers are used to break down the incoming power voltage to a voltage that can be easily used for control circuits.

Other control devices that we will discuss in this chapter include humidistats, oil safety switches, time-delay relays, time clocks, and solenoid valves.

In this chapter we will look at each of the control devices introduced in the preceding paragraphs and will describe how they operate in the system. In the next chapter we will examine heating control devices in more detail.

12.1 TRANSFORMERS

The **transformer** in a heating or cooling system provides the low-voltage power source for the control circuit. It transforms line voltage to the lower voltage needed for the control system. Most residences and small commercial installations use a 24-volt control system. The transformer for a residential unit, (Figure 12.1) is used to convert line voltage to 24 volts. Some commercial and industrial high-voltage equipment uses transformers that drop the line voltage to 240 or 120 volts (Figure 12.2).

Operation

Transformers are stationary inductive devices that transfer electric energy from one circuit to another by induction. The transformer has two windings,

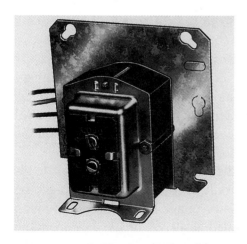

FIGURE 12.1 Transformer used in residential and small commercial systems *(Courtesy of Honeywell, Inc.)*

FIGURE 12.2 Large control transformer

primary and secondary. An alternating voltage is applied to the primary winding of a transformer and induces a current in the secondary winding.

Many different types of transformers are used in the industry. A step-down transformer induces a secondary voltage at a lower rate than the primary. This type of transformer is used for the power supply of a low-voltage system. A step-up transformer induces a secondary voltage at a rate higher than the primary. This type of transformer is used to boost the voltage.

Sizing Transformers

Transformers are like many other electric components. They are not 100% efficient. In other words, there is a loss between the primary and secondary windings. This loss must be considered when sizing transformers for a certain job. Transformers are rated by their primary voltage, secondary voltage, and voltamperes (VA). System equipment must be considered in transformer sizing along with the transformer rating.

The selection of a transformer vitally affects the performance and life of electric components in heating, cooling, and refrigeration equipment. A transformer too small for the control circuit will result in a lower-than-normal low voltage to the control circuit. This will result in improper operation of contactors or motor starters due to chattering or sticking contacts, burned holding coils, or the failure of contacts to close properly. All these conditions can cause system failure and possible damage to the compressor.

Even when transformers are sized correctly, care should be taken to avoid an excessive voltage drop in the low-voltage control circuit. When using a 24-volt control system with a remote thermostat, size the thermostat wire to carry sufficient current between the transformer and the thermostat.

The capacity of a transformer is described by its electric rating, primary voltage, frequency, secondary voltage, and the load rating in voltamperes. The maximum load of a transformer used in an air-conditioning control circuit is 100 VA. The maximum voltage is 30 V. Fuses can be used on the secondary side of the transformer for protection.

Transformers should be selected so they will operate all 24-volt loads without overheating. Transformers used only on furnaces will generally be rated 20 VA or less because of the light low-voltage loads. Air-conditioning systems usually require a transformer rated at 40 VA or larger, depending on the low-voltage loads. When replacing transformers in the field, use the following guidelines:

1. For replacement select a transformer the same size or larger than the one being replaced.
2. For new applications follow the manufacturer's recommendation.

Troubleshooting

Transformers can be easily diagnosed by two methods. An ohmmeter can be used to check the continuity of the windings of a transformer for an open, shorted, or good condition. However, it is difficult to diagnose a transformer with a spot burnout unless the second method of diagnosis is followed. A voltmeter can be used to check the secondary voltage of a transformer with the correct line voltage applied to the primary. When checking with a voltmeter, a load should be applied to the transformer. In some cases when a load is applied, the transformer secondary drops below an acceptable level, but this condition is rare.

When replacing a faulty transformer, the low-voltage control should be checked, because a short in the circuit will cause the transformer to burn out again. This is a common occurrence when a contactor coil or relay coil shorts out.

12.2 THERMOSTATS

The temperature in any structure, regardless of its age, location, or design, can be maintained at comfortable levels with a **thermostat.** Thermostats are designed and built in many different forms and sizes to meet the applications required in the industry. Thermostats play an important part in the total operation of most systems in the industry.

Applications

The basic function of a thermostat is to respond to a temperature change by opening or closing a set of electric contacts. Many different types of thermostats are used in the industry to perform a variety of switching actions. Figure 12.3 shows several common thermostats in use today.

Thermostats are used for many different purposes. An air-conditioning or heating thermostat would basically control the temperature of a given area for human comfort. Refrigeration thermostats are designed to maintain a specific temperature within a refrigerated space, such as in a domestic refrigerator, walk-in cooler, display case, and commercial freezer. There are many types of special application thermostats used in the industry, such as outdoor thermostats and safety thermostats. Whatever use a thermostat

FIGURE 12.3 Common low-voltage thermostats

is put to, it serves the same function: reacting to temperature with the opening and closing of a switch.

A heating thermostat closes on a decrease in temperature and opens on an increase in temperature. A cooling thermostat closes on an increase in temperature and opens on a decrease in temperature. This is a very important factor to consider when ordering or installing thermostats. A heating and cooling thermostat is available where heating and cooling are required. The heating and cooling thermostats are usually built so that the switch reacts either for heating or for cooling, with no intermediate position; in other words, a single-pole-double-throw switch. Some thermostats must isolate the heating and cooling contacts and therefore must use a separate set of contacts for heating and for cooling. Modern thermostats have a system switch that will determine whether the unit is heating or cooling.

The thermostat is merely a switching device that routes the voltage to the correct control for the operation prescribed by the thermostat setting. Figure 12.4 shows the schematic of a low-voltage thermostat with its routing. With the system selector in the cooling position, voltage flows through the system switch to the cooling thermostat, which operates the cooling equipment and the fan motor to maintain the temperature setting. If the system switch is in the heating position, the current flows through the switch to the heating thermostat, which operates the heating equipment. Many thermostats incorporate a fan switch that allows the fan to be operated manually or with the cooling equipment, since the fan on heating operates from a fan switch.

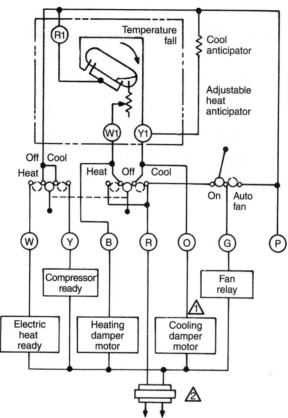

⚠1 With no O terminal load, thermostat current during heating cycle varies depending on whether fan switch is in the On or Auto position. Heater should be set for combined current level of heat ready and heat ready coils. With O terminal load, set thermostat heat anticipator to its maximum setting as cooling anticipator in series O terminal load provides heat anticipation in the heating cycle. (Limit the thermostat heating load current to 0.8 amps to assure good performance.)

⚠2 Power supply provides overload protection and disconnect means as required.

FIGURE 12.4 Schematic diagram of the routing of a low-voltage thermostat *(Courtesy of Honeywell, Inc.)*

Controlling Elements and Types of Thermostats

Two types of **thermostat controlling elements** are commonly used. The controlling element of a thermostat is the part that moves when a change in temperature is sensed. The bimetal thermostat shown in Figure 12.5 is commonly used to control the temperature of air in an air-conditioning or

FAN
AUTO
SYSTEM
COOL OFF HEAT

Mercury bulb

Bimetal

Temperature
adjustment
lever

FIGURE 12.5 Internal parts of a bimetal thermostat

heating application. The remote bulb thermostat is commonly used to control the temperature of any medium, whether liquid or vapor, in many applications.

Remote Bulb Thermostat. The remote bulb thermostat is shown in Figure 12.6. The power element, which is the bulb, and the diaphragm are connected with a section of small tubing. The bulb is filled with liquid and gas and then is sealed. The pressure exerted by the diaphragm on the mechanical linkage will open and close a set of contacts. As the bulb temperature changes, so will the pressure exerted on the diaphragm. If the temperature of the bulb increases, so will the pressure. If the temperature of the bulb decreases, so will the pressure. The increase or decrease in pressure causes the contacts to open or close, depending on the design of the thermostat.

Bimetal Thermostat. The heart of most other types of thermostats is a bimetal element. The element gets its name from the fact that it uses a bimetal to cause the movement that opens and closes a set of contacts. A bimetal is a combination of two pieces of metal, as shown in Figure 12.7(a), that are a given length at a certain temperature. The metals are welded together. Each metal has a different coefficient of expansion. If the temperature of these two metals is increased, one will become longer than the other because of the different expansion qualities. This causes the bimetal to arch,

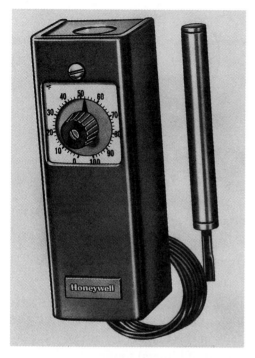

FIGURE 12.6 Remote bulb thermostat *(Courtesy of Honeywell, Inc.)*

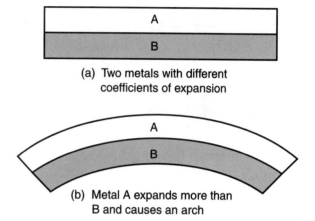

(a) Two metals with different coefficients of expansion

(b) Metal A expands more than B and causes an arch

FIGURE 12.7 A bimetal element for a thermostat

as shown in Figure 12.7(b). If the metal is anchored at one end, leaving the other end to move freely, it will move up and down according to the temperature surrounding it. In a low-voltage thermostat, the bimetal works better and gives better control if it is large.

The first type of bimetal thermostat that was produced is shown in Figure 12.8(a). It was unsatisfactory because of the unstable pressure from the bimetal that held the contacts together. It would react (close or open) to a relatively small change in the temperature around the room thermostat.

A thermostat must have a means of making a good connection with the contacts. This is accomplished by a **snap action** of the thermostat bimetal to the fixed contacts. The early type of thermostat was not snap-acting in the switching movement and thus caused problems because of its reaction to minor temperature changes. However, if a permanent magnet is placed near the bimetal arm, as shown in Figure 12.8(b), it will cause a snap action when the bimetal expands enough to move the contacts close together.

Two methods in common use enable thermostats to be snap-acting: a permanent magnet and a mercury bulb. A permanent magnet mounted near the

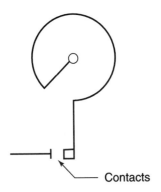
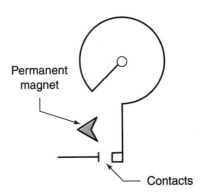

(a) Basic bimetal with no means of snap action to produce a stable set of contacts

(b) Bimetal with permanent magnet to produce snap action of the contacts

FIGURE 12.8 Basic bimetal elements

fixed contacts will cause the action of the bimetal to be snap-acting. The bimetal will make good contact once it is in the magnetic field of the permanent magnet. Figure 12.9 shows this type of bimetal thermostat.

The mercury bulb thermostat also provides snap action because of the globule of mercury moving between two probes sealed inside a glass tube, as shown in Figure 12.10. Figure 12.11 shows a bimetal thermostat with a mercury bulb that is used in the industry.

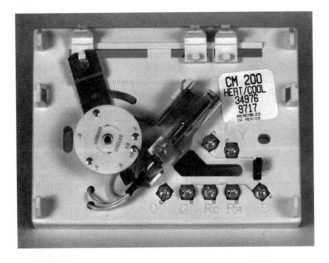

FIGURE 12.9 Bimetal thermostat with snap-action contacts

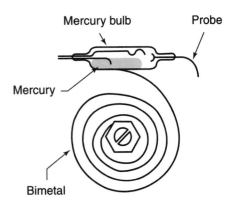

FIGURE 12.10 Bimetal with a mercury bulb used for the contacts

Mercury bulbs

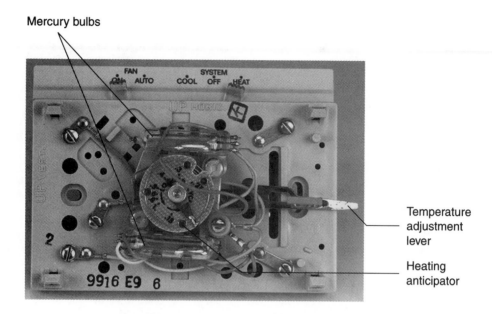

Temperature
adjustment
lever

Heating
anticipator

FIGURE 12.11 Bimetal thermostat with mercury bulb contacts

Line Voltage Thermostat.

The **line voltage thermostat** is designed to operate on line voltage, for example, 120 volts or 240 volts. The line voltage thermostat, shown in Figure 12.12, is used on many packaged air-conditioning units and refrigeration equipment of commercial design. This type of thermostat is used to open or close the voltage supply to a load in the system. Figure 12.13 shows the wiring diagram of a common 240-volt window air conditioner with a line voltage thermostat. Figure 12.14 shows a line voltage thermostat used to control an electric baseboard heater.

FIGURE 12.12 Line voltage thermostat

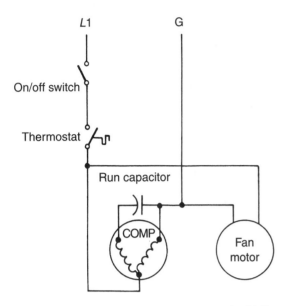

FIGURE 12.13 Wiring diagram of window unit with line voltage thermostat

FIGURE 12.14 Line voltage thermostat used to control an electric baseboard heater (*Courtesy of Honeywell, Inc.*)

The line voltage thermostat lacks many good qualities that are obtained with low-voltage thermostats. For example, depending on the application, the line voltage thermostat functions only to open or close a set of contacts. Line voltage thermostats are commonly used in the industry.

Low-Voltage Thermostat. The **low-voltage thermostat** is used on control systems with a 24-volt supply. The low-voltage thermostat is used on all residential heating and air-conditioning systems and many commercial and industrial systems. The low-voltage thermostat can be used for heating operation, cooling operation, automatic operation of fans, manual operation of fans, and automatic changeover from heating to cooling. The difference between a line voltage and a low-voltage thermostat is in the size of the bimetal element. With the larger line voltage contacts, more pressure is needed to close the contacts and therefore larger bimetal elements are required. Figure 12.15 shows a wiring schematic of a residential heating and cooling system with a low-voltage thermostat. The low-voltage thermostat is more accurate, less expensive, and requires smaller wiring than its counterpart, the line voltage thermostat. Figure 12.16 shows a low-voltage thermostat used to control the temperature in a residence. Figure 12.17 shows a digital low-voltage thermostat used in a residence.

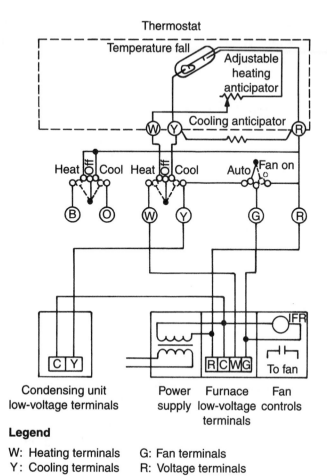

FIGURE 12.15 Wiring diagram of a residential heating and cooling system with a low-voltage thermostat

FIGURE 12.16 Low-voltage thermostat used in a residence *(Courtesy of Bill Johnson)*

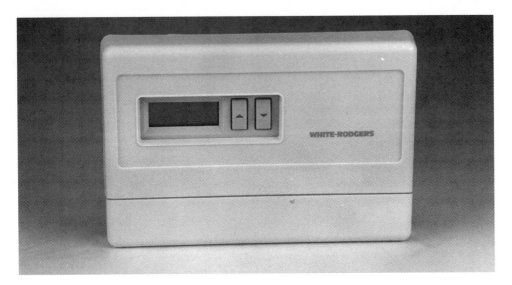

FIGURE 12.17 Low-voltage digital thermostat used in a residence

Anticipators

The success of any thermostat depends on the system being correctly sized, the air flow being balanced, and the thermostat being correctly placed. It is almost impossible for thermostats to maintain the exact temperature of any given area without spending an unreasonable amount for controls. Thus, anticipators are used to give a more evenly controlled temperature range. There are two types of anticipators: heating and cooling. We will discuss each type in turn.

Heating Anticipator.   A thermostat that has no means of heat anticipation will allow a wide swing from the desired temperature, especially on forced–warm-air systems. If the thermostat is set on 75°F, the furnace will come on for heating. But there is a delay before any warm air is carried to the conditioned space because a warm-air heating system must heat the furnace first. This delay will allow the air temperature to drop to 74°F or lower before the blower begins operation and heat is carried to the conditioned space.

The temperature difference between the closing of the thermostat and the time when warm air begins to reach the thermostat is called *system lag*. As long as the furnace remains on, the temperature will continue to rise. If the thermostat differential is 2°F, the thermostat will open at 77°F and stop the burner. However, the furnace is still warm and the blower must be operated

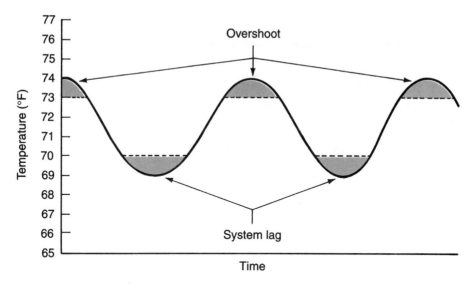

FIGURE 12.18 Effect of overshoot and system lag on room temperature

until the furnace cools, which will carry the temperature up to 78°F or higher.

The temperature difference between the opening of the thermostat and the time when the warm air is no longer being delivered to the room is called *overshoot*. The overshoot and system lag can produce an additional temperature swing of as much as 5°F. Figure 12.18 shows the effect of overshoot and system lag.

The wide difference in temperature can be controlled by using a heating anticipator in the thermostat. The heating anticipator is nothing more than a small source of heat close to the bimetal element. This allows the bimetal to be a little warmer than the surrounding air. The heating anticipator is placed in series with the contacts of the thermostat. When the contacts close and energize the burner, the current must also flow through the heating anticipator, which causes it to heat the bimetal to a certain temperature.

A heating anticipator anticipates the point at which the thermostat should open and provides a narrow thermostat differential. Suppose the thermostat is set at 75°F and the furnace burner is off, with the temperature dropping slowly. When the temperature falls below 75°F, the burner starts and at the same time the heat anticipator begins to heat the bimetal. The air temperature will continue to drop until the blower delivers warm air to the conditioned space. The room temperature begins to rise, but the bimetal temperature is slightly higher because of the heat from the heating anticipator. Thus, the thermostat anticipates the overshoot and cuts

FIGURE 12.19 Adjustable heating anticipator

off the burner. Due to the shorter length of time the burner operates, the furnace will cool more quickly, stopping the fan and preventing the overshoot. A heating anticipator does not eliminate system lag and overshoot. But these factors become negligible with a heating anticipator, producing a temperature swing of only 2°F to 3°F. The thermostat circuit shown in Figure 12.15 shows a heating anticipator connected in series with the contacts of the thermostat.

Two types of heating anticipators are used: fixed and adjustable. The fixed heating anticipator is nonadjustable and is not versatile. The adjustable heating anticipator, shown in Figure 12.19, can be matched to almost any control system. The adjustable heating anticipator should be set on the current draw of the primary control of the gas valve. For example, if a gas valve had a current draw of .2 amps, then the heating anticipator should be set on .2 amps. Figure 12.20 shows the procedure used to determine the current draw of the primary control by using a clamp-on ammeter. The ampere reading should be divided by the number of loops of wire that are made around the jaws of the meter. In most cases 10 loops is adequate for an accurate measurement. Many primary controls list the current draw directly on the control. Once the current draw of the primary control has been determined, the heat anticipator should be set at that current. Matching the adjustable heating anticipator to the current rating of the burner control assures the best possible heat anticipation.

Cooling Anticipator. The cooling anticipator operates somewhat differently from the heating anticipator. This type of anticipator is also known as an *off-cycle anticipator.* Figure 12.21 shows a cooling anticipator on a thermostat. The cooling anticipator is shown connected to a thermostat in

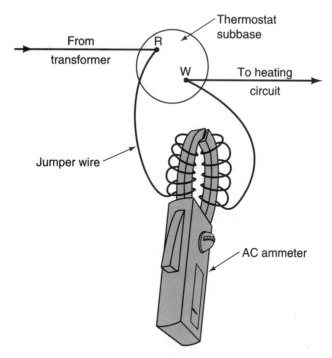

FIGURE 12.20 Procedure used to determine the current draw of primary control with clamp-on ammeter

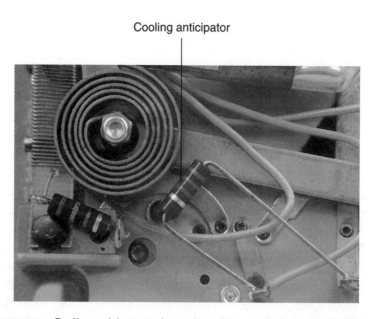

FIGURE 12.21 Cooling anticipator on low-voltage thermostat *(Courtesy of Bill Johnson)*

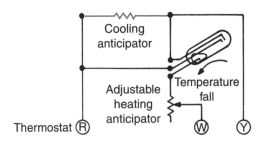

FIGURE 12.22 Schematic diagram of cooling anticipator connection in thermostat

Figure 12.22. The cooling anticipator adds heat to the bimetal on the "off" cycle of the equipment because of its parallel connection in the circuit. When the thermostat contacts close, the current takes the flow of least resistance, which is through the contacts rather than the cooling anticipator. On the "off" cycle the current passes through the anticipator and the contactor coil. The anticipator drops the voltage to a point that will not energize the contactor because of the series connection. The current flow through the anticipator heats the bimetal and causes the thermostat to anticipate the temperature rise. The cooling anticipator is not as important as the heating anticipator because cooling equipment operation is almost instant, while there is a delay in the heat delivery of a forced–warm-air system.

Thermostat Installation

The installation of a thermostat is simple because all thermostats are marked with identifiable letters, although the letter used are not consistent from manufacturer to manufacturer. Figure 12.23 shows a simple heating and cooling thermostat subbase. The terminals are shown with their letter designations (the thermostat attaches to the subbase). The letter designations to system functions are given in the chart in Figure 12.23. The letter designation should be followed for proper installation.

When checking thermostat operation, check the wiring connections and make sure that the selector switch and the temperature setting are properly selected.

When installing a thermostat, choose a location that will be most suitable for maintaining the correct temperature in the desired area. Thermostats installed in applications that require control of the temperature of some enclosed area, such as a refrigerator, walk-in cooler, or walk-in freezer, or some medium such as water in a chilled water system, should be installed

Letter Designation

Voltage to transformer	R–V
Heating	W–H
Cooling	Y–C
Fan	G–F
Heating damper (seldom used)	B
Cooling damper (seldom used)	O

FIGURE 12.23 Thermostat subbase with terminals and letter designations *(Courtesy of Honeywell, Inc.)*

where the correct temperature can be sensed by the thermostat without interference from some other source of heat or cold. Thermostat installations in commercial refrigeration applications should be where the thermostat can sense the temperature that reflects the average temperature of the box or case. When installing thermostats in applications where some medium such as water is being controlled, mount or insert the thermostat bulb or controlling element in the system so that the correct temperature is being sensed by the thermostat. When replacing a thermostat on a system, make certain that the thermostat is the correct type and that it is installed in the same way as the one being replaced. The thermostat should not be installed where there is a possibility of interference by some other installation.

The thermostat in an air-conditioning application should be located in the conditioned space on a solid inside wall where there is normal air circulation and where it will not be subject to any artificial heat or cold from lamps, televisions, appliances, or hot water pipes. The thermostat should be installed about five feet above the floor.

Troubleshooting

Troubleshooting thermostats will be a common job for installation and service technicians in the industry. Most thermostat manufacturers make available detailed installation and service procedures for their thermostats. When troubleshooting thermostats, the technician will be involved in three basic areas: calibration, diagnosis, and maintenance.

A common complaint is that the thermostat does not turn the equipment on and off at the proper temperature or that the temperature does not agree with the thermometer on the thermostat. Calibration is a means of resetting the thermostat so that the temperature more accurately reflects the temperature of the structure. When calibrating thermostats, follow the manufacturer's recommendations.

Common complaints about thermostats are excessive temperature swing, short cycling, the "on" cycle running too long, or the thermostat working incorrectly in regard to heating and cooling or not working at all. In many cases that involve temperature swing and cycling time, correctly setting the heat anticipator, if the thermostat has one, will correct the problem. Most line voltage thermostats do not have heat anticipators. Other problems often encountered by service technicians are that the thermostat is not doing what it is designed to do, or that the thermostat is defective and must be replaced.

Maintenance of thermostats covers checking the terminals for tightness, cleaning the contacts if possible, checking the calibration, and cleaning the sensing element.

12.3 STAGING THERMOSTATS

A **staging thermostat** is designed to operate equipment at different times with respect to the needs of the structure (Figure 12.24). The staging thermostat has more than one contact and opens and closes at different times with regard to the condition of the area being controlled.

The heating, cooling, and refrigeration industry has advanced in most phases, but in recent years there has been a demand for better control and more efficient operation with the modern systems. Staging thermostats have been designed to meet this need. The staging thermostat is becoming increasingly popular in the industry because of its versatility in system control.

Staging System

Many heating and cooling systems are operated in stages because the load in some structures fluctuates a great deal. A heating or cooling system that has been designed to operate on two different capacity levels is a *staged system*. A staged system is designed to operate at half of its capacity or more until the operating section of the equipment can no longer handle the heating or cooling needs of the structure. Then additional stages are called on as they are needed.

Staging systems offer many advantages because of their more efficient operation. For example, on a mild day the cooling load of a building is low and the full capacity of the equipment is not needed. The staging thermostat permits the equipment to operate at half of its capacity. If the day is extremely hot and the full capacity of the system is needed, then the staging thermostat makes that available. Staging can be used to many advantages in both heating and cooling.

Operation and Types

A staging thermostat is designed to be used on a system that has two stages of heating, cooling, or both. The staging thermostat operates on the differential in temperature between the stages. For example, a two-stage cooling thermostat could close one set of contacts at 75°F and the other set at 76.5°F to 78°F. A two-stage heating thermostat could close one set of contacts at 75°F and the other set at 73.5°F to 72°F. This allows systems to operate on partial capacity until the need arises for full capacity.

Staging thermostats can be obtained in a variety of stage configurations. Common staging thermostats used in the industry are the one-stage heating thermostat with two-stage cooling, the two-stage heating thermostat with one-stage cooling, and the two-stage cooling thermostat with two-stage heating. Figure 12.24 shows a two-stage heating and a two-stage cooling

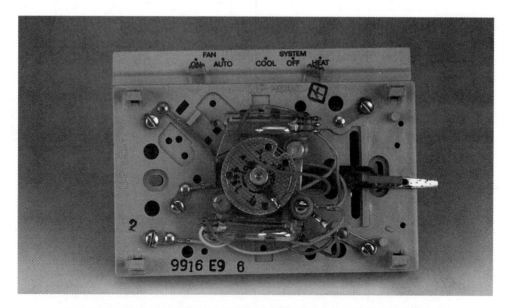

FIGURE 12.24 Low-voltage two-stage cooling and two-stage heating thermostat

thermostat. Note that four mercury tubes are used, with two mounted on the cooling control level and two mounted on the heating control level. Figure 12.25 shows the schematic diagram for the two-stage heating and two-stage cooling thermostat.

The letter designations are the same for staging thermostats as they are for regular thermostats except that heating or cooling stages would be lettered with a 1 or 2 after the main letter. The number 1 represents the first stage; the number 2 represents the second stage. When installing staging thermostats, pay careful attention to the letter designations and control hookup to avoid an improperly operating system.

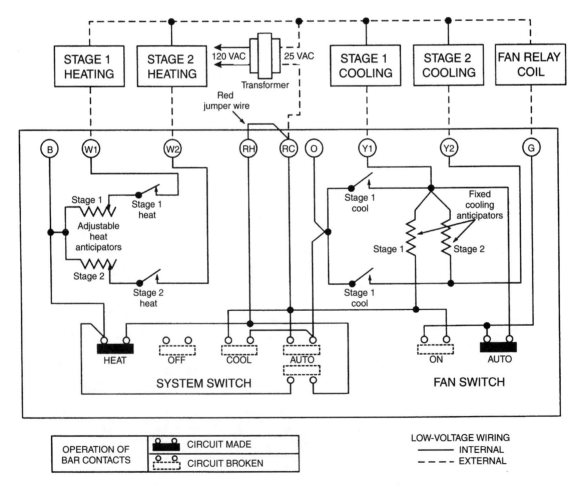

FIGURE 12.25 Schematic diagram for a two-stage cooling and two-stage heating thermostat *(Courtesy of Emerson Electric Company)*

Heat Pump

A heat pump is a refrigeration system that heats or cools by reversing the refrigerant cycle for the heating operation and then operating conventionally for cooling. Most heat pumps use a staging thermostat to operate a set of supplementary heaters. The first-stage thermostat operates the compressor. The second stage operates the supplementary heat when the first stage cannot handle the load. Figure 12.26 is a wiring diagram of a heat pump showing the two stages of heating and what they control.

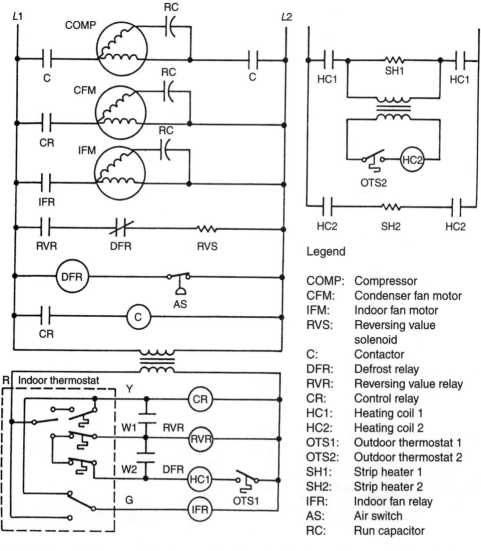

Legend

COMP:	Compressor
CFM:	Condenser fan motor
IFM:	Indoor fan motor
RVS:	Reversing value solenoid
C:	Contactor
DFR:	Defrost relay
RVR:	Reversing value relay
CR:	Control relay
HC1:	Heating coil 1
HC2:	Heating coil 2
OTS1:	Outdoor thermostat 1
OTS2:	Outdoor thermostat 2
SH1:	Strip heater 1
SH2:	Strip heater 2
IFR:	Indoor fan relay
AS:	Air switch
RC:	Run capacitor

FIGURE 12.26 Schematic diagram of a heat pump using a two-stage heating and one-stage cooling thermostat to control the heat pump

12.4 CLOCK THERMOSTATS

A **clock thermostat** is used to control the temperature of a structure, allowing the customer to lower or raise the temperature control point for at least one period in 24 hours. Figure 12.27(a) shows a clock thermostat and Figure 12.27(b) shows a digital programmable thermostat that are used in the industry. With energy conservation and high utility bills playing a larger part in air-conditioning and heating systems, clock and programmable thermostats are becoming increasingly popular. These thermostats give the customer the option of raising or lowering the temperature for a specified period of time to a more energy-efficient level and then, at a time that is appropriate, bringing the temperature back to a more comfortable setting.

Applications

Clock and programmable thermostats are widely used to allow a homeowner to set the temperature down or up for at least one period in 24 hours. The clock and programmable thermostats are used in a structure to allow for a setback temperature. This setback operation is usually done at a period of time that would least affect the homeowner. In most cases, the temperature would be set back when all occupants are sleeping or away from home and would be automatically reset to the desired level before the occupants needed the more comfortable temperature.

Clock and programmable thermostats are available for heating, cooling, and heating and cooling conditioned air systems. There are many different types of clock thermostats being produced today, and there is almost no limit to the functions they can perform.

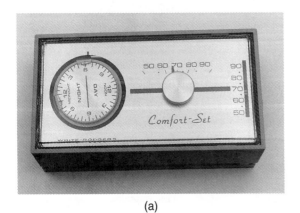

(a)

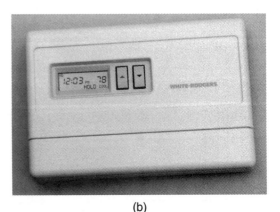

(b)

FIGURE 12.27   (a) Electromechanical clock thermostat (b) Electronic digital programmable thermostat

Several different types of clock and programmable thermostats are used today. There are many single-setback and multi-setback clock and programmable thermostats available at a wide variety of prices. Most clock thermostats use a battery or batteries as the power supply to operate the clock and are easily accessible for changing. An electronic digital programmable thermostat is available that is completely programmable with the homeowner selecting the setback temperatures and times.

The digital programmable thermostats are the most popular clock or setback thermostats because of their versatility and the many functions they offer to the homeowner. Figure 12.28(a) shows a digital programmable thermostat with Figure 12.28(b) showing the thermostat without the cover. Figure 12.29(a) shows a deluxe digital programmable thermostat with Figure 12.29(b) showing the thermostat without the cover. Most digital programmable thermostats have separate five-day (weekday) and two-day (weekend) programming, allowing four separate time/temperature settings per 24-hour period. Most have a continuous LCD display that shows the set point and alternately displays time and temperature. These thermostats must be programmed by the homeowner. The homeowner should follow the manufacturer's instructions for programming the thermostat. Once these thermostats have been programmed, the homeowner has the option of running the program in order to check the times and temperatures. Most programmable thermostats have a feature that allows the homeowner to override the program if the home is occupied at a time that was previously set to call for a setback.

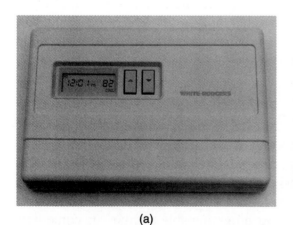

(a)

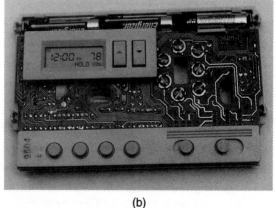

(b)

FIGURE 12.28 (a) Electronic digital programmable thermostat (b) Electronic digital programmable thermostat with cover removed

(a)   (b)

FIGURE 12.29 (a) Deluxe electronic digital programmable thermostat (b) Deluxe electronic digital programmable thermostat with cover removed

Installation

The installation of a clock thermostat with regard to the temperature function is the same as that of any low-voltage thermostat. Most clock or programmable thermostats today use a battery to power the thermostat. The letter designations are also the same as for other low-voltage thermostats. Figure 12.30 shows the diagram of an electromechanical clock thermostat with the letter designations shown. Figure 12.31 shows a typical installation wiring diagram for a programmable thermostat. Most clock and programmable thermostats are easy to install. The technician should make certain that the homeowner is instructed on how to set or program the new thermostat if the homeowner is available. The operating instructions or guide should be left with the homeowner at the time of installation.

Programming the Electronic Digital Programmable Thermostat

The programming of the thermostat shown in Figure 12.28 is similar to that of most programmable thermostats manufactured today. Most programmable thermostats are designed so that the average homeowner can accurately set and program them. However, each manufacturer will use its own method of programming. Most programmable thermostats come with detailed instructions on the programming. Figure 12.32 shows some of the programming instructions for the thermostat shown in Figure 12.28.

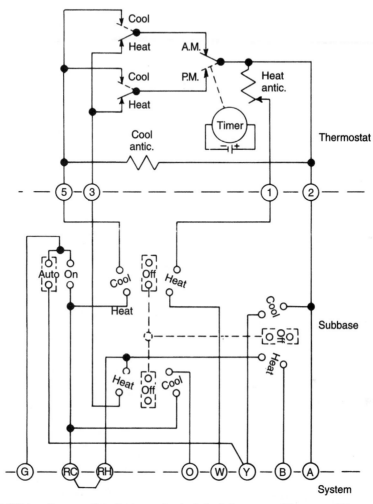

FIGURE 12.30 Wiring diagram of an electromechanical clock thermostat *(Courtesy of Emerson Electric Co.)*

PROGRAMMABLE ELECTRONIC THERMOSTAT

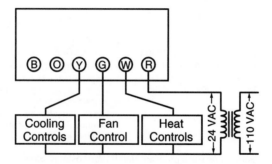

FIGURE 12.31 Installation wiring diagram of an electronic programmable thermostat

Before you begin programming your thermostat, you should be familiar with its features and with the display and the location and operation of the thermostat buttons. Your thermostat consists of two parts: the **thermostat cover** and the **base**. To remove the cover, gently pull it straight out from the base. To replace the cover, line up the cover with the base and press gently until the cover snaps onto the base.

THE THERMOSTAT BASE

Other than ⏶ and ⏷, the following buttons and switches are located behind the door on the bottom of the thermostat cover (see fig. 6). Pull the door down to open it.

The Thermostat Buttons and Switches

① (Red arrow) Raises temperature setting.

② (Blue arrow) Lowers temperature setting.

③ SET TIME button.

④ VIEW PRGM (program) button.

⑤ RUN PRGM (program) button.

⑥ HOLD TEMPerature button.

⑦ FAN switch (**ON, AUTO**).

⑧ SYSTEM switch (**COOL, OFF, HEAT, EMER**).

The Display

⑨ Indicates day of the week.

⑩ **HEAT** is displayed when the SYSTEM switch is in the HEAT position. **COOL** is displayed when the SYSTEM switch is in the COOL position. **COOL** or **HEAT** is displayed (flashing) when the compressor is in lockout mode.

⑪ Alternately displays current time and temperature.

⑫ **EMERGENCY** is displayed when the system switch is in the EMER position.

⑬ Displays currently programmed set temperature (this is blank when SYSTEM switch is in the OFF position).

⑭ The word **HOLD** is displayed when the thermostat is in the HOLD mode.

OPERATING FEATURES

Now that you are familiar with the thermostat buttons and display, read the following information to learn about the many features of the thermostat.

• **SIMULTANEOUS HEATING/COOLING PROGRAM STORAGE** — When programming, you can enter both your heating and cooling programs at the same time. There is no need to reprogram the thermostat at the beginning of each season.

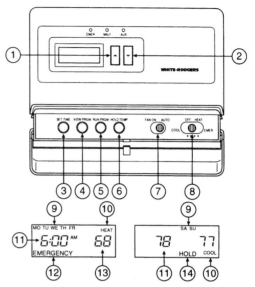

Figure 6. Thermostat display, buttons, and switches

• **TEMPERATURE OVERRIDE** — Press ⏶ or ⏷ until the display shows the temperature you want. The thermostat will override current programming and keep the room temperature at the selected temperature until the next program period begins. Then the thermostat will automatically revert to the program.

• **HOLD TEMPERATURE** — The thermostat can hold any temperature within its range for an indefinite period, without reverting to the programmed temperature. Press HOLD TEMP button. **HOLD** will be displayed. Then choose the desired hold temperature by pressing ⏶ or ⏷. The thermostat will hold the room temperature at the selected setting until you press RUN PRGM button to start program operation again.

• **°F/°C CONVERTIBILITY** — Press SET TIME and HOLD TEMP buttons until the temperature display is in Celsius (°**C**). To display Fahrenheit (°**F**), repeat the process.

• **TEMPERATURE DISPLAY ADJUSTMENT** — Your new thermostat has been accurately set in our factory. However, if you wish, you may adjust your new thermostat temperature display to match your old thermostat. This can be accomplished (within a ±4°F range) as follows:

1. Press VIEW PRGM and HOLD TEMP buttons at the same time.

2. Press ⏶ or ⏷ to adjust the displayed temperature to your desired setting.

3. Press RUN PRGM to resume normal program operation.

• **RESET BUTTON** - (see fig. 2) resets the thermostat program to the factory setting. This button can be used if you do not like the program you have entered or if you wish to start over in the programming procedure. The reset button can also be used to reset the program if the thermostat has been subjected to a voltage spike and the program has become scrambled or frozen.

FIGURE 12.32 Manufacturer's instructions for a programmable thermostat *(Courtesy of White-Rodgers Division, Emerson Electric Co.)*

PROGRAMMING YOUR THERMOSTAT

Now you are ready to program your thermostat. This section will help you plan your thermostat's program to meet your needs. For maximum comfort and efficiency, keep the following guidelines in mind when planning your program.

- When heating (cooling) your building, program the temperatures to be cooler (warmer) when the building is vacant or during periods of low activity.
- During early morning hours, the need for cooling is usually minimal.

Look at the factory preprogrammed times and temperatures shown below. If this program will suit your needs, simply press the RUN PRGM button to begin running the factory preset program.

FACTORY PREPROGRAMMING

Heating Program for ALL days of the Week:			Cooling Program for ALL Days of the Week:		
PERIOD	TIME	TEMP	PERIOD	TIME	TEMP
1st	6:00 AM	68°F	1st	6:00 AM	78°F
2nd	8:00 AM	68°F	2nd	8:00 AM	82°F
3rd	5:00 PM	68°F	3rd	5:00 PM	78°F
4th	10:00 PM	64°F	4th	10:00 PM	78°F

If you want to change the preprogrammed times and temperatures, follow these steps.

Determine the time periods and heating and cooling temperatures for your weekday and weekend programs. You must program four periods for both the weekday and weekend program. However, you may use the same heating and cooling temperatures for consecutive time periods. You can choose start times, heating temperatures, and cooling temperatures independently for both weekday and weekend programs (for example, you may select 5:00 AM and 70° as the weekday **1st period heating** start time and temperature, and also choose 7:00 AM and 76° as the weekday **1st period cooling** start time and temperature).

Use the table at the bottom of the page to plan your program time periods, and the temperatures you want during each period. You may also want to look at the sample program table to get an idea of how the thermostat can be programmed.

Entering Your Program

Follow these steps to enter the heating and cooling programs you have selected.

Set Current Time and Day

1. Press SET TIME button once. The display will show the hour only.

EXAMPLE: `12: PM`

2. Press and hold either ▲ or ▼ until you reach the correct hour and AM/PM designation (**AM** begins at midnight; **PM** begins at noon).

3. Press SET TIME once. The display window will show the minutes only.

EXAMPLE: `:00`

4. Press and hold either ▲ or ▼ until you reach the correct minutes.

5. Press SET TIME once. The display will show the day of the week.

6. Press ▲ or ▼ until you reach the current day of the week.

7. Press RUN PRGM once. The display will show the correct time and room temperature alternately.

Enter Heating Program

1. If you want to change the display from Fahrenheit to Celsius (or vice-versa), press SET TIME and HOLD TEMP at the same time.

2. Move the SYSTEM switch to **HEAT**.

Heating/Cooling Schedule Plan

Period	WEEKDAY (5 DAY)		WEEKEND (2 DAY)	
	Start Time*	Temperature	Start Time*	Temperature
1ST HEAT				
2ND HEAT				
3RD HEAT				
4TH HEAT				
1ST COOL				
2ND COOL				
3RD COOL				
4TH COOL				

* We suggest that you program the thermostat so that the cooling system has about **30 minutes** to reach the desired temperature (for example, if the first person gets up at 6:00 AM, program the **first** period to begin at 5:30 AM).

SAMPLE Heating/Cooling Schedule Plan

Period	WEEKDAY (5 DAY)		WEEKEND (2 DAY)	
	Start Time*	Temperature	Start Time*	Temperature
1ST HEAT	5:30 AM	68°F	7:00 AM	68°F
2ND HEAT	8:00 AM	65°F	11:00 AM	70°F
3RD HEAT	5:00 PM	70°F	6:00 PM	70°F
4TH HEAT	10:30 PM	65°F	11:30 PM	65°F
1ST COOL	6:30 AM	76°F	7:00 AM	76°F
2ND COOL	2:00 PM	78°F	12:30 PM	74°F
3RD COOL	5:00 PM	72°F	6:00 PM	72°F
4TH COOL	10:30 PM	78°F	11:30 PM	78°F

FIGURE 12.32 Continued.

3. Press VIEW PRGM once. "**MO TU WE TH FR**" (indicating weekday program) will appear in the display. Also displayed are the currently programmed start time for the **1st heating** period and the currently programmed temperature (flashing).

EXAMPLE:

```
MO TU WE TH FR
6:00 AM    68
```

This display window shows that for the 1st weekday period, the start time is 6:00 AM, and 68° is the programmed temperature (this example reflects factory preprogramming).

4. Press ▲ or ▼ to change the displayed temperature to your selected temperature for the 1st heating program period.

5. Press SET TIME once (the programmed time will flash). Press ▲ or ▼ until your selected time appears. The time will change in 15 minute increments. When your selected time is displayed, press SET TIME again to return to the change temperature mode.

6. Press VIEW PRGM once. The currently programmed start time and setpoint temperature for the **2nd heating** program period will appear.

7. Repeat steps 4 and 5 to select the start time and heating temperature for the 2nd heating program period.

8. Repeat steps 4 through 6 for the 3rd and 4th heating program periods. Weekday heating programs are now complete.

9. Press VIEW PRGM once. "**SA SU**" (indicating weekend program) will appear in the display, along with the start time for the 1st heating period and the currently programmed temperature.

10. Repeat steps 4 through 8 to complete weekend heating programming.

11. When you have completed entering your heating program, press RUN PRGM.

Enter Cooling Program

> ⚠ **CAUTION**
>
> If the outside temperature is below 50°F, disconnect power to the cooling system before programming. Energizing the air conditioner compressor during cold weather may cause personal injury or property damage.

1. Move SYSTEM switch to **COOL** position.

2. Follow the procedure for entering your heating program, using your selected cooling times and temperatures.

CHECK YOUR PROGRAMMING

Follow these steps to check your thermostat programming one final time before beginning thermostat operation.

1. Move SYSTEM switch to **HEAT** position.

2. Press VIEW PRGM to view the 1st weekday heating period time and temperature. Each time you press VIEW PRGM, the next heating period time and temperature will be displayed in sequence for weekday, then weekend program periods (you may change any time or temperature during this procedure).

3. Press RUN PRGM.

4. Move SYSTEM switch to **COOL** position.

5. Repeat step 2 to check cooling temperatures.

6. Press RUN PRGM to begin program operation.

YOUR THERMOSTAT IS NOW COMPLETELY PROGRAMMED AND READY TO AUTOMATICALLY PROVIDE MAXIMUM COMFORT AND EFFICIENCY!

FIGURE 12.32 Continued.

12.5 PRESSURE SWITCHES

A **pressure switch** is a device that opens or closes a set of contacts when a certain pressure is applied to the diaphragm of the switch. A high-pressure switch is connected to the discharge side of the system to sense discharge pressure. A low-pressure switch is connected to the suction side of the system to sense suction pressure. Pressure switches can be used as safety devices, as main operating controls, or to operate other parts of the system. Figure 12.33(a) shows a common pressure switch used in the industry.

Two types of pressure switches are used in the industry today. A nonadjustable pressure switch is used by many manufacturers to prevent the pressure setting from changing. An adjustable pressure switch can be adjusted to meet any specific need that might arise. Adjustable pressure switches are usually obtained for field replacement of a pressure switch. Pressure switches can also be classified as high pressure or low pressure.

(a)

(b)

FIGURE 12.33 Pressure switch (a) with enclosure *(Courtesy of Johnson Controls)* (b) without enclosure *(Courtesy of Bill Johnson)*

These switches open or close on a rise or fall of pressure. Pressure switches are manufactured with a case enclosing the switch, as shown in Figure 12.33(a) and for mounting in a control panel without the case, as shown in Figure 12.33(b). A dual-pressure switch contains both a high-pressure and a low-pressure switch, as shown in Figure 12.34. Pressure switches can be obtained for most any purpose that would arise in the industry.

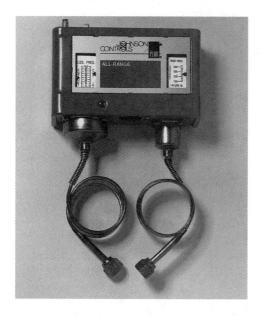

FIGURE 12.34 Dual-pressure switch (combination of high and low pressure) *(Courtesy of Johnson Controls)*

High-Pressure Switches

A **high-pressure switch** is usually used as a safety device to protect the compressor and system from excessively high discharge pressure. A high-pressure switch used as a safety control must open on a rise in pressure to shut the equipment down.

The high-pressure setting of the switch must correspond to the type of refrigerant in the system. The setting of a high-pressure switch used with Freon 134A would be different from the setting of a high-pressure switch used with Freon 22. Figure 12.35 lists the common setting points of a high-pressure switch for the common refrigerants used in the industry today. High-pressure switches may also close on a rise in pressure to operate a device that would control the discharge pressure.

Low-Pressure Switches

Low-pressure switches are used as safety devices, as operating controls, and as devices to operate any component by the suction pressure of the system. All low-pressure switches are connected to the suction side of a refrigeration system. Low suction pressure can damage the compressor. Therefore, low-pressure switches are often used as safety devices to prevent damage to the system when the suction pressure drops below a predetermined point.

Figure 12.36 lists the common setting points of a low-pressure switch for different types of Freon.

If the low-pressure switch opens, it should break the control circuit that operates the compressor. Low-pressure switches can also be used as oper-

Refrigerant	Condenser Type	High-Pressure Setting	
		Cut-out	Cut-in
12	Air cooled	225	145
	Water cooled	170	90
134A	Air cooled	245	85
	Water cooled	190	95
22	Air cooled	380	300
	Water cooled	210	130
404A	Air cooled	450	354
	Water cooled	250	160

FIGURE 12.35 Table of approximate setting points for a high-pressure switch used as a safety control

Refrigerant	Low-Pressure Setting	
	Cut-out	Cut-in
12	15	35
134A	15	35
22	38	68
404A	50	85

FIGURE 12.36 Table of approximate setting points for a low-pressure switch used as a safety control

| | **REFRIGERANT** | | | | | |
| | **12** | | **22** | | **404A** | |
Application	**In**	**Out**	**In**	**Out**	**In**	**Out**
Walk-in Cooler	32	15	60	35	72	40
Dairy Case—Open	35	12	64	24	75	30
Reach-in Display Case—Open	36	14	66	33	78	40
Meat Display Case—Closed	35	15	65	34	78	45
Meat Display Case—Open	30	12	55	30	65	35
Vegetable Display—Open	40	15	80	36	90	45
Beverage Cooler—Wet Type	30	20	58	38	68	58
Beverage Cooler—Dry Type	35	15	65	35	76	43
Florist Box	45	28	80	55	85	63
Frozen Food—Open	6	6	16	4	25	10
Frozen Food—Closed	10	2	22	12	30	15
Wall-in Freezer—(−10°F)	12	1	32	10	25	15

Caution: The figures are approximate and should be used only as a guide. Special applications should not be considered.

FIGURE 12.37 Table of approximate pressure control settings of a low-pressure switch used as an operating control

ating controls to operate the system by a pressure setting that corresponds to a temperature setting. Therefore, by properly setting a low-pressure switch, you can also control the temperature. Figure 12.37 lists the proper pressure control settings for specific applications.

Notation and Terms

You need to understand many terms to successfully maintain and set pressure switches. The **differential** of a pressure switch is the difference between the cut-in and cut-out pressure of the switch. The cut-in is the pressure of the system when the pressure switch closes. The cut-out is the pressure of the system when the pressure switch opens. The **range** of a control is the operating range of the system—for example, the overall pressure—over which the switch can operate. These terms are used frequently when setting pressure to obtain the correct operation of the system. Many pressure controls can be set by carefully reading the pressure dial of a pressure switch. Care should be taken when setting or replacing switches.

Troubleshooting

There are many commonly encountered problems with pressure switches. The contacts of pressure switches often cause problems because of pitting, wear, mechanical linkage faults, and sticking. The contacts should be checked to see if they are closed or open in relation to the setting of the control.

If the contacts show continuity, the resistance should also be checked. If the resistance is above allowable limits, or if the contacts prove to be faulty, the pressure switch should be replaced. If any resistance is read through a set of contacts, voltage is being lost. A reading of 2 or 3 ohms of resistance in a set of switches indicates that the switch should be replaced or at least that the voltage should be checked across the contacts.

The pressure connections of a pressure switch can leak and cause a faulty control or no control at all. To check for a leak, use an acceptable method for locating Freon leaks. Never replace a pressure switch before its correct position and purpose have been determined.

12.6 MISCELLANEOUS ELECTRIC COMPONENTS

Humidistats

In certain air-conditioning applications, it is essential to control the humidity. **Humidistats** are used to control the humidity of a structure. The humidistat uses a moisture-sensitive element to control a mechanical linkage that will open or close an electric switch according to the humidity. Figure 12.38 shows a humidistat.

In certain fiber and cotton industries, the humidity must be accurately controlled by a humidistat, which controls the operation of air washers to prevent damage to the finished product. Humidity control is also advisable in the winter in many structures to produce a more comfortable area by adding more moisture to the air.

FIGURE 12.38 Humidistat *(Courtesy of Honeywell Inc.)*

Oil Safety Switch

Oil safety controls are essential on commercial and industrial equipment to provide adequate protection for the compressor in case the oil pressure drops. Large compressors use a pressure type of lubrication system that must be maintained at a given pressure to ensure proper lubrication of the compressor.

To correctly read the oil pressure of a compressor, you must subtract the suction pressure from the oil pressure to get the net oil pressure, because the suction pressure exerts pressure on the oil in the crankcase. Therefore, the pressure connections of an **oil safety switch** should be connected to the oil pressure port on the compressor oil pump and to the suction or crankcase pressure. These connections of pressure to the oil safety switch will transfer the net oil pressure to the control by some type of mechanical linkage to open and close a set of contacts, depending on the oil pressure setting. Figure 12.39 shows an oil safety switch and its schematic diagram.

The oil safety switch is designed to allow a certain time delay so that the oil pressure will build up in the compressor after the start-up of the compressor. When the compressor is energized, the time-delay switch in the oil

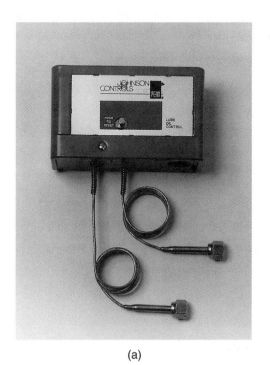

(a)

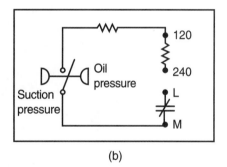

(b)

FIGURE 12.39 Oil safety switch (a) photo *(Courtesy of Johnson Controls)* (b) schematic

safety control is also energized. If the oil pressure does not reach a certain level within the time-delay period, the control circuit will be de-energized. However, if the oil pressure reaches the desired level, the time-delay switch is removed from the circuit and the compressor continues to operate. The time-delay switch is nothing more than a heater that opens a bimetal switch after the specified period of time delay.

Oil safety switches are rated for pilot duty and in most cases must be manually reset. Oil safety controls are essential on large, expensive compressors to prevent undue damage because of inadequate lubrication.

Time-Delay Relay

Time-delay relays are used in the industry to delay the starting of some load for a designated period of time. Time-delay relays usually use some type of heating element that closes a bimetal element. The time-delay function is brought about by the period of time that it takes the heating element to open or close the bimetal. Two types of time-delay relays are shown in Figure 12.40. The relay contacts are usually rated as pilot duty. The coil or heater of the relay can be rated at 24 volts, 120 volts, or 240 volts.

Time-delay relays can be used to prevent two heavy loads of a system or two units connected in parallel from starting at the same time by putting the contacts of the time-delay relay in series with the load or controlling element of the unit. Some commercial and industrial units use a part winding motor, which uses a time-delay relay to energize the windings at different times. The delay period of a time-delay relay can vary from 15 seconds to 30 seconds or longer.

FIGURE 12.40 Time-delay relay

Time Clock

Time clocks are devices that open and close a set of contacts at a selected time by a mechanical linkage between the contacts and a clock. Time clocks can operate on a one-day basis or a seven-day basis. The one-day time clock, shown in Figure 12.41, operates on a 24-hour period without regard to the day of the week. A seven-day time clock operates on an hourly or a daily basis during a week.

Time clocks are set by attaching clips to the clock wheel or dial, as shown in Figure 12.41. There are two clips that attach to the wheel or dial. On movement of the dial, one clip closes the contacts at the point in time where it is attached to the dial. The other clip opens the contacts at the point in time where it is set on the dial.

Time clocks are used to operate equipment or sections of equipment on a time basis. One common use of a time clock is to stop and start the defrosting cycle of a refrigeration system. Time clocks are also used to cut systems off when a structure is to be unoccupied and to cut systems in before building occupants arrive. Buildings such as churches, offices, or manufacturing plants that close and open at specific times often use time clocks. Time clocks can be used effectively to save the operating cost of equipment when it is not needed. A seven-day time clock is shown in Figure 12.42.

FIGURE 12.41 Twenty-four-hour time clock
(Courtesy of Paragon Electric Co., Inc.)

FIGURE 12.42 Seven-day time clock
(Courtesy of Paragon Electric Co., Inc.)

FIGURE 12.43 Solenoid valve *(Courtesy of Sporlan Valve Company)*

Solenoid Valves

Solenoid valves are valves that open and close due to a magnetic or solenoid coil being energized, which pulls a steel core into the magnetic field
of the solenoid. A common solenoid is shown in Figure 12.43. Solenoid
valves stop or start the flow of a fluid such as water, air, or refrigerant.
Solenoids may be normally open or normally closed and care should be
taken to identify the normal position of the valve.

In this section only a few miscellaneous electric controls have been covered, but many others are used in the industry for special purposes. Most
controls not covered can be understood by studying the device and reading
the manufacturer's instructions on the device. Service technicians should
understand all basic electric components and their operation to correctly
diagnose problems in components. Some devices are difficult and complex,
but in most cases manufacturers will assist technicians in becoming familiar with special devices by their literature and through service schools.

12.7 SERVICE CALLS

Service Call 1

Application: Residential conditioned air system

Type of Equipment: Gas furnace and air-cooled condensing unit

Complaint: No heat

Service Procedure:

1. The technician reviews the work order from the dispatcher for available information. The work order information reveals that nothing about the system is operating. The gas furnace is located in the basement.
2. The technician informs the homeowner of his or her presence and obtains any additional information about the system problem.
3. Upon entering the residence, the technician should make certain that no dirt or foreign material is carried into the structure. The technician also takes care not to mar or damage interior walls.
4. The technician sets the thermostat fan switch to the "on" position and there is no system action. Next the technician sets the thermostat to call for heat and there is no system action. At this point, the technician should suspect that the transformer is bad because there appears to be no control voltage.
5. The technician proceeds to the location of the furnace to determine if control voltage is present.
6. The voltage being supplied to the transformer must be checked along with the transformer output voltage.
7. The technician measures 120 volts input voltage to the transformer and 0 volts output voltage and determines that the transformer is bad. It must be replaced with one of the correct VA rating.
8. The technician replaces the transformer and checks the operation of the conditioned air system.
9. The technician informs the homeowner of the problem and that it has been corrected.

Service Call 2

Application: Domestic refrigerator

Type of Equipment: Domestic frost-free refrigerator

Complaint: Temperature too cold (freezing some food)

Service Procedure:

1. The technician reviews the work order from the dispatcher for available information. The work order information reveals that the produce and liquids are being frozen while stored in the fresh food compartment. The homeowner will leave the back door open, and the technician is to lock the door when leaving the home.
2. Upon entering the residence, the technician makes certain that no dirt or foreign materials is carried into the structure. The technician also takes care not to mar or damage interior walls.

3. Upon examnination of the refrigerator, the technician notices that the milk is almost frozen solid.
4. The technician measures the temperature of the fresh food compartment of the refrigerator and finds that the temperature is 15°F.
5. The technician locates and rotates the thermostat to a higher setting and observes that the compressor continues to operate. When the technician turns the thermostat to the "off" position, the compressor stops. A resistance check of the thermostat should be made to ensure that the thermostat will not open at the desired temperature. The thermostat remains closed in any position except the "off" position.
6. The technician replaces the thermostat with the correct replacement model.
7. The technician leaves a note for the homeowner explaining the problem and the actions that have been taken to correct it.
8. The technician should make certain that the residence is secure before leaving.
9. A call later in the day when the customer is at home to check on how the refrigerator is operating builds confidence and good will between the homeowner and service company.

Service Call 3

Application: Residential conditioned air system

Type of Equipment: Gas furnace with air-cooled condensing unit

Complaint: No cooling

Service Procedure:

1. The technician reviews the work order from the dispatcher for available information. The work order information reveals that the system is not cooling. The homeowner can turn the fan switch to the "on" position and the indoor fan will operate normally.
2. The tecnhnician informs the homeowner of his or her presence and obtains any additional information about the system.
3. Upon entering the residence, the technician makes certain that no dirt or foreign material is carried into the structure. The technician also takes care not to mar or damage interior walls.
4. The technician sets the thermostat to the cooling mode of operation and nothing happens.
5. The technician sets the fan switch to the "on" position and the indoor fan runs. The technician can safely assume that 24 volts are available for this action to occur.

6. The next step for the technician is to determine if the thermostat is closing, allowing 24 volts to pass to the indoor fan relay and the outdoor condensing unit. This check can be made at the terminal board on the furnace, at the outdoor condensing unit low-voltage connections, or the indoor fan relay coil connections. If 24 volts are not read at any of these locations, then the thermostat is open and must be replaced.

7. The thermostat must be replaced with one that is designed for the system. The technician sets the heat anticipator to the correct setting.

8. The technician checks the operation of the system and make certain that the system is operating properly.

9. The technician informs the homeowner of the problem and that it has been corrected.

Service Call 4

Application: Commercial refrigeration

Type of Equipment: Walk-in cooler with air-cooled condensing unit

Complaint: Stored product too warm

Service Procedure:

1. The technician reviews the work order from the dispatcher for available information. The work order information reveals that the temperature of the walk-in cooler is higher than acceptable for the product being stored.

2. The technician informs the store manager of his or her presence and obtains any additional information about the problem. The store manager informs the technician that the cooler is doing some cooling, but is not maintaining the required 35°F.

3. The technician measures a temperature of 50°F in the cooler. The evaporator fan motor is operating normally and the air-cooled condensing unit is running for a considerable period of time and cutting off before the desired temperature is reached. The technician must determine what is cutting the condensing unit off.

4. The technician examines the controls of the walk-in cooler and determines that the operating control is a pressure switch.

5. The technician must determine if the pressure switch is out of adjustment or faulty.

6. The technician installs a gauge manifold set on the refrigeration system to determine the operating pressure of the refrigeration system and if the pressure switch is opening at the actual setting on the switch. The technician determines that the pressure switch is operating correctly.

The pressure switch must be adjusted to maintain the correct temperature for the product.

7. The technician sets the pressure switch at the estimated setting.

8. The technician checks the operation of the pressure switch to ensure that the temperature is maintained in the right range.

9. Once the technician has determined that the correct temperature is being maintained, the store manager is informed of the problem and that it has been corrected by adjusting the setting on the pressure switch. The technician advises the store manager not to allow anyone to change the setting of the pressure switch.

Service Call 5

Application: Commercial and industrial conditioned air system

Type of Equipment: Water chiller with one air-handling unit

Complaint: System not operating

Service Procedure:

1. The technician reviews the work order from the dispatcher for available information. The entire system is not operating. The technician is to communicate with the manager upon arrival.

2. The technician informs the manager of his or her presence and obtains any additional information. The manager relays the importance of a fast remedy of the problem because of a meeting in the structure in the afternoon.

3. The technician checks the wiring diagram of the HVAC system controls and determines that the only two possible causes could be a power failure or the failure of the time clock to bring the equipment on at the desired time.

4. The technician measures the control system voltage and determines that it is correct.

5. The technician locates a seven-day time clock that is used to start and stop the equipment when the structure is going to be occupied. The time clock is at the Monday at 2 am setting indicating that the power is not available to the clock motor, or the time clock motor is faulty. The technician measures and finds 120 volts available to the time clock motor and determines that the motor is bad.

6. The technician removes the on and off pins from the clock dial and sets the time clock to the "on" position. The time clock will stay in the "on" position until the time clock is manually turned off.

7. The manager is informed of the action taken by the technician and told that when a new time clock is obtained, the faulty one will be replaced.

8. The technician replaces and sets the time clock to the manager's specifications.

SUMMARY

In this chapter we discussed several different types of control devices, among them thermostats, pressure switches, and transformers.

Thermostats are designed to open and close a set of electric contacts. They are used to control temperature in a structure or the temperature of a medium such as a liquid or gas. There are two types of thermostats used in the industry today. The remote bulb thermostat is used for the control of any medium and is usually used in commercial and industrial applications. The bimetal thermostat is used in all residences and in many cases when air temperature is controlled in commercial and industrial applications. Thermostats can also be of the line voltage or low-voltage type. The line voltage thermostat is used to open or close the voltage supply to a load in a system. The low-voltage thermostat is often used as a switching device to control low-voltage loads.

The heating thermostat that does not have a heating anticipator will allow the temperature of the controlled area to have a large temperature differential. Heating anticipators, when incorporated with a heating thermostat, will create a small temperature differential. Heating anticipators are connected in series with the contacts of the thermostat and set to match the burner control. Cooling anticipators are connected in parallel and operate when the equipment is off. They anticipate the need of the equipment to operate sooner because of system lag. Thermostats play an important part in the industry because of their temperature control function.

Staging thermostats are used to provide a means of controlling two stages of heating or cooling with a set differential. The staging of equipment gives better comfort and more efficient operation of equipment. Staging is used exclusively in commercial and industrial applications.

A clock thermostat is used to control the temperature of a structure. It allows the customer to lower or raise the temperature control point for at least one period in 24 hours. The clock thermostat gives the homeowner the flexibility of setting back the temperature in his or her home when the family is sleeping, for example, or when the family is away from home.

Pressure switches are widely used devices in the industry. They are often used to control a load in a system. Pressure switches are designed to open or close on a rise or fall in pressure, depending on their application.

Low-pressure switches are often used in the industry to protect refrigeration components from low suction pressure. Low-pressure switches are used to operate systems or components and to act as safety devices to protect the refrigeration components of the system.

High-pressure switches are used as safety devices to protect the refrigeration system from excess pressure or to operate a component for head pressure control. High head pressure can damage a compressor, metering device, and high-side system piping. High-pressure switches can also be used to control loads that must be started from the discharge pressure. Careful attention should be paid to the type of system and refrigerant when checking the pressure setting of a control.

Transformers decrease or increase the applied voltage to the desired voltage by use of induction. Transformer ratings include primary voltage and frequency, secondary voltage, and voltamperes. Capacity is rated in voltamperes, that is, the voltage times the current of the control circuit loads. The more low-voltage loads in a control circuit, the larger the transformer that must be used.

REVIEW QUESTIONS

1. What is the purpose of a thermostat in a control system?

2. Which of the following are the two main types of thermostats?
   a. low-voltage and pilot duty
   b. line voltage and high-voltage
   c. low-voltage and line voltage
   d. low-voltage and high-voltage.

3. A heating thermostat opens on a
   _____ .
   a. rise in temperature
   b. fall in temperature

4. A cooling thermostat opens on a
   _____ .

   a. rise in temperature
   b. fall in temperature

5. Explain the operation and application of a remote bulb thermostat.

6. True or False: A bimetal is a combination of two pieces of metal that are welded together, each with a different coefficient of expansion.

7. Why should a thermostat be snap-acting?

8. Which of the following methods are used to ensure snap action of low-voltage thermostat elements?

a. mercury bulbs and small permanent magnets
b. small permanent magnets and springs
c. springs and solenoids
d. solenoids and mercury bulbs

9. True or False: The line voltage thermostat is more common in commercial applications than the low-voltage thermostat.

10. Low-voltage thermostats would be used for which of the following applications?
a. residential air-conditioning control systems
b. refrigerator thermostats
c. window unit thermostats
d. all of the above

11. What are the major differences between the line and low-voltage thermostats?

12. What mode of operation do the letter designations R, Y, Y1, Y2, W, W1, W2, and G represent in a low-voltage thermostat?

13. What is the purpose of a heating anticipator?

14. What is the proper procedure for troubleshooting thermostats?

15. The temperature difference between the closing of the thermostat and the time when warm air begins to reach the thermostat is called _____.
a. overshoot
b. heat anticipation

c. system lag
d. system lead

16. The temperature difference between the closing of the thermostat and the time when the warm air is no longer delivered to the room is called _____.
a. overshoot
b. heat anticipation
c. system lag
d. system lead

17. The overshoot and system lag can produce an additional temperature swing of as much as _____.
a. 2°F
b. 5°F
c. 10°F
d. 20°F

18. What is the purpose of staging heating and cooling equipment?

19. Why are two-stage heating thermostats used on heat pumps?

20. What is the purpose of a low-pressure switch?

21. What is the purpose of a high-pressure switch?

22. How can a low-pressure switch be used as an operating and safety control on a refrigeration system?

23. Define the following terms:
a. cut-in
b. cut-out
c. range
d. differential

24. What is the purpose of a transformer and how does it work?

25. True or False: A step-down transformer is used to boost voltage.

26. What is the correct procedure for checking transformers?

27. How is a transformer sized or rated?
    a. primary volts
    b. secondary volts
    c. voltamperes
    d. all of the above

28. What precaution should be taken when replacing transformers?

29. What is the purpose of a humidistat?

30. The oil safety control is a differential pressure switch and operates on what two pressures?
    a. oil pressure and suction pressure
    b. oil pressure and discharge pressure
    c. suction pressure and discharge pressure
    d. suction pressure, oil pressure, and discharge

31. Time-delay relays are used in heating, cooling, and refrigeration systems to _____.

32. Give several reasons for using a time clock in the heating and cooling industry.

33. Why is it important for service technicians to understand the operation of any electric component in the system?

34. What is the purpose of a clock thermostat?

35. What is a multi-setback clock thermostat?

36. How is the clock portion of a clock thermostat powered?

37. Explain the difference between a single-setback and multi-setback clock thermostat.

38. What routine maintenance should be performed on a thermostat controlling residential heating and cooling?

39. What is calibration with respect to a thermostat?

40. What is the purpose of the fan switch on a residential low-voltage thermostat?

PRACTICE SERVICE CALLS

Determine the problem and recommend a solution for the following service calls. Be specific; do not list components as good or bad.

Practice Service Call 1

Application: Residential conditioned air system

Type of Equipment: Packaged air conditioner

Complaint: No cooling

Symptoms:

1. Line voltage and control voltage available to unit.
2. Condensing unit and evaporator blower will not operate.
3. Thermostat is set in cool mode.
4. Evaporator fan motor operated when fan switch is set to "on" position.

Practice Service Call 2

Application: Commercial refrigeration

Type of Equipment: Display refrigerator with air-cooled condensing unit

Complaint: Unit not cooling

Symptoms:

1. Evaporator fan operating normally.
2. Compressor and condenser fan motor in good condition, but not operating.
3. Pressure switch is used for safety control.
4. Pressure switch contacts closed.
5. Thermostat used for operating control.

Practice Service Call 3

Application: Residential conditioned air system

Type of Equipment: Split-system heat pump

Complaint: No heating

Symptoms:

1. Equipment is inoperative no matter where thermostat is placed.
2. Line voltage to indoor and outdoor units correct.
3. No open safety controls in system.
4. Thermostat in good condition.

Practice Service Call 4

Application: Commercial refrigeration system

Type of Equipment: Walk-in freezer with air-cooled condensing unit with R-12

Complaint: No cooling

Symptoms:
1. Compressor and condenser fan motor are in good condition.
2. Thermostat is calling for cooling.
3. System is equipped with a high-pressure and a low-pressure switch for safety controls.
4. Discharge pressure is 450 psig.
5. Suction pressure is 15 psig.

Practice Service Call 5

Application: Domestic refrigerator

Type of Equipment: Frost-free refrigerator equipped with a defrost timer to initiate defrost cycle

Complaint: Temperature too high in fresh food compartment

Symptoms:
1. Excess frost in frozen food compartment.
2. Thermostat in good condition.
3. Compressor, evaporator fan motor, and condenser fan motor are operating correctly.
4. Refrigerator runs continually.
5. Defrost heaters in good condition.

LAB MANUAL REFERENCE

For experiments and activities dealing with material covered in this chapter, refer to Chapter 12 in the Lab Manual.

13

Heating Control Devices

OBJECTIVES

After completing this chapter, you should be able to

- Explain the purpose of the electrical controls in warm-air and hydronic heating applications that are necessary to safely operate and maintain the desired temperature in a conditioned space.
- Describe the pilot safety controls and the methods of ignition of the burners in a gas furnace.
- Describe the operation of primary controls used to supervise the operation of an oil burner.
- Draw a wiring diagram of an oil-fired, warm-air furnace.
- Draw a wiring diagram of a gas-fired, warm-air furnace.
- Explain the operation of an electric furnace or electric resistance duct heaters and the methods of control that are in common use.
- Draw a wiring diagram of an electric furnace.
- Troubleshoot a gas furnace.
- Troubleshoot an oil furnace.
- Troubleshoot an electric furnace or electric resistance duct heaters.

KEY TERMS

Cad cell
Electrical resistance heater
Fan switch
Gas valve
Hot surface ignition
Ignition module
Pilot

Pilot assembly
Primary control
Sequencer
Spark ignition
Stack switch
Thermocouple

INTRODUCTION

The air conditioning of structures during the winter months requires a source of heat to maintain the desired temperature level. The three major sources of energy used to supply heat to a structure are gas (natural or liquefied petroleum), oil, and electricity. These sources of energy can be used to heat air in a warm-air heating system, heat water in a hydronic heating system, or produce steam for a steam heating system. In this chapter, the electrical components will be given major emphasis while any gas principles will only be briefly discussed to ensure the understanding of the material. The electric heat pump will not be specifically discussed in this chapter with the exception of the method of supplementary heat that is used.

Many of the controls in warm-air furnaces are similar regardless of what type of energy is used as the heating source. There must be a method to start the air supply when warm air is available to the conditioned space and to stop the air supply when the supply of warm air is no longer available. The safety of the heating appliance must also be considered to prevent the furnace from operating under unsafe conditions such as high temperatures, flame rollout, inoperable combustion fan, or open blower door.

In this chapter the controls used for each type of energy source will be discussed. Gas heating equipment requires one of the following methods to ensure that the **pilot** is available or that there is a method of ignition for the main burner: a continuously burning pilot (standing pilot), a method of igniting the pilot when needed (intermittent pilot), or directly igniting the main burner (direct ignition). The intermittent pilot system and the direct ignition system have gained popularity over the past 10 years while the old standing pilot system has decreased in popularity because of the waste of energy and advanced technology that has allowed for the production of inexpensive ignition modules and direct ignition controllers.

There are two types of oil burners in use in the industry today: the vaporizing burner and the atomizing burner. In general, before fuel oil can be ignited it must be vaporized or broken into a fine mist and mixed with air. The vaporizing oil burner depends on natural evaporation facilitated by the heating of the fuel oil to provide oil vapor for combustion. A good example of this method is the old carburetor used on older model oil furnaces. The atomizing oil burner uses a pump and fan connected to the burner motor to accomplish this purpose. The oil pump supplies oil from the storage tank and also raises the pressure of the oil entering the burner, causing the oil to be supplied to the combustion chamber as a fine mist. The fan causes the air to mix with the oil and is ignited by a spark. There are two types of combustion controls commonly used on residential oil-fired heating equipment, the stack switch and the cad cell primary control.

Heating appliances that use electrical energy for the heating source use many of the same controls that have already been discussed in this text. Controls that are essential to electrical resistance heat will be covered again with reference to the control systems used on electric furnaces, supplementary heaters for heat pumps, and electric heaters installed in air ducts.

Supplementary heat used for heat pump application is usually electric resistance heat, but in some cases gas and oil furnaces are being used for this purpose. It is not the intent of this chapter to specifically cover the control systems used in fossil fuel applications that provide supplementary heat for heat pump systems. The controls for the gas and oil appliances used as heat pump supplementary heat will be the same as those in primary heating systems, but the way in which they are connected together will require a different control system.

Troubleshooting gas, oil, and electric heating controls will be covered in this chapter. Troubleshooting the electronic ignition devices that are used in gas heating appliances is generally accomplished using manufacturers' troubleshooting guides along with necessary technical skills of the technician. Troubleshooting the combustion controls of a residential oil heating system is accomplished by knowing what input to the control is required to prove that combustion has been made. Troubleshooting the air supply to the conditioned space in the heating operation of the equipment will be similar no matter what fuel is used.

The controls that are necessary for hydronic and steam heating systems will be briefly covered to give an overview of each application. Hydronic and steam control systems are beyond the scope of this text because of the size and complexity of these types of systems.

13.1 HEATING FUNDAMENTALS

The basic heating appliance used in the heating and cooling industry usually heats air or water, or produces steam. Air is the most popular method of transferring heat from the appliance to the structure, and there are many different styles and designs of warm-air furnaces. Water is also a popular method of transferring heat from the appliance to the conditioned space;

this type of heating is accomplished by using a hot water boiler to heat water, which is then pumped to a location where a supply of hot water is needed. Steam is also used as a heating source and requires a steam boiler and connecting pipe to transfer the steam to the desired location in a structure. The hot water and steam boilers are not as popular in the residential market as the warm-air furnaces, but as the size of the structure increases, their popularity also increases. Hot water and steam boilers are used mainly in the light commercial and commercial and industrial structures.

The warm-air furnace, whether its source of energy is gas, oil, or electricity, will require a source of heat and a method to supply the heat to the desired conditioned space. The gas and oil furnace requires a heat exchanger to transfer the heat from the fuel source to the air while preventing the products of combustion from entering the conditioned space. The heat source, either gas or oil, must be supplied within the heat exchanger. The heat source is supplied to the heat exchanger by the use of a gas or oil burner. There must be some method of moving air across the heat exchanger to facilitate the heat transfer from the heat exchanger to the air. A typical upflow gas furnace is shown in Figure 13.1 and a typical oil furnace is shown in Figure 13.2.

Fossil fuel heating appliances must have a method to remove the products of combustion from the heat exchanger. This can be accomplished by natural draft (gravity) or by forced draft (fan) through a vent to the outside.

The electric furnace requires no heat exchanger because the heat source is from **electrical resistance heater**. A typical electric furnace is shown in Figure 13.3.

The hot water heating system must have some means of heating water or producing steam. The hot water system heats water in a large heat exchanger called a *hot water boiler,* as shown in Figure 13.4. This water is then pumped through hot water lines to the area of the structure that requires heat. The steam boiler shown in Figure 13.5 produces steam that flows by pressure to the areas where heat is required in the structure. The hot water and steam boilers are merely heat exchangers where heat is transferred from the source to the water. In the case of the hot water boiler, water is merely heated to its desired temperature. Hot water boilers usually used in residences and small commercial applications are known as *low-pressure boilers* and operate at a maximum of 15 psig. In large commercial and industrial applications, high-pressure boilers are used and operate at pressures higher than 15 psig.

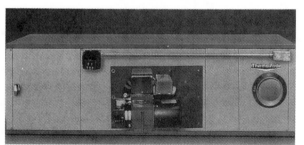

FIGURE 13.2 Typical horizontal oil furnace *(Courtesy of ThermoPride Williamson Co.)*

FIGURE 13.1 Typical upflow gas furnace

FIGURE 13.3 Typical electric furnace

FIGURE 13.4 Typical hot water boiler (*Courtesy of Dunkirk Radiator Corporation*)

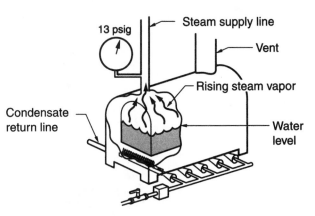

FIGURE 13.5 Drawing of typical steam boiler

13.2 BASIC HEATING CONTROLS

In the basic warm-air furnace there are many controls that are applicable to warm-air furnaces regardless of the type of energy that is being used to supply the heat to the structure. Several factors have to be considered in the operation and function of a warm-air heating system such as the fan operation at the desired time to effectively supply the warm air to the structure, and safety controls that ensure safe operation of the warm-air heating appliance. There are other basic considerations but these two are by far the most important, with the exception of gas and oil combustion controls which will be discussed later in this chapter.

Fan Controls

In all types of forced-air heating equipment there must be some method of controlling the fan motor in order that the warm air is delivered to the conditioned space at the correct temperature. If the temperature of the air is not warm enough, it has a tendency to create a cold draft; this is extremely uncomfortable when this draft is coming directly in contact with the occu-

FIGURE 13.6 Combination fan and limit switch

pants of the structure. At the same time, the air temperature cannot be high enough to cause damage to the equipment or structure or to harm the occupants. The fan must be started at the correct time to make sure that the air is warm enough so as not to be uncomfortable, but it also can't be so hot that it damages equipment or is uncomfortably hot to the occupants. Fan controls must be correctly set in order to supply the conditioned air to the structure at the right temperature. There are several types of **fan switches** used in the industry: temperature controlled, time and temperature controlled, and time controlled.

The temperature-controlled fan switch is nothing more than a thermostat that closes on a rise in temperature to start the fan motor when the furnace can supply warm air to the structure. The fan switch element is inserted into the air cavity of the furnace close to the heat exchanger in order to pick up the temperature of the heat exchanger. This type of fan switch is available with different element lengths to meet specific applications. This type of fan switch is oftentimes incorporated with a limit switch in the same enclosure, allowing both the fan and limit switch to use the same thermal element to control both switches. Limit switches will be covered later in this section. Figure 13.6 shows a combination fan and limit switch often used on furnaces. The wiring diagram of a furnace using the temperature-controlled fan switch is shown in Figure 13.7.

The temperature-controlled fan switch must be set correctly in order to maintain the temperature of the air delivered to the structure and prevent overheating the combustion chamber. The setting of the fan switch must be

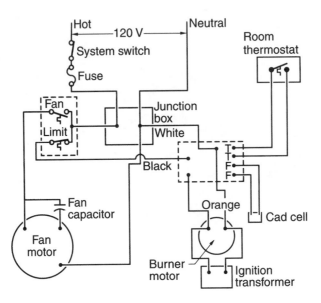

FIGURE 13.7 Typical wiring diagram of furnace with temperature-controlled fan switch

such that the air supplied to the structure is warm enough to prevent uncomfortable drafts and not so hot that it is uncomfortable to technicians or dangerous to the equipment or structure. Figure 13.8 shows the dial of a combination fan and limit switch. This type of fan switch uses a bimetal element to provide quick and accurate response to the air temperature changes. The temperature range of this type of fan switch is approximately

FIGURE 13.8 Dial of a combination fan and limit switch

Adj

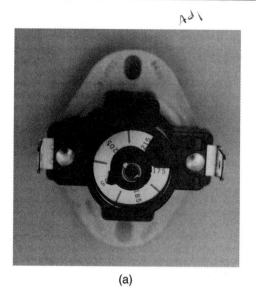

NON-Adj

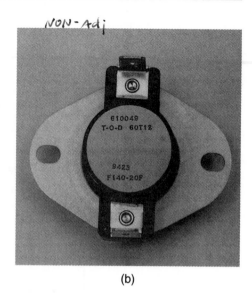

(a)   (b)

FIGURE 13.9 Adjustable (a) and nonadjustable (b) temperature-controlled fan switches

50°F to 200°F with a minimum differential of 15°F. Figure 13.9 shows adjustable and nonadjustable temperature-controlled fan switches. The adjustable type has only a pointer that can be set to turn the fan on with a set differential.

Some furnaces in the industry use a time-controlled fan switch that simply uses the time element to cut the fan on and off. This type of control is nothing more than a time-delay relay that delays the starting of the fan for a certain period of time. The time-delay element is usually connected in parallel to a 24-volt circuit that would be energized with a call for heat. This type of fan switch is usually nonadjustable and is shown in Figure 13.10. This type

• Holds fan off on a start for 45 sec

• Holds fan op. for 75 sec. on a stop.

FIGURE 13.10 Time-controlled fan switch

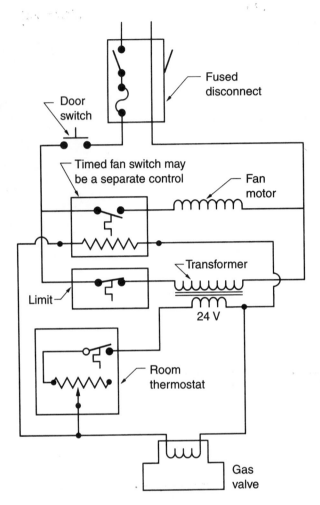

FIGURE 13.11 Typical diagram of a circuit using a timed-on fan switch

of fan switch is available with different on and off timing. For example, one model closes its contacts in 45 seconds and opens the contacts in 75 seconds after the fan has been on for two minutes if de-energized. A diagram of the fan circuit using a timed-on fan switch is shown in Figure 13.11.

● The time/temperature-controlled fan switch provides starting of the fan within a certain period of time (usually 25 to 60 seconds) and/or by temperature. This time function comes after the room thermostat has called for heat. The temperature function would come into effect if the furnace

FIGURE 13.12 Time/temperature-controlled fan switch *(Courtesy of Therm-O-Disc)*

FIGURE 13.13 Flame rollout limit switch

warmed faster than the setting of the time function. The off mode of this type of control is temperature and can be set for 80°F to 110°F. This type of fan switch is shown in Figure 13.12. It can include a limit switch.

Limit Switches

Limit switches on heating appliances are basically thermostats that open when an unsafe condition exists in the furnace, such as high furnace temperatures. This type of unsafe condition could exist if the fan switch failed to start the fan at the proper time or the fan motor was faulty, thus allowing the temperature of the furnace to reach an unsafe condition. In most cases, this temperature is sensed from within the air cavity of the warm-air furnace. Other temperature limit switches are used in various places in warm-air systems to prevent unsafe conditions from damaging the equipment or the structure. Limit switches are used in case of flame rollout, which is flame extending outside the combustion chamber or heat exchanger. A limit switch used for this purpose is shown in Figure 13.13. Figure 13.14 shows the circuitry of a line voltage limit switch and Figure 13.15 shows the circuitry of a low-voltage limit switch.

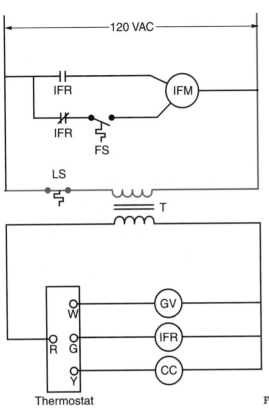

IFM – Indoor fan motor
IFR – Indoor fan relay
FS – Fan switch
LS – Limit switch
GV – Gas valve
CC – Cooling controls
T – Transformer

FIGURE 13.14 Control circuitry of a line voltage limit switch

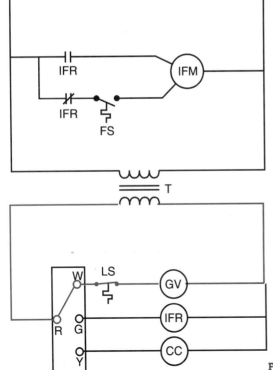

IFM – Indoor fan motor
IFR – Indoor fan relay
FS – Fan switch
LS – Limit switch
GV – Gas valve
CC – Cooling controls
T – Transformer

FIGURE 13.15 Control circuitry of a low-voltage limit switch

13.3 GAS HEATING CONTROLS

The basic control cycle of a gas heating appliance is initiated when a switch, usually a thermostat, closes to call for heat. In a gas heating system this call for heat completes the heating control circuit, starting a chain reaction that results in lighting the burner. In a gas heating appliance, once there is a call for heat, the function of the gas burner control is to ensure safe ignition of the main burner. There are three basic types of gas burner controls: standing pilots, where the pilot burns continuously; intermittent pilots, where the pilot is automatically lit on a call for heat; and direct ignition, where some method is used to light the main burner upon a call for heat. Some type of automatic **gas valve** is used to allow the flow of gas to reach the burner at the proper time ensuring ignition.

Standing Pilot Burner Control System

The standing pilot is often called a *continuous ignition system.* The pilot has to be lit by hand and burns continuously until shut off by hand or until a pilot outage occurs. This standing pilot lights the main burner when there is a call for heat from the thermostat. This type of gas burner control system must ensure that the pilot is lit when the main gas valve opens. If the pilot is not lit, the control system must lock out the opening of the main gas valve to prevent the combustion chamber from filling with un-ignited gas, causing an unsafe condition.

In almost all cases, this burner control system uses a **thermocouple** to supply up to 30 mV of power to the safety shutoff pilot solenoid in the gas valve. This thermocouple is made of two dissimilar metals that when heated will produce a small voltage. A thermocouple is used to supply a millivoltage to the pilot solenoid in the gas valve, holding the pilot safety valve open and indicating that a pilot is available to light the main burner. Figure 13.16 shows a thermocouple and a pilot burner. The pilot burner has three primary functions: to direct the pilot flame for proper burner ignition, to provide a mount for the thermocouple, and to heat the thermocouple to provide the voltage required to hold in the pilot solenoid valve. Figure 13.17 shows the proper pilot flame adjustment.

The gas valve is used to supply gas to the main burner when a pilot has been proved. Figure 13.18 shows a drawing of a typical gas valve with a pilot safety valve and a main gas valve; both of these valves have to be open for the gas supply to reach the main burner. A photograph of a gas valve that incorporates a pilot safety solenoid with the connection for the thermocouple is shown in Figure 13.19. A typical schematic diagram of a

FIGURE 13.16 (a) Thermocouple and (b) pilot burner

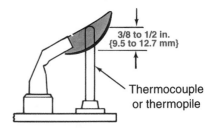

FIGURE 13.17 Proper pilot flame *(Courtesy of Honeywell, Inc.)*

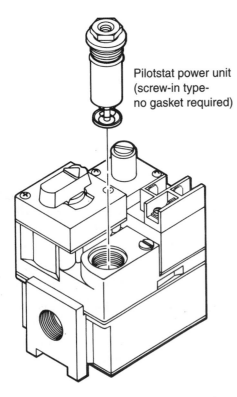

FIGURE 13.18 Drawing of a typical gas valve using thermocouple as pilot safety *(Courtesy of Honeywell, Inc.)*

FIGURE 13.19 Photograph of a gas valve

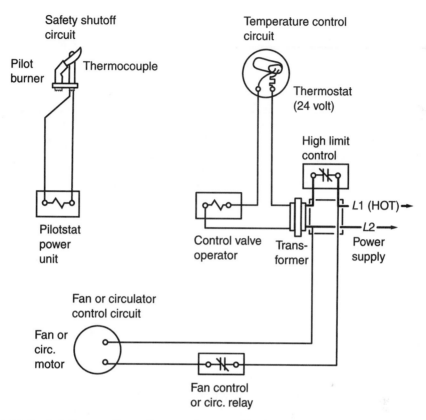

FIGURE 13.20 Typical diagram of a gas furnace using a combination gas valve *(Courtesy of Honeywell, Inc.)*

furnace with a combination gas valve is shown in Figure 13.20. Gas valves used in this application are available in 24 and 120 volts.

The pilot solenoid of this type of gas valve is not strong enough to open the pilot valve of the main gas valve. The thermocouple only produces a voltage source that is capable of holding the pilot solenoid open. The pilot valve must be manually opened for the initial lighting of the pilot until the pilot has adequate time to heat the thermocouple, thus allowing the production of a voltage source that is capable of holding the pilot valve open once it has been manually depressed.

Figure 13.21 shows a flowchart of the sequence of operation of a standing pilot type of ignition system.

Intermittent Pilot Burner Control System

The intermittent pilot burner control system must light the pilot and control the main gas valve. The intermittent pilot burns only when there is a call

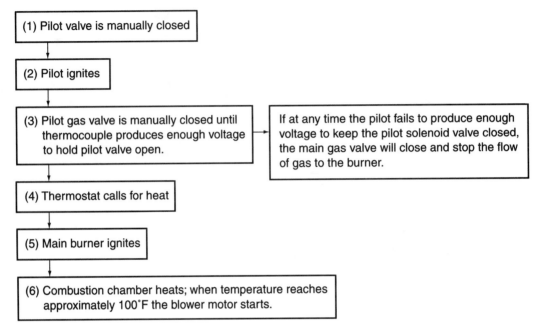

FIGURE 13.21 Flowchart of the sequence of operation of a standing pilot ignition system

for heating and remains off when there is no call for heat. This saves operating cost because less gas is being consumed in the furnace operation. However, a method of lighting the pilot on the call for heat and proving the lighting of the pilot is necessary. Once the pilot has been proved, the pilot flame lights the main burner and remains on until the main burner goes off at the end of the heating cycle.

There must be some method of igniting the pilot burner in the intermittent pilot control system. The function of the igniter is to smoothly light the pilot burner. Once the pilot has been ignited, there must be some method of proving that the pilot is lit. This can be accomplished by using a liquid-filled pilot sensor, a temperature sensor, a "fire eye" sensor, or a flame rod. A liquid-filled pilot sensor is nothing more than a probe that is inserted in the lit pilot. As the probe is heated, the pressure increases, which exerts pressure on a diaphragm to close a set of contacts signaling that the pilot is lit. The "fire eye" or "cad cell" type sensor is a device that changes its resistance in the presence of light. When this type of sensor is directed toward light its resistance will decrease, and when it is directed toward darkness the resistance will increase. This sensor must be connected to an electronic module to control the operation of the pilot and main burner, and it is used mainly on commercial and industrial gas burn-

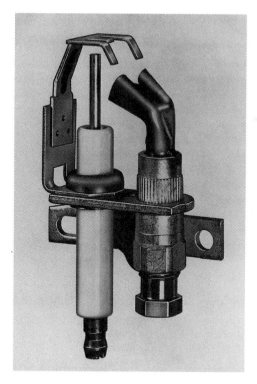

FIGURE 13.22 Photograph of a pilot assembly with ignition components and flame rod *(Courtesy of Honeywell, Inc.)*

ers. The flame rod is a sensor that is used to detect a flame at the pilot or the main burner. A gas flame conducts electricity, and a path is made from the flame rod back to the main or pilot burner head. A **pilot assembly**, which includes the flame rod and ignition components, is shown in Figure 13.22. Oftentimes the ignition sensor is used as a flame rod to send the correct signal back to the ignition or burner control. An **ignition module** is shown in Figure 13.23. For proper operation, this ignition module requires a combination pilot burner/igniter-sensor as shown in Figure 13.22, the proper gas valve, and an ignition cable.

This type of intermittent pilot burner control system requires a special gas valve, which is shown in Figure 13.24. Most of the gas valves used in this type of system are the redundant type or dual-type valves. In most cases, these types of valves actually have three valves within each valve. One is a manual shut-off valve that mechanically blocks all gas flow when turned off. There are two additional main gas valves that open on a call for heat. Essentially the main gas flow is stopped twice with a redundant-type gas valve.

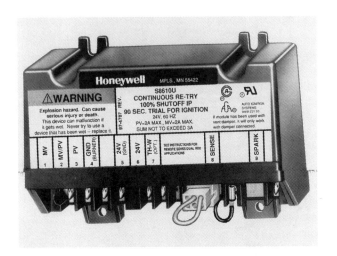

FIGURE 13.23 Photograph of an ignition module *(Courtesy of Honeywell, Inc.)*

FIGURE 13.24 Photograph of redundant gas valve

The operational sequence of an intermittent pilot burner control system is as follows:

1. On a call for heat, some modules have a pre-purge cycle that occurs before the spark starts. During this pre-purge cycle, the combustion blower runs to clear the heat exchanger of any unburned gas. This cycle usually lasts 30 to 45 seconds.

2. On a call for heat, the ignition module does a self-check, and if a failure is shown, the ignition won't start. If the checks are good, the module begins a safety lockout timing, powers the spark igniter, and opens the solenoid valve so gas can flow to the pilot. The pilot must light within a certain period of time or the module closes the valve.

3. When the pilot lights, current flows from the ignition sensor through the pilot flame to the burner head and then to ground; the ionized pilot flame provides a current path between the rod and burner head, rectifying the current. Because of the difference in size of the sensor and burner, current flows in only one direction. The current is a pulsating direct or rectified current, and it tells the module that a flame has been established. Ignition stops and the second main gas valve opens, allowing gas to flow to the main burner.

4. As long as this rectified flame current remains above the minimum, the module keeps the main gas valves open. If the current drops below

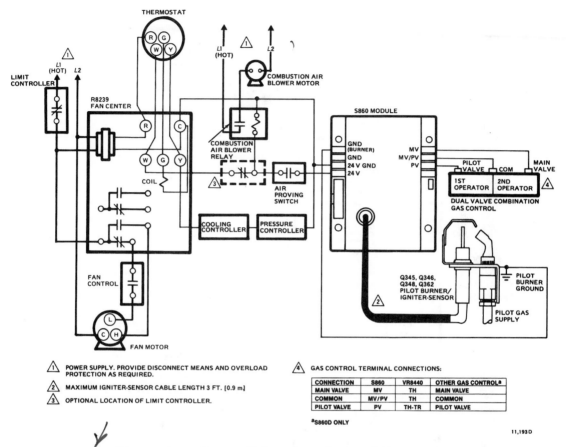

FIGURE 13.25 Diagram of a gas furnace with intermittent pilot control *(Courtesy of Honeywell, Inc.)*

the minimum or becomes unsteady, the module closes the main gas valves. The module then performs the start safety check and, if it's safe, the module attempts ignition again.

A diagram of this type of control system is shown in Figure 13.25. Figure 13.26 shows a flowchart of the sequence of operation of an intermittent ignition system.

Direct Ignition Burner Control System

The direct ignition systems use a spark igniter (direct **spark ignition**) or a silicon carbide igniter (**hot surface ignition**) to light the main gas burner directly. Ignition stops after a designated time or when the main burner flame ignition has been properly proved. The typical components of a

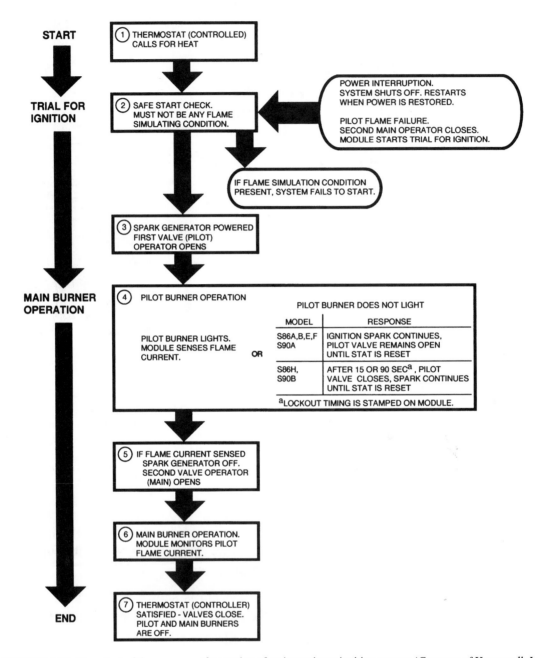

FIGURE 13.26 Flowchart of the sequence of operation of an intermittent ignition system *(Courtesy of Honeywell, Inc.)*

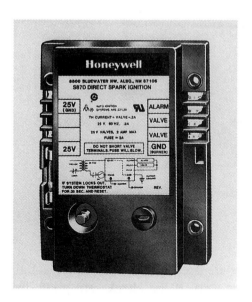

FIGURE 13.27 Photograph of an ignition module used on direct ignition control system *(Courtesy of Honeywell, Inc.)*

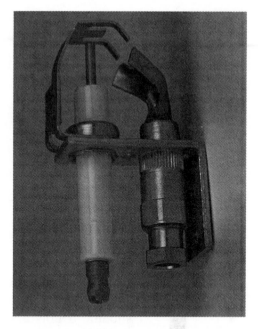

FIGURE 13.28 Photograph of a spark igniter

direct ignition burner control system are the ignition module, igniter, sensor, gas control, and other common controls used on any type of gas furnace. The ignition module is an electronic device that supervises the ignition of the flame and monitors the flame during the "on" cycle. It is shown in Figure 13.27. The igniter is the device that lights the main burner. There are basically two types used in the industry: the spark igniter as shown in Figure 13.28 and the hot surface ignition device as shown in Figure 13.29. The gas valve is the device that controls the flow of gas to the burner and is similar to that shown in Figure 13.24. All other controls are common to most gas furnaces used in the industry.

The operational sequence of the direct ignition burner control system is as follows:

1. On a call for heat, most modules have a pre-purge cycle that occurs before ignition. During this pre-purge cycle, the combustion blower runs to clear the heat exchanger of unburned gas. The cycle usually lasts 30 to 45 seconds.

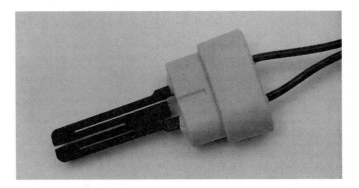

FIGURE 13.29 Photo of a hot surface igniter

2. On a call for heat, the ignition module does a self-check, and if a failure is shown, the ignition will not start. If the checks are good, the module begins a safety lockout, powers the igniter, and opens the gas valve.
3. Once ignition starts, the burner must light and ignition must be proved within the safety lockout timing. If the burner does not light, then the ignition stops and the gas valve closes. On a lockout, the system must be manually reset. Many modules allow for several attempts at ignition before locking out.
4. When the flame lights, current flows from the sensor through the ionized pilot flame to the burner head and then to ground. The current is a pulsating, direct, or rectified current, and it tells the module that flame has been established. Ignition stops and the burner continues to run.
5. As long as this rectified flame current remains above the minimum, the module keeps the gas valve open. If the current drops below the minimum or becomes unsteady, the module interrupts power to the gas valve, closing the valve and stopping gas flow. The module then performs the start safety check and, if it's safe, the module attempts ignition again. Figure 13.30 shows a flowchart of the operation of a direct ignition system.

A diagram of the direct spark ignition burner control system is shown in Figure 13.31, and the hot surface ignition burner control system is shown in Figure 13.32. Figure 13.33 shows a typical diagram of a gas-fired, warm-air furnace using a direct ignition burner control.

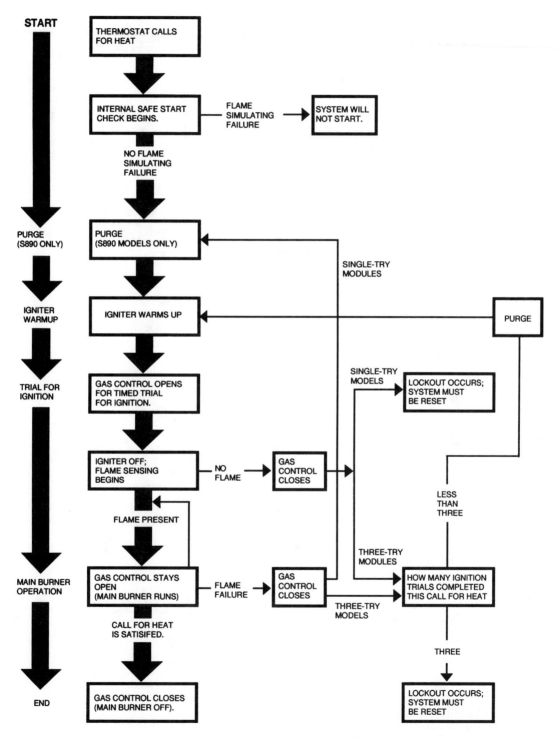

FIGURE 13.30 Flowchart of the sequence of operation of a direct ignition system *(Courtesy of Honeywell, Inc.)*

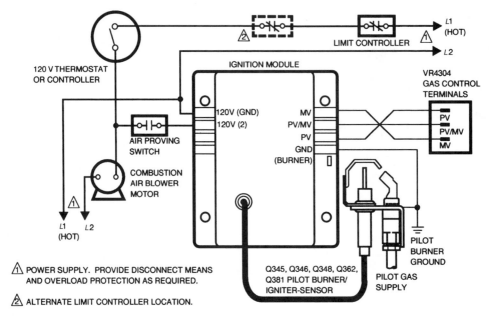

FIGURE 13.31 Diagram of direct spark ignition module *(Courtesy of Honeywell, Inc.)*

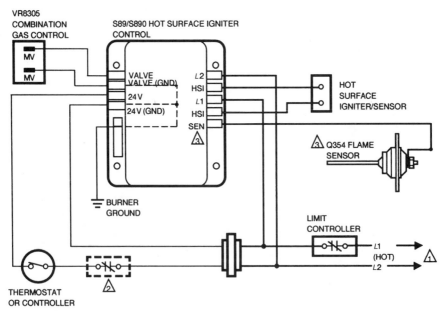

FIGURE 13.32 Diagram of hot surface ignition module *(Courtesy of Honeywell, Inc.)*

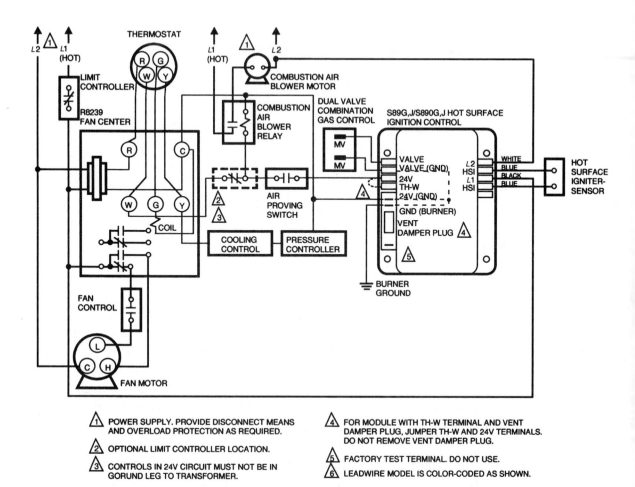

① POWER SUPPLY. PROVIDE DISCONNECT MEANS AND OVERLOAD PROTECTION AS REQUIRED.

② OPTIONAL LIMIT CONTROLLER LOCATION.

③ CONTROLS IN 24V CIRCUIT MUST NOT BE IN GORUND LEG TO TRANSFORMER.

④ FOR MODULE WITH TH-W TERMINAL AND VENT DAMPER PLUG, JUMPER TH-W AND 24V TERMINALS. DO NOT REMOVE VENT DAMPER PLUG.

⑤ FACTORY TEST TERMINAL. DO NOT USE.

⑥ LEADWIRE MODEL IS COLOR-CODED AS SHOWN.

FIGURE 13.33 Typical diagram of gas furnace with direct ignition controller *(Courtesy of Honeywell, Inc.)*

13.4 OIL HEATING CONTROLS

The function of an oil burner control system is to turn the heating system on and off in response to the needs of the conditioned space. The control must also safeguard the operation of the heating appliance and oil burner. Various thermostats in the control system are used to maintain the temperature in the conditioned space while others are used as limits that respond to unsafe conditions in the heating system. The **primary control** is the heart of an oil burner control system and supervises the operation of the oil burner. The primary control must control the oil burner motor, ignition transformer, and oil solenoid valve, if used, upon a call for heat.

The gun-type atomizing oil burner will be discussed briefly to provide a basic understanding of its operation. Figure 13.34 shows a typical oil

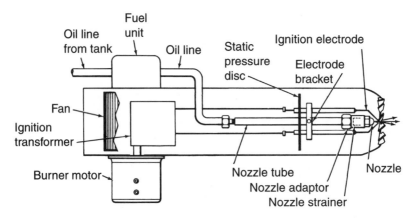

FIGURE 13.34 Typical oil burner with components labeled *(Courtesy of Honeywell, Inc.)*

burner with its components labeled. The oil burner is made up of an oil pump and fan driven by an electric motor. In many cases, this oil pump is responsible for transferring oil from the oil supply and for delivering the oil to the nozzle under high pressure. As the oil is forced through the nozzle under high pressure, the electrodes of the oil burner create a high-voltage spark from the ignition transformer. The ignition electrodes should be behind the extreme outer edge of the oil spray pattern that is created by the oil nozzle, and they will ignite the oil in the combustion chamber. If so equipped, the oil valve will open at the correct time by the supervision of the primary control. The oil burner directs the flame into a heat exchanger that is isolated from the conditioned space and has a vent that allows products of combustion to exit the heat exchanger through a chimney or vent stack.

The primary control must safely control the operation of the oil burner. The primary control must ensure that the burner has lit and that the flame has been proved. If the burner is allowed to run without a flame being established, a large amount of oil will be pumped into the heat exchanger, creating an unsafe condition and problems when the burner does light. The primary control senses the heat from the oil burner flame through a thermal sensor known as a **stack switch**, shown in Figure 13.35, or a light-sensitive sensor know as a **cad cell**, shown in Figure 13.36. The oil burner primary control can interrupt the transformer operation once a flame has been established. However, not all oil burner primary controls are equipped with this function. Most primary controls will have to be manually reset once a flame failure has occurred in the oil burner.

FIGURE 13.35 Photograph of a stack switch *(Courtesy of Honeywell, Inc.)*

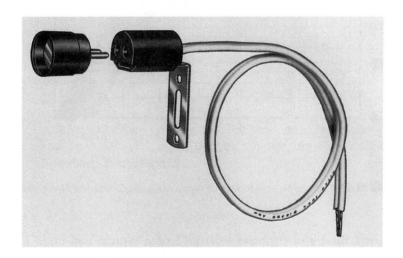

FIGURE 13.36 Photograph of a cad cell *(Courtesy of Honeywell, Inc.)*

Cad Cell Oil Burner Primary Controls

This primary control device consists of a primary control and a light-sensitive cad cell mounted so that it views the oil burner flame. The cad cell changes its resistance according to the intensity of the light. The resistance of the cad cell decreases as the intensity of the light increases. Figure 13.37

Resistance (ohms)

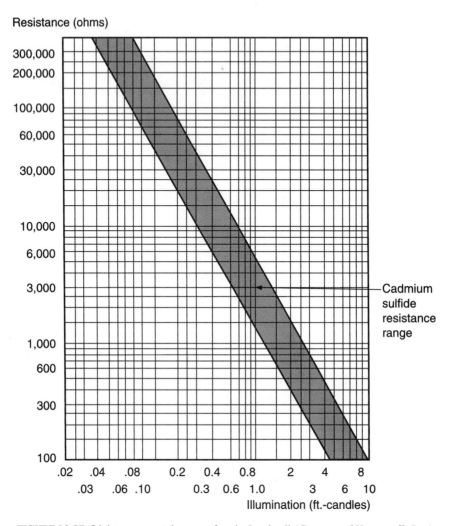

FIGURE 13.37 Light response tolerance of typical cad cell *(Courtesy of Honeywell, Inc.)*

shows the light response tolerance of a typical cad cell. The resistance of the cad cell is used to determine if there is a flame present in the combustion chamber. The resistance of the cad cell is the input to the primary control detection circuit, which determines whether the control senses the flame, thus allowing the oil burner to continue operation or to lock out on safety. One of the major advantages of the cad cell is the rapid response time to light, making the control fast acting.

The cad cell is made from a ceramic disc coated with cadmium sulfide and overlaid with a conductive grid. Electrodes are attached to the disc that transmit a resistance to the primary control. The entire assembly is sealed

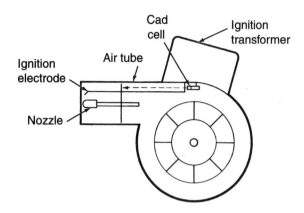

FIGURE 13.38 Correct placement of a cad cell *(Courtesy of Honeywell, Inc.)*

to prevent the cad cell from deteriorating. If an oil burner is properly adjusted, the cad cell resistance will be approximately 300 to 1000 ohms when the burner is operating. To continue operation, the cad cell resistance should remain below 1600 ohms. The placement of the cad cell should be so that the cad cell has a good view of the flame with adequate light reaching the cell. The cad cell must be protected from external light sources. Figure 13.38 shows the correct placement for a cad cell in an oil burner.

When the primary control starts the burner, a bimetal-operated safety switch in the primary control starts to heat. The primary control will open the circuit to the oil burner motor and ignition transformer unless a flame has been established. If a flame is established, the cad cell resistance drops and the current through the cad cell actuates a relay or an electronic network in the primary control. The circuit to the safety switch is broken and the electrical flow to the bimetal is interrupted, thus removing the heat and allowing the burner to continue operation. If the flame goes out, the cad cell resistance goes up, causing a signal to be sent to the primary control to energize the safety switch heater. The heater being energized will, in a matter of seconds, break the electrical circuits to the oil burner. Figure 13.39 shows a typical diagram of an oil-fired, forced-air furnace. Figure 13.40 shows the wiring diagram of a cad cell primary control.

Stack Switch Oil Burner Primary Controls

The stack switch is a heat-actuated control that uses the stack temperature to indicate that the oil burner has or has not established a flame. A bimetal element inserted into the stack actuates a push rod when the bimetal senses heat, signaling that the flame has been established and breaking the circuit

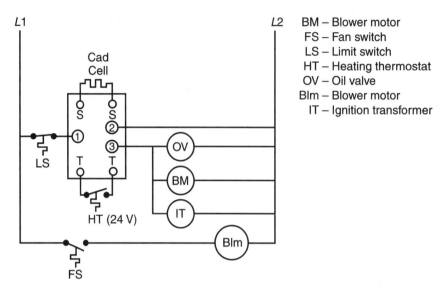

FIGURE 13.39 Typical diagram of an oil-fired furnace with cad cell primary control

FIGURE 13.40 Cad cell primary control with electronic components *(Courtesy of Honeywell, Inc.)*

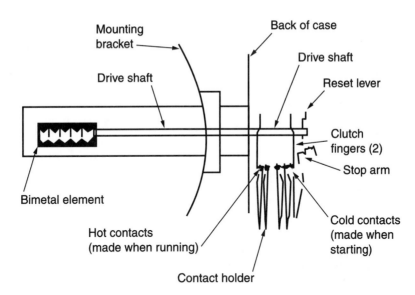

FIGURE 13.41 Components of a typical stack switch flame detector *(Courtesy of Honeywell, Inc.)*

to the safety switch. At the same time, another circuit is established to allow continued operation of the oil burner. The components of the flame detector are shown in Figure 13.41. The correct location and mounting of the stack switch is in the center of the stack or vent in the direct path of the hot flue gases.

The stack switch primary control starts the burner and supervises burner operation. When the thermostat calls for heat, the stack switch closes a relay, which starts the burner motor and ignition transformer and opens the oil solenoid, if used. At the same time, the safety switch heater starts to heat. If the oil burner establishes a flame and heat is felt in the stack, the bimetal in the stack switch will open a set of contacts, thus de-energizing the safety switch heater. The oil burner will run until the thermostat has been satisfied. If the stack switch does not sense a temperature rise in the stack, thus determining that no flame has been established, the safety switch heater will remain energized and the bimetal will be heated sufficiently to de-energize the relay in the primary control, thus interrupting the electrical path to the burner motor, ignition transformer, and oil solenoid, if so equipped. Figure 13.42 shows a typical diagram of an oil-fired, forced-air furnace with a stack switch. Figure 13.43 shows a typical diagram of the internal wiring of the stack switch.

Stack switches are available with intermittent ignition, which stops the ignition transformer when the flame has been proven.

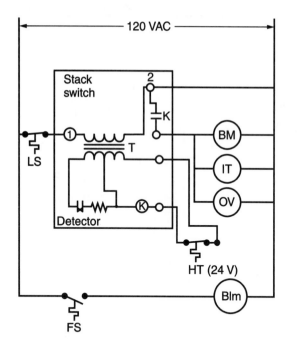

BM –	Blower motor
LS –	Limit switch
HT –	Heating thermostat
IT –	Ignition transformer
OV –	Oil valve
Blm –	Blower motor
FS –	Fan switch

FIGURE 13.42 Typical diagram of oil-fired furnace with stack switch

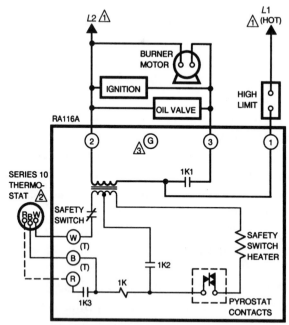

⚠ POWER SUPPLY. PROVIDE DISCONNECT MEANS AND OVERLOAD
PROTECTION AS REQUIRED.

⚠ IF USING TWO-WIRE THERMOSTAT, TAPE LOOSE END OF RED
WIRE IF NECESSARY.

⚠ USE GREEN TERMINAL TO CONNECT CASE TO GROUND.

FIGURE 13.43 Typical wiring of stack switch *(Courtesy of Honeywell, Inc.)*

13.5 ELECTRIC HEATING CONTROLS

Forced-air electric furnaces utilize electric resistance heaters to provide the heat to the conditioned area with the assistance of a blower motor to facilitate the movement of air. Electric resistance heaters are also used as duct heaters with an external means of supplying the air flow. Some other types of primary heating systems are manufactured so that electric resistance heaters can be mounted in the equipment.

The electric heating control system will be similar whether the application is an electric furnace or a duct heater. An electric furnace is shown in Figure 13.44. An electric resistance heater is shown in Figure 13.45. Most electric heaters are similar whether they are mounted in an electric furnace or straight into the air flow as a duct heater.

There are several methods of controlling the electric heaters in an electric heating system. One of the most popular methods is a **sequencer**, which is nothing more than a time-delay relay. Sequencers can have as many as

FIGURE 13.44 An electric furnace

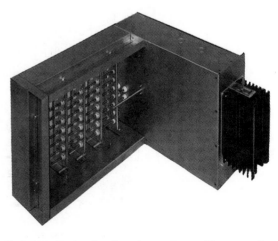

FIGURE 13.45 An electric resistance heater *(Courtesy of Indeeco, St Louis, MO)*

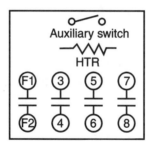

FIGURE 13.46 Diagram of sequencer *(Courtesy of Honeywell, Inc.)*

five sets of contacts that close at different intervals after the control has been energized. This function allows the control system to sequence the heaters and fan motor in over a period of time. The diagram of a sequencer is shown in Figure 13.46 with a picture of a sequencer in Figure 13.47. A typical diagram of a sequencer used to control the fan and heaters in an electric furnace is shown in Figure 13.48. Care should be taken that the fan is brought on before any overheating occurs. Contactors are sometimes used to control electric resistance heaters. When contactors are used for this purpose, care should be taken that the fan is interlocked so that the heaters cannot be on without the fan operating. Thermostats are used as safety overloads in the case of an electric heater overheating.

FIGURE 13.47 Photograph of sequencer *(Courtesy of Honeywell, Inc.)*

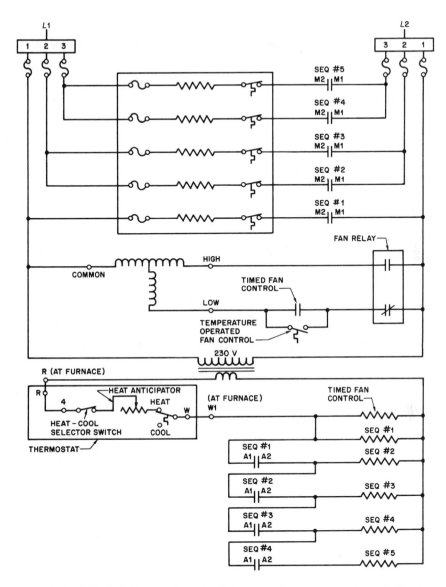

FIGURE 13.48 Typical diagram of an electric furnace using a sequencer to control heaters

13.6 HYDRONIC AND STEAM CONTROLS

Most controls used in hydronic and steam heating systems have been discussed in the thermostat and the pressure switch sections of Chapter 12 in this text. Hydronic and steam heating systems require some type of heat exchanger, usually a boiler, that is used to heat water or produce steam.

There are additional controls on many of these systems that are used basically for safety, such as low water cutoffs, water temperature limits, pressure limit switches, and flow switches. The supervision of the gas or oil burner will be the responsibility of the gas or oil burner controls. Gas and oil burner controls for the smaller boilers will be much the same as those discussed in earlier sections of this chapter. However, the larger heating plants used in the commercial and industrial sectors usually require a quicker-acting control because of the amount of fuel that could enter the combustion chamber in a short period of time.

The controls of a hydronic system are mostly immersion thermostats that are inserted into the water in the boiler or piping. These thermostats are used as operating controls, which control the temperature of the water in the system, and safety controls, which are used as limits to disrupt power in case of unsafe temperatures within the system. Thermostats that are immersed in water are sometimes called *aquastats*. Many hydronic systems have a low water cutoff that will interrupt the power to the burner if the level of water in the boiler gets too low. A typical diagram of the control system of a hydronic heating system is shown in Figure 13.49.

Hydronic heating systems must have some method of circulating the water throughout the structure in order to deliver heat to the conditioned space or spaces. Hydronic heating systems commonly use centrifugal water pumps to circulate water through the hot water piping of the building. There must be some method of controlling the flow of water to the building in order to maintain the desired temperature. In small structures with only one heating zone, it would be necessary to cut the flow of water off by stopping the pump or closing a hot water valve that is installed in

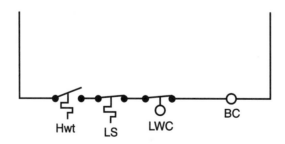

Hwt – Hot water thermostat
LS – Limit switch
LWC – Low water cutoff
BC – Fuel control

FIGURE 13.49 Simplified diagram of a control system for a hydronic heating system

the hot water supply lines. A thermostat would be used to control the hot water valve or hot water pump. In larger structures that have more than one heating zone, a valve must be installed in the hot water line supplying each zone in order to control the flow of water to each zone. Hot water valves used in this application generally regulate the flow proportionally, supplying whatever flow is needed for the zone being supplied to maintain the desired temperature. Some hot water valves are being used that have only two positions, open and closed.

Some larger residences now use a type of zoned hydronic system that uses a zone control system connected to the pumps and valves to supply the hot water needed to condition each zone. In commercial and industrial control systems the hot water zone valves can be electronic or pneumatic, but the same principle is applied for each zone.

There are two main types of hot water heating units used to supply heat to the conditioned space: the natural convection units and the forced convection units. Natural convection units use the natural flow of warm air to heat the desired space and are commonly found as baseboard heaters, radiators, and panel heaters. The forced convection type of heating unit uses a fan to circulate air across a coil to supply the desired heat to the conditioned space. These types of units are available as room forced-air convectors, forced-air unit heaters, or hot water coils placed in an air supply.

A steam system would use a pressure switch as an operating control and a limit switch. The system would use a low water cutoff to stop the burner if the water level is low. The steam system uses a condensate pump to return water to the boiler and is controlled by a flow switch. A simplified control system for a steam boiler is shown in Figure 13.50.

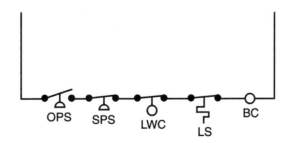

OPS – Operating pressure switch
SPS – Safety pressure switch
LWC – Low water cutoff
LS – Limit switch
BC – Burner control

FIGURE 13.50 Simplified control system for a steam boiler

Steam flows naturally because of the pressure in the system. As the steam is cooled at the heating transfer unit it will condense back to a liquid and will have to be delivered back to the boiler. Steam systems have devices called steam traps that allow the condensate to flow back through the lines to the boiler or to a reservoir where it is pumped back into the boiler. A condensate pump is used to pump the condensate water back to the boiler. These pumps are controlled by a flow switch which, when the water level is high enough in the reservoir, closes to energize the pump. Various types of control valves are used to regulate the flow of steam to devices that are controlling the temperature of a conditioned space or devices that need a steam source in a manufacturing process.

The hydronic and steam control systems become increasingly complex as the size increases with more sophisticated combustion controls and system controls. The intent of this material on hydronic and steam heating systems has been only to provide an overview of what a technician might come in contact with in the industry.

13.7 SERVICE CALLS

Service Call 1

Application: Residential conditioned air system

Type of Equipment: Gas furnace with standing pilot ignition

Complaint: No heat

Service Procedure:

1. The technician reviews the work order from the dispatcher for available information. The work order reveals that there is no pilot light.
2. The technician informs the homeowner of his or her presence and obtains any additional information about the problem. The furnace is located in the crawl space.
3. The technician asks the homeowner to set the thermostat to the heating mode of operation.
4. The technician proceeds to the furnace in the crawl space and determines that there is no pilot burning. The technician attempts to light the pilot, but every time the pilot safety valve is released, the pilot flame goes out.
5. The technician must determine if the problem exists with the thermocouple or the gas valve.

6. To check the thermocouple, the technician uses an adapter that allows a millivolt reading of what the thermocouple is producing. The voltage reading of the thermocouple is 4 mV, which is an indication that the thermocouple is at fault.
7. The thermocouple must be replaced with a new thermocouple of adequate length.
8. The technician runs the furnace through at least one complete cycle to make certain that the system is operating properly.
9. The technician informs the homeowner of the problem and that it has been corrected.

Service Call 2

Application: Residential conditioned air system

Type of Equipment: Gas furnace with intermittent pilot control

Complaint: No heat

Service Procedure:

1. The technician reviews the work order from the dispatcher for available information. The work order information reveals that the furnace is in the basement. The homeowner will leave the basement door open, which is located on the north end of the house. The thermostat will be set in the heating mode.
2. The gas supply is turned off by the technician.
3. The technician measures the line voltage available to the furnace and reads 120 volts.
4. The technician determines that limit switches and other safety devices are in good condition.
5. The technician measures the voltage at the ignition module and reads 24 volts.
6. The technician resets the ignition module by resetting the thermostat or breaking 24 volts at the terminal board of the furnace.
7. After the ignition module has been reset, the module should produce a spark across the igniter/sensor gap. The ignition lead should be pulled off and the spark checked at that location. If no spark exists, the technician should check the fuse on the ignition module. If the technician finds the fuse is good, the ignition module should be replaced. The fuse is good.
8. The technician replaces the ignition module with the correct replacement.

9. The technician leaves a note for the homeowner explaining the problem and that the furnace has been repaired.
10. The technician secures the basement door upon leaving the residence.

Service Call 3

Application: Residential conditioned air system

Type of Equipment: Gas furnace with direct spark ignition

Complaint: No heat

Service Procedure:
1. The technician reviews the work order from the dispatcher for available information. The work order reveals that the furnace is an upflow gas-fired, forced-air furnace located in a closet within the living space of the house.
2. The technician informs the homeowner of his or her presence and obtains any additional information about the system problem. The homeowner informs the technician that the furnace goes off before the desired temperature within the structure is reached.
3. Upon entering the structure, the technician makes certain that no dirt or foreign material is carried into the structure. The technician also takes care not to damage any interior walls.
4. The technician locates the furnace and sets the thermostat to the heating mode. The technician observes that the furnace burner is igniting, but only stays lit for approximately five minutes.
5. The lighting of the burner eliminates some of the possible problems that the technician needs to check. If the burner lights, the technician can eliminate checking for a call for heat, voltage to the ignition module, and correct operation of the spark across the igniter/sensor gap.
6. The module does not lock out, so there is no need to reset the control. The technician checks to see if flame current is correct, and for excessive heat at the sensor insulator. The technician checks these items and both are in acceptable ranges or in good condition. The ignition module is faulty and needs to be replaced.
7. The technician replaces the ignition module with an acceptable replacement.
8. The technician checks the operation of the furnace and makes certain that the problem has been corrected.
9. The technician informs the homeowner of the problem and that it has been corrected.

Service Call 4

Application: Residential conditioned air system

Type of Equipment: Oil furnace with a cad cell primary control

Complaint: No heat

Service Procedure:

1. The technician reviews the work order from the dispatcher for available information. The work order reveals that the oil furnace is located in the basement, which is accessed through the house. The housekeeper will be at the residence and will let the technician in.
2. The technician informs the housekeeper of his or her presence and obtains any additional information about the operation of the furnace. The housekeeper informs the technician that they can hear the burner start and light but it goes off almost immediately.
3. Upon entering the residence, the technician makes certain that no dirt or foreign material is carried into the structure and that no interior walls are marred.
4. The technician makes a visual inspection of the oil burner and resets the safety switch. The oil burner starts, ignites, and cuts off on safety control.
5. The cad cell lead wires are disconnected and connected to an ohmmeter. The safety switch is again reset and while the burner is operating, the technician checks the ohm reading of the cad cell, which should be between 300 and 1000 ohms. The ohm reading is 1000 ohms, proving that the cad cell is good.
6. Once the technician has proved that the cad cell is in good working condition, this indicates that the primary control is bad and should be replaced.
7. The technician changes the primary control and checks the operation of the furnace.
8. The technician informs the housekeeper of the problem and that it has been corrected.

Service Call 5

Application: Residential conditioned air system

Type of Equipment: Oil furnace with stack switch control

Complaint: No heat

Service Procedure:

1. The technician reviews the work order from the dispatcher for available information. The work order reveals that the furnace is in the crawl space which is unlocked. No one will be available to allow the technician access to the home; the key will be located under the mat at the front door. The thermostat will be in the heat position and will be set for the desired temperature. No additional information is available.
2. The technician decides that there is no need to enter the home if the system is receiving a call for heat from the thermostat.
3. The technician proceeds to the crawl space and makes a visual inspection of the furnace and sees nothing out of the ordinary.
4. The easiest way to determine if power is available and the thermostat is calling for heat is to reset the stack switch safety switch and observe the operation of the burner. If the technician resets the switch and the burner starts, it confirms that both power and a call for heat are available to the burner. When the safety switch is reset, the burner starts and lights but cuts off on a safety control.
5. The technician cleans the contacts of the primary control relay(s) and also cleans the bimetal element that is inserted into the stack. The technician again tries the oil burner controls to see if the burner operates properly.
6. The technician again resets the safety switch and the burner operates correctly. By cleaning the contacts and bimetal element of the control, the technician has corrected the problem.
7. The technician operates the furnace through a complete cycle of starting and stopping. (This can be accomplished by disconnecting the thermostat leads at the primary control, which the technician makes certain to reconnect when completed.)
8. The technician leaves a note for the homeowner explaining the problem and what was done to correct it.

SUMMARY

There are basically two mediums that are used to heat structures, air and water. The forced-air furnace heats and delivers air to the desired location in the structure. Hydronic boilers heat water that is pumped through pipes to the area of the structure where heat is required. The steam boiler produces steam with piping directing the steam to the desired location. Warm-air furnaces have many controls that are the same whether the fuel being used is gas, oil, or electricity. There are also some differences in the control of the fossil fuels; for example, oil and gas burners are controlled in a different manner. The hydronic control system uses thermostats as operating and safety controls. Steam systems are controlled by pressure, which requires the use of a pressure switch for an operating control.

In most types of heating appliances using fossil fuels a heat exchanger is required to transfer the heat to the medium being cooled. In a warm-air furnace the fuel source heats a combustion chamber and the furnace forces air across the surface thus heating the air and forcing it to the conditioned space. The hot water or hydronic and steam boiler are nothing more than heat exchangers that heat water or produce steam. In an application where fossil fuel is used, there must be some method to vent the products of combustion.

There are some common controls to all types of forced-air furnaces such as fan switches and limit switches. The fan switch is used to stop and start the fan when the air is at the proper temperature for delivery to the structure. The air delivered to the structure must be warm enough to prevent cold drafts coming in contact with occupants of the structure and causing discomfort. If the temperature of the air is excessively hot, it could damage the structure and/or equipment. The three types of fan switches most commonly used in warm-air furnaces are temperature, time and temperature, and time controlled. The temperature-controlled fan switch is controlled by the temperature of the ambient air surrounding the heat exchanger. The time- and temperature-controlled fan switch is controlled by the timing of the fan switch and/or the ambient temperature of the heat exchanger. The time-controlled fan switch is nothing more than a time-delay relay and the only function is time. All forced-air heating appliances have some form of safety control that opens if the furnace, or a section of the furnace, gets too hot. These limit switches open on a temperature rise and are used at strategic locations in the furnace that might be prone to overheating.

The basic function of gas heating controls are to safely operate the heating appliance and to ensure safe ignition of the main burner. The three basic types of gas burner controls are standing pilot, intermittent, and direct ignition. These three methods have to do with how the main burner ignition is accomplished. The standing pilot is a continuous flame that is located near the burner and can easily ignite the main burner. The intermittent pilot is lit when there is a call for heat from the thermostat and is extinguished when the call has been satisfied. The direct ignition burner control lights the main burner without a pilot. The intermittent pilot and direct ignition burner controls use electronic ignition modules to light pilots or main burners. The ignition module uses a spark or a hot surface element to light the pilot or main burner.

The function of the oil burner control is to turn the oil burner on and off in response to the needs of the conditioned space. The oil burner control supervises the operation of the oil burner, maintaining the level of safe operation that is required by the industry. The two types of oil burner controls currently used in the industry are the stack switch and cad cell controls. The stack switch primary control uses the heat available in the stack or vent to determine if the burner has ignited the fuel oil. The cad cell primary control visually inspects the combustion chamber to determine if the oil burner has properly ignited the oil. The cad cell primary control has a much faster response time than does the stack switch. The resistance of the cad cell is determined by the intensity of the light that the cad cell is viewing. If an oil burner is properly adjusted, the approximate resistance of the cad cell will be 300 to 1000 ohms. Most oil burner controls are available with continuous or intermittent ignition.

Burner controls used for hydronic and steam heating systems, with the exception of the burner controls on the larger applications, would be similar to those discussed in this chapter. Hydronic systems use thermostats as operating and safety controls. Steam heating systems use pressure switches as their operating and safety controls.

REVIEW QUESTIONS

1. Which of the following is the most popular medium used to transfer heat from the heating appliance to the conditioned area of a residential structure?

   a. air
   b. water
   c. steam
   d. none of the above

2. Which of the following types of heating systems is most popular in the commercial and industrial section of the industry?

   a. air
   b. water
   c. steam
   d. either b or c

3. How is the heat transfer from the heating appliance to the conditioned space accomplished in a residential warm-air furnace?

4. What methods are used to remove the products of combustion from the combustion chamber of a heating appliance using a fossil fuel source?

5. True or False: An electric furnace requires a vent from the heat exchanger.

6. Which of the following is *not* a common method used to start the blower motor in a forced-air furnace?

   a. temperature
   b. time and temperature

c. time

d. pressure

7. What is the purpose of the fan switch in a warm-air furnace?

8. Explain the operation of the three types of fan switches used in the industry.

9. What is the result of the following conditions in regards to the operation of a fan switch in a furnace?

a. fan switch setting too low

b. fan switch setting too high

10. What is the purpose of a limit switch in a warm-air furnace?

11. True or False: The standing pilot is a continuous flame that is located near the burner and can easily ignite the main burner.

12. An intermittent pilot ignites when the _____.

a. thermostat is set to "heat"

b. thermostat calls for heat

c. manual pilot valve is opened

d. main gas valve opens

13. Which component is most commonly used to prove that a standing pilot is lit?

a. bimetal element

b. thermostat

c. thermocouple

d. none of the above

14. How are most intermittent pilots lit?

a. manually

b. glow coils

c. heat

d. spark

15. What is a typical operational sequence of an intermittent pilot burner control system?

16. True or False: A thermocouple is strong enough to close the pilot safety solenoid in a gas valve.

17. What is the purpose of a gas valve?

18. What is a redundant gas valve?

19. How many valves could be incorporated in a redundant-type gas valve?

a. one

b. two

c. three

d. four

20. Which of the following are the two types of ignition used with direct ignition?

a. spark and glow coil

b. spark and hot surface

c. spark and thermocouple

d. thermocouple and glow coil

21. Most hot surface ignition devices are made of _____.

a. silicon

b. crystallized sulfur

c. silicon-coated steel

d. silicon sulfide

22. What is a typical operational sequence of a direct ignition pilot control system?

23. What is the function of an oil burner primary control?

24. Which of the following is *not* a component of a gun type atomizing oil burner?

    a. burner motor
    b. oil pump
    c. ignition electrodes
    d. thermocouple

25. What are the two types of primary controls used to supervise the operation of an oil burner?

26. Which of the following devices uses light to prove the flame of an oil burner?

    a. silicon sulfide cell
    b. cad cell
    c. photo cell
    d. none of the above

27. True or False: A stack switch uses pressure to prove the flame of an oil burner.

28. What is the approximate resistance reading of a cad cell viewing a properly adjusted oil burner flame?

    a. 300 to 1000 ohms
    b. 1000 to 1300 ohms

    c. 1500 to 2000 ohms
    d. over 2000 ohms

29. Where does the stack switch primary control pick up the signal that a flame has been proven?

30. Explain the operation of a cad cell primary control.

31. Explain the operation of a stack switch primary control.

32. True or False: The ignition transformer is always energized as long as the thermostat is calling for heat.

33. Where is a stack switch mounted in the oil burner system?

34. What is the correct location of a cad cell?

35. What type of control is commonly used as an operating control in a hydronic system?

36. What type of control is commonly used as an operating control for a steam boiler?

PRACTICE SERVICE CALLS

Determine the problem and recommend a solution for the following service calls. (Be specific; do not list components as good or bad.)

Practice Service call 1

Application: Residential conditioned air system

Type of Equipment: Oil furnace with stack switch control

Complaint: No heat

Symptoms:
1. Correct line voltage to primary control.
2. Thermostat calling for heat.
3. Contacts and element of stack switch have been cleaned.
4. Contacts have been put in step.

Practice Service Call 2

Application: Residential conditioned air system

Type of Equipment: Gas furnace with standing pilot

Complaint: No heat

Symptoms:
1. Correct line voltage available to primary control.
2. Thermostat calling for heat.
3. Pilot is lit.
4. Gas valve is supplied with 24 volts.

Practice Service Call 3

Application: Residential conditioned air system

Type of Equipment: Gas furnace with direct spark ignition

Complaint: No heat

Symptoms:
1. Correct line voltage available to ignition module.
2. Thermostat calling for heat.
3. No spark at igniter or sensor gap.

Practice Service Call 4

Application: Residential conditioned air system

Type of Equipment: Gas furnace with intermittent pilot

Complaint: Constant clinking noise when the furnace is operating

Symptoms:
1. Correct line voltage available to ignition module.
2. Thermostat calling for heat.
3. Spark continues when pilot lights.

Practice Service Call 5

Application: Residential conditioned air system

Type of Equipment: Oil furnace with cad cell primary control

Complaint: Oil burner ignites but cuts off almost immediately

Symptoms:
1. Correct line voltage to primary control.
2. Thermostat calling for heat.
3. Safety switch has been reset with same results.
4. Cad cell reads 5000 ohms when the burner is ignited.

LAB MANUAL REFERENCE

For experiments and activities dealing with material covered in the chapter, refer to Chapter 13 in the Lab Manual.

14

Troubleshooting Electric Control Devices

OBJECTIVES

After completing this chapter, you should be able to

- Troubleshoot electric motors.
- Troubleshoot contactors and relays.
- Troubleshoot overloads.
- Troubleshoot thermostats.
- Troubleshoot pressure switches.
- Troubleshoot transformers.
- Troubleshoot electric heating controls.
- Troubleshoot gas heating controls.
- Troubleshoot oil heating controls.

KEY TERMS

Contactor	Relay
Ignition module	Thermostat
Motor	Transformer
Pressure switch	

INTRODUCTION

Most troubleshooting in a system involves a specific problem that the customer is encountering. Most problems in a system stem from one source—an electric component that is not functioning properly or is faulty. It is the responsibility of the service technician to locate the component that is not functioning correctly and to replace or repair it. Sometimes it is difficult to locate the exact trouble in the entire system, but the task should be relatively simple once the problem has been narrowed down to a single or several components. In this chapter we will discuss troubleshooting for most of the basic electric control components.

The first step in troubleshooting any component is to understand its operation and function. If the operation of an electric component is not understood, it is impossible to effectively check the component. Electric meters will usually be needed in the diagnosis of the component. Thus, it is essential for service technicians to understand the use of electric meters. Service technicians and other personnel must also understand the proper procedures for checking electric components and be able to correctly diagnose the condition of the component. In the following sections, we will discuss some guidelines to use in checking electric components and in diagnosing problems of components in modern heating, air-conditioning, and refrigeration systems.

Caution should be used when servicing any electrical device that is supplied with electrical energy.

14.1 ELECTRIC MOTORS

Motors are the most important loads in any heating, cooling, or refrigeration system. They are used almost exclusively to cause the rotating motion of fans, compressors, pumps, and dampers. Many different types of electric motors are used in the industry. However, the type of motor used will have no effect on diagnosing the condition of the motor but will have a great effect on the selection of a replacement motor if needed.

Always stand clear of the shaft end of an electric motor when it is mechanically connected to a device that requires rotation when it is started.

Open-type electric motors will fail in three different areas: windings, centrifugal switch, and bearings. The windings can be checked for opens, shorts, and grounds. One important element in the diagnosis of an electric motor winding is to know the type of motor because of the winding layout. The condition of the centrifugal switch can best be determined by visual inspection after disassembling the motor; the operating characteristics can also tell much about the condition and operation of the centrifugal switch. The bearings can usually be checked by turning the motor by hand to determine if hard or rough spots exist in the rotating movement of the motor.

CAUTION Make sure that the electrical power is turned off and locked before attempting to turn a motor by hand. Fans and pumps are generally easy to turn by hand, while compressors may require a wrench.

The only part of a sealed motor, such as a hermetic compressor, that can be checked is the windings because there are no internal parts other than the bearings. The winding can be checked as in any other motor for opens, shorts, and grounds. Diagnosing bearing failure in a hermetic motor is often difficult because no visual inspection or feeling of the rotation can be accomplished. Bearing failure on hermetic motors must be diagnosed through amperage readings of the motors along with running characteristics. However, several different types of starting apparatus need to be checked. It is imperative that service technicians be able to diagnose the condition of the many types of starting relays and start assist devices used in the industry. These starting components were discussed in Chapter 10.

CAUTION When motor changes are necessary, the technician should make certain that all electrical characteristics and the mounting are correct.

CAUTION Motors should be mounted in the method intended by the manufacturer.

Motor replacement procedure is important to the technician in the field. Many times the exact replacement motor is not available, and the technician must adapt a different motor to meet the specifications. Replacement motors can sometimes be identified by a manufacturer's part number or by the model number of the unit, but in many cases identification must be made by the motor and its nameplate. The technician must have enough information to identify a motor that will replace the defective one. Many unknown elements such as type of motor, rotation, number of motor speeds, and horsepower can be identified by the technician when removing the faulty motor from the unit. The technician must be sure the replacement motor selected will operate properly for the application.

Troubleshooting motors was covered in detail in Chapter 9.

14.2 CONTACTORS AND RELAYS

Contactors and **relays** are used on most heating, cooling, and refrigeration equipment for the operation of loads in the system. Contactors and relays are similar in their operation because both contain sets of contacts and a coil used to open or close the contacts. The contactor is larger and capable of carrying more amperage than the relay.

The same procedure can be used to check both contactors and relays. Three types of problems are encountered with contactors and relays: contacts, coil, and mechanical linkage. Any one of the three areas can cause a contactor or relay to malfunction.

Contacts

The contacts of a relay or contactor must make good direct contact when energized for the device to function properly. One problem often encountered with contactors and relays is the contacts' inability to make a good contact. The contacts can be burned, pitted, or stuck together. A set of burned or pitted contacts can cause a voltage drop across the contacts.

There are several methods of checking a set of contacts to determine if they are burned or pitted enough to warrant changing the device. The easiest method is to make a visual inspection. Figure 14.1 shows a contactor with a severely damaged set of contacts. Most contactors have movable covers, which allow easy visual inspection. Most relays are sealed and visual inspection is impossible.

A resistance check can also determine the condition of a set of contacts. The device must be energized to check normally open contacts. Normally closed contacts must be checked with the device de-energized. If the resistance is greater than 1 ohm, the contact should be considered faulty.

A voltage check can also determine the condition of a set of contacts. When a voltage check is made, the contactor or relay should be energized. To make a voltage check on a set of contacts, take a voltage reading from one side of the contacts to the other, as shown in Figure 14.2. The reading will show how much voltage is being lost. The load must be energized when a voltage test is being performed. If the voltage loss across the con-

FIGURE 14.1 Contactor with damaged contacts *(Courtesy Square D Company)*

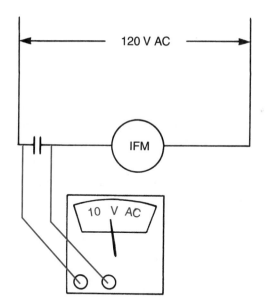

FIGURE 14.2 Voltage test of a set of contacts;10 volts AC are lost across contacts

tacts exceeds 5% of the line voltage, the contacts are faulty and the contactor or relay should be replaced.

The contact alignment can also cause a contactor or relay to malfunction. Contacts should close directly in line with each other and seat directly in line with good, firm contact. The major cause of contact misalignment is a faulty mechanical linkage. If the contacts are out of alignment, the contactor or relay must be rebuilt or replaced.

Coil

The coil of a relay or contactor is used to close the contacts by creating a magnetic field that will pull the plunger into the magnetic field. If the coil of a relay or contactor is faulty, the device will not close the contacts. A contactor or relay coil should be checked for opens, shorts, or a measurable resistance. If a coil is shorted, the resistance will be 0 ohms and the coil should be replaced. An open coil will give a resistance reading of infinity, and this coil should also be replaced. A measurable resistance almost always indicates that the coil is good. Almost any measurable resistance indicates a good coil because of the variance in coil voltages. A shorted contactor coil will cause a transformer to burn out, and the service technician should take caution not to allow this to happen.

Mechanical Linkage

The mechanical linkage of a contactor or relay can cause malfunctions in many different forms, such as sticking contacts, contacts that will not close due to excess friction, contacts that do not make good direct contact, and misalignment of contacts. The best method for detecting a faulty mechanical linkage is by visual inspection. On contactors and some relays, this can be done by removing the device and merely looking it over. However, most relays are sealed and inspection is impossible. A sealed relay must be checked by determining if the contacts open and close when the relay coil is energized or de-energized.

A mechanical linkage problem can cause a contactor or relay to stick open or closed or cause misalignment of the contacts. If a contactor or serviceable relay has a mechanical linkage problem, it should be replaced (unless it can be easily repaired). A sticky armature can cause a relay, contactor coil, or transformer to burn out.

14.3 OVERLOADS

Most major loads used in heating, cooling, and refrigeration equipment have some type of overload protection. Overloads are often overlooked as being a problem in the system but they may be faulty. A faulty overload can cause the equipment to run without protection or not operate at all. The high cost of the major loads in a system makes it necessary to protect all major loads.

CAUTION When replacing overload devices, make sure they are correctly sized.

Fuse

The fuse is the easiest type of overload to check because of its simplicity. A fuse can easily be checked with an ohmmeter in most cases. If a 0-ohm resistance is shown, the fuse is good. No continuity indicates a bad fuse.

A fuse on rare occasions will not completely blow or break but will partially burn out. In this case, the fuse will show 0 ohms but will not allow enough current through it to operate the load. A voltage check across each fuse while power is applied to the load will show a partially burned-out fuse. The voltage check is done by placing the leads of a voltmeter across each end of the fuse, as shown in Figure 14.3. If line voltage is read on the meter, the fuse is bad and should be replaced. If no voltage is read, the fuse is good. This method of checking fuses would not always be accurate on three-phase circuitry.

Circuit Breaker

The circuit breaker is another type of overload device used by some equipment manufacturers and in many electric panels that are commonly used in

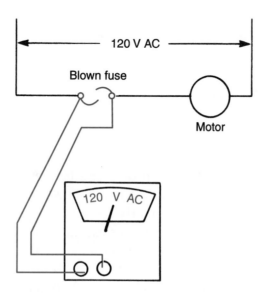

FIGURE 14.3 Voltage check of a bad fuse

FIGURE 14.4 Checking a circuit breaker with a voltmeter

the industry today. The circuit breaker is a device that will trip or open on an overload and must be manually reset. The circuit breaker is checked by taking a voltage reading on the load side of the circuit breaker, as shown in Figure 14.4. If line voltage is read on the load side of the circuit breaker, it is probably good.

A circuit breaker can also cause trouble if it trips at an amperage lower than the rated amperage of the breaker. (A clamp-on ammeter is used to check the amperage of a circuit breaker.) If this occurs, replace the breaker. A circuit breaker can also cause nuisance trippings if it is unable to handle its rated amperage.

Line Voltage Overload

A line voltage overload installed on a load device is the easiest type of overload to check. It is used on small hermetic compressors and motors and is connected directly to the line voltage supply. The line voltage overload can be open, permanently closed, or open on a lower-than-rated ampere draw. A line voltage overload has only two or three terminals to check. An ohmmeter across the terminals will indicate whether the overload is open or closed.

Care should be taken not to condemn an overload when, in fact, it is open because of a malfunction of the load it is controlling. If a line voltage overload is weak, an amperage check should be made to see what amperage is causing the overload to open. If the dropout amperage is lower than the overload rating, the device should be replaced.

Pilot Duty Overload

A pilot duty overload has a set of contacts that will open if an overload occurs in the line voltage side of the overload. These overloads are arranged so that the line voltage feeds directly through them and then on to the load. The line voltage section of a pilot duty overload can be controlled by heat, amperage, or magnetism—all three are in common use in the industry today. This type of overload is harder to check than the overloads discussed previously because of its complexity. The pilot duty contacts and the controlling line voltage element must be checked.

The pilot duty contacts on a pilot duty overload are easy to check by using an ohmmeter. The pilot contacts are usually easy to distinguish from the line voltage components of the overload because of their small size in relation to the large size of the line voltage connections, as shown in Figure 14.5. The contacts will either be open or closed. If the contacts are open, the overload is bad (or there is an overload in the circuit). If the contacts are closed, the overload is good.

The line voltage part of the pilot duty overload indicates an overload by several methods: heat, current, and magnetism. All three methods determine the current flow, but they use different elements to determine an

FIGURE 14.5 Pilot duty and line voltage connections of an overload

(a) Heat (thermal) type of pilot duty overload

(b) Magnetic type of pilot duty overload

FIGURE 14.6 Pilot duty overloads

overload. The heat (thermal) type of pilot duty overload shown in Figure 14.6(a) actually transfers current to heat. The current type of overload is similar in design to the magnetic. The current type of pilot duty overload, shown in Figure 14.5, uses the current through a coil to indicate an overload and open the pilot duty contacts. The magnetic overload shown in Figure 14.6(b) uses the strength of the magnetic field to open and close the pilot duty contacts. The magnetic overload gives a certain amount of built-in time delay due to its makeup.

To check the thermal element of a thermal overload, take a resistance reading across the thermal element. If the resistance is above 0 ohms, the element is bad and the overload should be replaced. The magnetic and current types of pilot duty overload can be checked in the same way. However, the magnetic and current overloads use a coil that is larger than the element of a thermal overload. This coil is connected in series with the load and therefore indicates the current being used by the load. The coil in the magnetic or current types of pilot duty overload can be easily checked with an ohmmeter. The resistance of the coil should be 0 ohms because it is part of the conductor going to the load. The overload should be replaced if the coil reads any resistance.

FIGURE 14.7 Early type of internal overload on a hermetic compressor

Internal Overloads

Another type of overload is the internal overload used in hermetic compressors. This type of overload is actually embedded in the windings of the hermetic compressor motor, which gives it a faster response to overloads.

The early type of internal overload used separate terminals extending from inside the compressor to the outside of the compressor terminal box, as shown in Figure 14.7. This type of internal overload can be simply and easily checked with an ohmmeter to determine if it is open or closed. It is embedded in the winding but makes no electric connections to the windings.

The type of internal overload currently used is hard to check because it has no external connections. It is in series with the common terminal of the compressor motor. This type of overload must be checked as part of the windings of the motor, which makes it extremely hard to diagnose for troubles. The only way to check this type of overload is with an ohmmeter, just as you would check a motor. If an open is present, it could be due to the windings or the overload. Service technicians should never condemn a hermetic compressor that is temporarily overloaded. The compressor should be given ample time to cool.

CAUTION When troubleshooting electric motors or hermetic compressor motors that are extremely hot, make sure they have ample time to cool before condemning them.

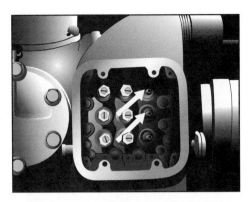

FIGURE 14.8 Semihermetic overload terminals

Many semihermetic compressors have an internal thermostat embedded in the motor windings. The thermostat has a separate connection to the compressor terminals as shown in Figure 14.8. An internal thermostat can easily be checked with an ohmmeter. If any resistance is read, the device is faulty and should be replaced.

Electronic overloads are becoming increasingly popular for the protection of large compressors in the industry. Electronic overloads use sensors with a certain resistance that are placed in the motor windings. The resistances of these sensors change with the temperature of the motor winding: the higher the temperature of the motor winding, the higher the resistances of the sensors. The sensors are connected to an electronic module that amplifies the resistances of the sensors. If the module senses an overloaded condition due to the resistances of the sensors, then the control contacts of the module open and stop the compressor. A more detailed explanation of electronic overloads is given in Chapter 17. Electronic overloads are easy to troubleshoot because most manufacturers give detailed resistance charts on the sensors at certain temperatures. The technician only needs to measure the resistances of the sensors to determine if an overloaded condition exists. If an overloaded condition exists, then the pilot duty contacts should be open; but if the motor windings are at normal temperature, the pilot duty contacts should be closed.

CAUTION Safety devices should be connected in series to ensure that unsafe conditions will cut off the load being protected.

At no time should a technician jump out safety controls and leave them jumped out.

14.4 THERMOSTATS

Some type of **thermostat** is used on most heating, cooling, and refrigeration equipment. Therefore, it is essential to know how to correctly diagnose the condition of thermostats.

There are two basic types of thermostats used in the industry today: the line voltage thermostat and the low-voltage thermostat. The line voltage thermostat is used to make or break line voltage to a load. Its only function is to open or close a set of contacts on a temperature rise or fall. Thus, the line voltage thermostat is usually simpler than the low-voltage thermostat because it does not have as many functions as the low-voltage thermostat. The low-voltage thermostat is used when a voltage lower than 120 volts—usually 24 volts—is used to operate a control system. The low-voltage thermostat can have many functions. It can stop and start a fan motor, operate a fan motor independently of other parts of the system, and do many other functions sometimes required in control systems. The line voltage thermostat is not as accurate as the low-voltage thermostat due to the contacts' larger size, which is necessary to carry the higher voltage. Low-voltage thermostats are usually used on residential heating and cooling control systems and on many commercial and industrial systems. The line voltage thermostat is used on window air conditioners and commercial and industrial air-conditioning, heating, and refrigeration equipment.

Line Voltage Thermostat

The line voltage thermostat is easy to troubleshoot because of its simplicity. A line voltage thermostat could have two, three, or four terminals. A typical thermostat is shown in Figure 14.9. The most important element of checking line voltage thermostats is to be sure the contacts are closed in the correct temperature range. Once it has been determined that the thermostat should be opened or closed, it can be checked with an ohmmeter. Or the control voltage can be checked at the equipment. The wires must be removed from the thermostat to check it with an ohmmeter. The ohmmeter will read 0 ohms if the thermostat is closed, and infinity if it is open. The

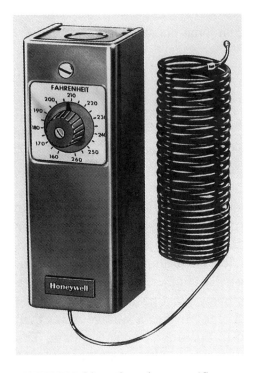

FIGURE 14.9 Line voltage thermostat *(Courtesy of Honeywell, Inc.)*

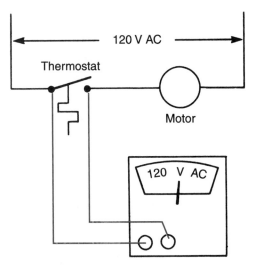

FIGURE 14.10 Voltage check of open thermostat

service technician must determine the correct terminals to check on the thermostat. This information can be found on the unit's wiring diagram. In many cases, it is difficult to remove the wires of a thermostat, so a voltage check is made, as shown in Figure 14.10. When voltage is read across the contacts of a thermostat, the thermostat is open. If no voltage is read, the thermostat is closed. The voltage check must be made with power applied to the unit and the switches placed in an operating position.

Low-Voltage Thermostat

The low-voltage thermostat is more difficult to troubleshoot than the line voltage thermostat because of the many functions of the low-voltage thermostat. The low-voltage thermostat operates the heating and cooling of the system, operates the fan motor with the heating and cooling opera-

FIGURE 14.11 Common low-voltage thermostat subbase *(Courtesy of Honeywell, Inc.)*

tions, operates the fan motor independently, often operates two-stage systems, operates damper motors, and operates a pilot function of a gas heating system. A common subbase of a low-voltage thermostat is shown in Figure 14.11. All the letter designations, which indicate the different functions of the low-voltage thermostat, are also shown in the figure.

In troubleshooting a system, the low-voltage thermostat and the subbase may be at fault. The low-voltage thermostat and subbase can be checked with an ohmmeter. This can be done at the equipment or at the junction point of the thermostat wires. Service technicians seldom need to remove the thermostat and install a set of wires on it to check it. A low-voltage thermostat can also be checked by taking a voltage check at the equipment to ensure that the thermostat is functioning properly. The low-voltage thermostat and subbase are merely a point in the control system that is fed with low voltage. The thermostat sends a voltage signal to the equipment, which must then operate to meet the conditions called for by the thermostat.

The chart shown in Figure 14.12 can often be used in troubleshooting thermostats.

Condition:

Possible causes:

NOTE: T/S indicates room thermostat.

T/S jumpered; system won't work.	T/S jumpered; system works.	Room temp. overshoots setting; too cold.	Room temp. doesn't reach setting; too warm.	System cycles too often.	System doesn't cycle often enough.	Room temp. swings excessively.	
●							T/S not at fault; check elsewhere.
			●				T/S wiring hole not plugged; drafts.
					●	●	T/S not exposed to circulating air.
		●	●				T/S not mounted level (mercury switch types).
		●	●				T/S not properly calibrated.
		●		●			T/S exposed to sun, source of heat.
	●				●		T/S contacts dirty.
	●		●				T/S set point too high.
		●					T/S set point too low.
	●		●				T/S damaged.
			●				T/S located too near cold air register.
	●						Break in T/S circuit.
		●	●		●	●	System sized improperly.

FIGURE 14.12 Troubleshooting chart for thermostats *(Courtesy of Honeywell, Inc.)*

14.5 PRESSURE SWITCHES

Pressure switches are used on heating, cooling, and refrigeration systems to start or stop some electric load in the system when the pressure in the system dictates this action. Pressure switches are used as safety devices or as operating controls. A pressure switch used as a safety device will stop an electric load when the pressure in a system reaches an unsafe condition. A pressure switch used as a safety device can be used to protect a refrigeration system from excessive discharge pressure or low suction pressure. It can be used in a gas heating system to protect the equipment from low or high gas pressure. It can be used to protect air-moving equipment from low air pressure.

An operating-control pressure switch is used to operate some load in the system. The most common use of an operating-control pressure switch is on a commercial refrigeration system to control the temperature of a walk-in cooler or freezer. Some pressure switches are also used to operate unloading devices on large compressors, to operate condenser fan motors to control the discharge pressure of a refrigeration system, and to operate pumps and cooling tower fans on water-cooled condensers.

The most important aspect of checking a pressure switch is to understand what it is used for in the system. The service technician must also determine if the pressure switch should be opened or closed. Once the service technician has determined what the pressure switch is used for and whether it should be opened or closed, it is easy to check the pressure switch. A resistance or voltage check should be made to determine if the pressure switch is opened or closed.

In troubleshooting pressure switches, it may be that the pressure switch is faulty or that the system is malfunctioning. Figure 14.13(a) shows a pressure switch in the control circuit malfunctioning by opening at a higher setting than that on the pressure switch. Figure 14.13(b) shows a pressure switch operating correctly and opening due to a low suction pressure.

The pressure switch can be stuck open or closed, it can open or close on the wrong pressure, or there can be a mechanical problem with the switch itself. If a pressure switch is stuck open or closed, then the pressure in the system will have no effect on the pressure switch. The service technician will have no trouble detecting a faulty pressure switch that should be opened or closed and is keeping a load from operating as it should. The technician should make sure the pressure switch is not stuck in a position that will result in damaged loads. If a pressure switch is not opening and closing on the right pressure, it should be correctly set or replaced.

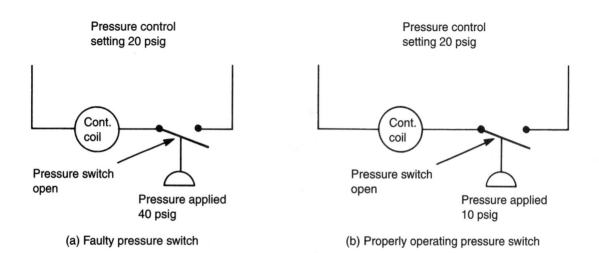

FIGURE 14.13 Pressure switch operation

On some occasions, a pressure switch has been opened and closed so many times that the switch is worn out. Other mechanical linkage problems include broken springs, leaking bellows, corroded linkages, and broken linkages. Any condition that occurs in the mechanical linkage of a pressure switch will usually call for replacement of the pressure switch.

In most cases when a pressure switch opens, this is due to a malfunction in the system. But occasionally the pressure switch itself is at fault. The service technician must determine which condition is occurring. A pressure switch used as an operating control will often have to be reset but seldom replaced. The same troubleshooting procedure is used regardless of the function of the pressure switch.

14.6 TRANSFORMERS

A **transformer** is a device used to raise or lower the incoming voltage by induction to a more usable voltage for the control system. Some types of transformers are used to buck (lower) or boost (raise) the incoming voltage to an air-conditioning unit. A buck-and-boost transformer is used in conjunction with a voltage system that is too high or too low to supply the correct voltage to a system. Figure 14.14 shows a typical buck-and-boost transformer.

FIGURE 14.14 Buck-and boost transformer *(Courtesy of Acme Transformer)*

A transformer can be checked in two ways: a resistance check or a voltage check. An ohmmeter can be used to check the condition of the windings of a transformer. If the ohmmeter reads 0 ohms, the windings of the transformer are shorted. A reading of infinity indicates an open transformer. A measurable resistance indicates that the transformer is probably good. The number of ohms measured would be determined by the voltage of the transformer.

A transformer can also be checked by reading the output voltage if the correct primary voltage is applied. In some cases, the transformer might check out at good if the load is not put across the secondary. But when the load is inserted in the line, the transformer voltage will not be enough to energize the load. In this case, the transformer has a spot burnout and should be replaced. Often transformers are burned out because other devices in the circuit are being shorted. The service technician should take every precaution to prevent this from happening.

14.7 ELECTRIC HEATING CONTROLS

The sequencer and contactor are commonly used to control the operation of electric resistance heaters in electric furnaces and other applications such as duct heaters or supplementary heat for heat pumps. There are many other electrical components that are used in electric heating control circuits for effective and safe operation of the electric heating appliance. Thermostats are used as operating controls and limits switches in these appliances. Electric motors are used to move air through the heaters to the conditioned space. Controls and devices that have been covered previously will not be covered again. When troubleshooting these components, the technician should focus on the function of the component in the circuit. For example, a thermostat could be used as an operating control to control the temperature of conditioned space or a safety control to interrupt the power to an electrical load if an unsafe condition occurs. The technician must know the function of the control in the control system in order to effectively troubleshoot electrical devices.

In most cases, sequencers are used to control the operation of the electrical resistance heaters in an electric furnace and other applications. The sequencer is an electrical switch that acts much like a time-delay relay. The sequencer has an electric heater that heats the bimetal element, causing the contacts of the sequencer to close. Sequencers can have as many as five sets of contacts that close in sequential order. The contacts of a sequencer close and open in a sequential order, thus spreading out the switching times

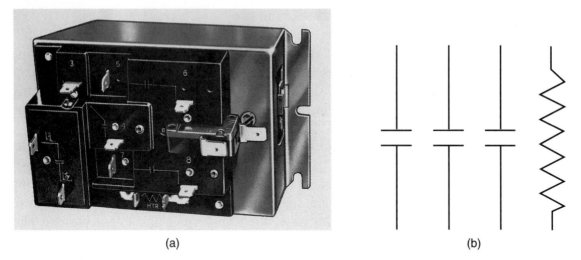

(a)   (b)

FIGURE 14.15 (a) Photograph of sequencer *(Courtesy of Honeywell, Inc.)* (b) Diagram of a sequencer

to avoid large electrical loads at one time. The closing of the contacts of a currently used sequencer are 18, 30, and 45 seconds. This allows each resistance heater to be staged into operation, decreasing large loads being placed on the electrical system at one time. Figure 14.15(a) shows a sequencer used in an electric furnace and Figure 14.15(b) shows the diagram of that sequencer. Figure 14.16 shows a diagram of an electric furnace with three electric resistance heaters and the blower motor being controlled by a single sequencer.

The first check on any electrical circuit would be to determine if the correct voltage is available to the device controlling the load. In checking the electric heating components, the technician should first check the power supply to the sequencer, both line voltage for the heaters and low voltage for the control element. If the control voltage is not available to the sequencer, then the technician must check the control circuits, but if the sequencer element is receiving 24 volts and not closing the contacts, then the sequencer is faulty and should be replaced. The condition of the sequencer element can be checked to determine if the element is open, shorted, or good (has continuity). With the heavy electrical load placed on contacts with electrical resistance heaters, oftentimes the contacts merely burn out because of the heat. Each set of contacts of a sequencer must be checked if there is reason to believe that not all heaters are operating; it is possible for any one set of contacts of a sequencer to be bad even though the remaining contacts may be operating properly. The condition of the contacts can be determined by a voltage check as shown in Figure 14.17. If

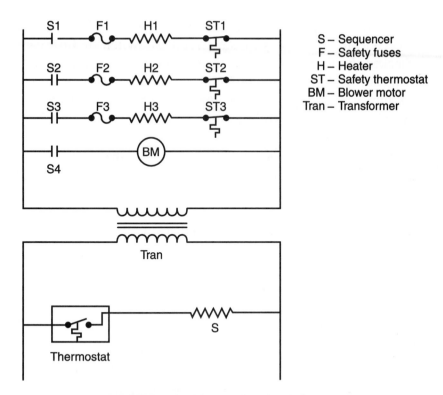

FIGURE 14.16 Diagram of an electric furnace

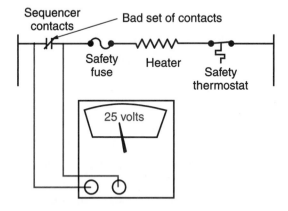

FIGURE 14.17 Voltage test of a set of contacts

a technician reads a voltage across the contacts of a sequencer, contactor, or relay, it shows that voltage is being lost across that set of contacts. If the contacts are open, then the voltage reading will be line voltage because of feedback through the circuit. A closed set of good contacts should read 0 volts because it is the same leg of power, but if the sets read line voltage, then it is reading one line and the other reading is from feedback through other electrical devices in the circuit. A switch or fuse can usually be checked in this manner. Of course, control voltage can be supplied to the sequencer and then a resistance check can be made of each set of contacts, but this method requires more time than the voltage check. The technician must know the timing of the contacts' closing in order to effectively troubleshoot a sequencer.

Other elements of electric resistance heaters that often give problems are the limits used for overcurrent or overtemperature protection. The check of these devices can be accomplished like any thermostat or switch. Electrical resistance heaters can burn in two the call on this type of problem would be "insufficient heat" and the technician would only have to check the continuity of the heater.

14.8 GAS HEATING CONTROLS

There are many thermostats used in gas heating control circuits. Thermostats that close on a rise in temperature are used in gas heating circuits to start the fan once the furnace heat exchanger has been heated. Thermostats that open on a rise in temperature are used as limits to prevent dangerous conditions from occurring in or around the furnace. The operating thermostat is usually low voltage, but line voltage thermostats can also be used for this function. Limit switches can also be line or low voltage, but will generally have only one function in the control circuit. The technician must know the function of the thermostat in the circuit. Oftentimes wiring diagrams will help in determining this.

Gas Valves

Gas valves are available in many different designs. Some valves have only one function, which is to open or close, while many have a number of different functions. The two commonly used gas valves today are the combination and redundant type gas valves, both having more than one function. The combination gas valve has a pilot solenoid that is held open by the millivoltage produced by a thermocouple. The thermocouple produces a small voltage when heated. The source of heat is the pilot flame. Figure 14.18 shows a thermocouple with the proper flame. The thermocouple is attached to the pilot solenoid of the gas valve; the small voltage applied to the solenoid will not open the valve without assistance but will hold the valve open once it has been manually opened. The thermocouple produces a maximum voltage of 30 mV with the operating range being between

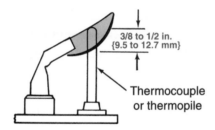

FIGURE 14.18 Proper pilot flame *(Courtesy of Honeywell, Inc.)*

10 mV and 20 mV when loaded. Along with pilot solenoids, combination gas valves house an on and off gas flow manual valve and an automatic valve that opens when a call for heat occurs, provided that the manual on and off valve is open and the pilot solenoid valve is open. The redundant gas valve has three valves incorporated in one housing: a manually operated shutoff valve that mechanically blocks all gas flow when turned off; an electrically operated solenoid valve that opens to allow gas to flow to the second valve; and a pilot duty electrically operated servo valve that opens to allow gas to enter the gas burners. There are many types of gas valves that have only one function and are operated by various types of elements such as the heat motor valve, the diaphragm valve, electrically operated solenoid valves, and others.

Troubleshooting gas valves is accomplished first by knowing the type of gas valve used and its operation. The technician must determine the type of gas valve used in the control system. New model gas warm-air furnaces will almost always be equipped with redundant gas valves. The gas valves used on older model heating appliances will be a combination of different types of gas valves. The most common of these will be the combination gas valve with one manual valve, one semiautomatic valve, and one automatic valve. Other types of gas valves have only one valve that opens and closes by energizing whatever element controls the valve. The technician must know how the gas valve operates in order to correctly troubleshoot it.

The easiest gas valves to troubleshoot will be valves that only open and close on a signal from other controls, generally thermostats. The technician has several options when checking these types of valves. A voltage check can be made to determine if the valve is receiving the correct voltage. If the valve is receiving the correct voltage and is still not operating, the valve should be changed. The technician should give the valve adequate time to open because in many cases, the controlling element of the valve does not operate instantly, but has a delay in opening because of the design of the controlling element. Others operate instantly. A resistance check with an ohmmeter is another method of checking the controlling element of a gas valve. The resistance reading on the controlling element of a gas valve should be a measurable resistance. The gas valve could be electrically sound, but faulty if any internal ports within the valve are restricted.

The combination gas valve has several valve operating elements that must be checked. The combination gas valve has a manually operated valve that interrupts the gas flow at the entrance of the gas valve. The next valve in a combination gas valve is the pilot solenoid, which opens the first valve, allowing gas to the pilot and to the second valve. This pilot solenoid is not strong enough to open the valve without assistance. The valve must be physically opened and then the pilot solenoid will keep the valve open as long as the signal from the thermocouple is adequate. Figure 14.19 shows the unloaded and loaded millivolt output of a good thermocouple. The pilot solenoid or the thermocouple could be bad and prevent the pilot valve from remaining open. A special thermocouple adapter used to measure the voltage produced by the thermocouple is shown in Figure 14.20. The millivoltage of the thermocouple must be adequate to keep the pilot solenoid energized with a correctly burning pilot flame. The pilot solenoid could be open, shorted, or grounded, which would prevent the solenoid from holding the valve open. To determine the condition of a pilot solenoid, it can be checked with an ohmmeter. If the thermocouple is not producing adequate millivoltage to hold the pilot solenoid open, it must be

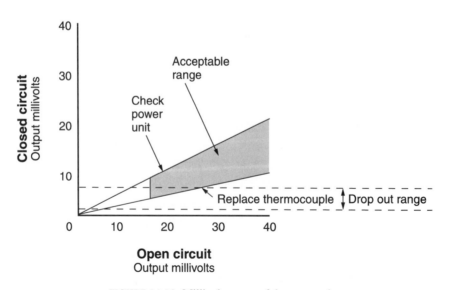

FIGURE 14.19 Millivolt output of thermocouple

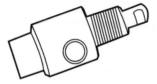

FIGURE 14.20 Adapter to check millivolt production by thermocouple

replaced. The pilot solenoid or gas valve must be replaced if the pilot solenoid is faulty. Oftentimes pilot solenoids are available and can be replaced,
but many technicians choose to change the entire valve. Figure 14.21
shows a safety troubleshooting chart for a standing pilot control system
using a thermocouple.

Redundant type gas valves have two valves built into the main body. The
first valve acts as a pilot valve and on a call for heat, the igniter and the pilot
valve open; if the pilot is established, then the main valve will open, allowing

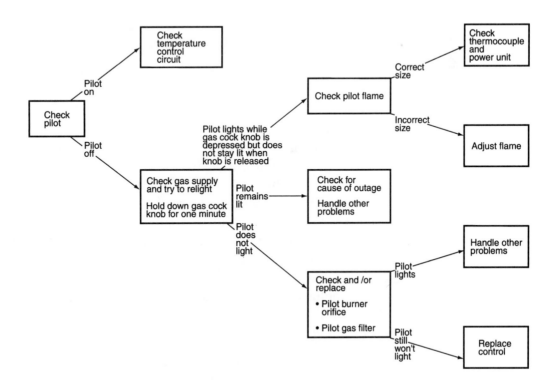

FIGURE 14.21 Safety shutoff troubleshooting chart *(Courtesy of Honeywell, Inc.)*

gas to pass to the burner for ignition. On the gas valve, the electrical elements operating these internal valves are PV and MV; PV is the pilot valve and MV is the main valve. This type of valve operates by supervising the ignition module; on a call for heat, the pilot must be established before the main valve is energized. Troubleshooting of the ignition modules will be covered in the next section. The technician could perform an electrical check or continuity check of the pilot and main valve elements to determine their condition. If the correct voltage is available to the valve and the valve being checked is not opening, then the valve is faulty and should be replaced.

A gas valve used on a direct ignition system houses two valves, but the electrical connection is for both valves. There is no need for a pilot valve because no pilot flame is used on a direct ignition burner control. On some valves the connections for both valves are made externally and can be checked by using the voltage or resistance methods. If only one connection is available for both valves, the technician must have an idea of what the resistance reading should be for a resistance check, while the voltage check is merely for the correct voltage to the valve. If the valve is receiving voltage and not opening, the valve is bad.

Troubleshooting Intermittent Pilot Systems

An intermittent pilot control system supervises the operation of the heating source in a gas warm-air furnace and other gas heating appliances. This type of control system lights the pilot and ensures that it is established before allowing the main gas valve to open and supply gas to the main burners. When troubleshooting or checking the operation of an intermittent ignition module there is always a fire or explosion hazard that can cause personal injury or property damage. If a technician suspects a gas leak or smells gas, the manual gas valve should be turned off until the condition is corrected. Do not try to light any pilot or appliances while spilled gas remains. A gas leak test should be performed at the time of installation and any time new connections are made or components changed.

The technician often will have to check out the operation of the gas burner of a heating appliance. This procedure could be performed during a preseason start-up call or a heating check call. The following steps could be used when performing this type of call.

1. Visually inspect the heating appliance, making sure that all electrical connections are tight and clean. Open gas valves and make sure there are no leaks.

2. Review the normal operating sequence and the timing of the ignition control system. The normal operating sequence of this type of control is shown in Figure 14.22.

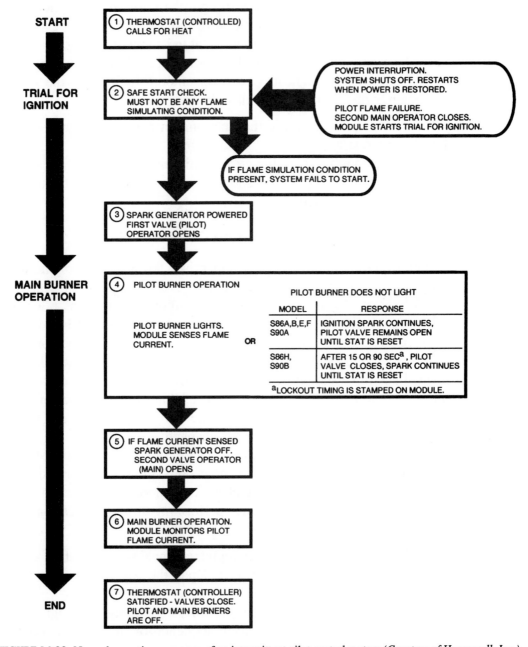

FIGURE 14.22 Normal operating sequence of an intermittent pilot control system *(Courtesy of Honeywell, Inc.)*

3. Reset the module by turning the thermostat to its lowest setting. Wait one minute and proceed to step 4.

4. Check the safety operation of the ignition module by closing the manual gas valve. Set the thermostat to a call for heat. The ignition control system should spark and try to light the pilot. When a pilot can't be established, check the operation of the pilot or safety lockout. Open the gas supply and set the thermostat to its lowest setting.

5. Check the ignition module for normal operation by setting the thermostat to call for heat. Check for smooth ignition of the pilot and main burner without flame disturbances.

This procedure will ensure that the ignition control system is operating safely and properly.

A troubleshooting chart for an intermittent ignition control system is shown in Figure 14.23. Troubleshooting charts are available for most ignition modules in production.

Troubleshooting Direct Ignition Control Systems

A direct ignition control system supervises the operation of the heating source in a gas warm-air furnace and other heating appliances. This type of control system lights the main burner without any type of pilot flame. When troubleshooting or checking the operation of a direct **ignition module** there is always a fire or explosion hazard that can cause personal injury or property damage. If the technician suspects a gas leak or smells gas, the manual gas valve should be turned off until the condition is corrected. Do not attempt to light any pilot or appliance while spilled gas remains. A gas leak test should be performed at the time of installation and any time new connections are made or components changed.

The technician often will have to check out the operation of the gas burner of a heating appliance. This procedure could be performed during a preseason start-up call, a heating check call, or an initial start-up call. The following steps could be used when performing this type of call on a direct ignition system.

1. Visually inspect the heating appliance, making sure that all electrical connections are clean and tight. Open the gas valve and make sure there are no leaks.

2. Review the normal operating sequence and timing of the ignition control system. Normal operating sequence of this type of control is shown in Figure 14.24.

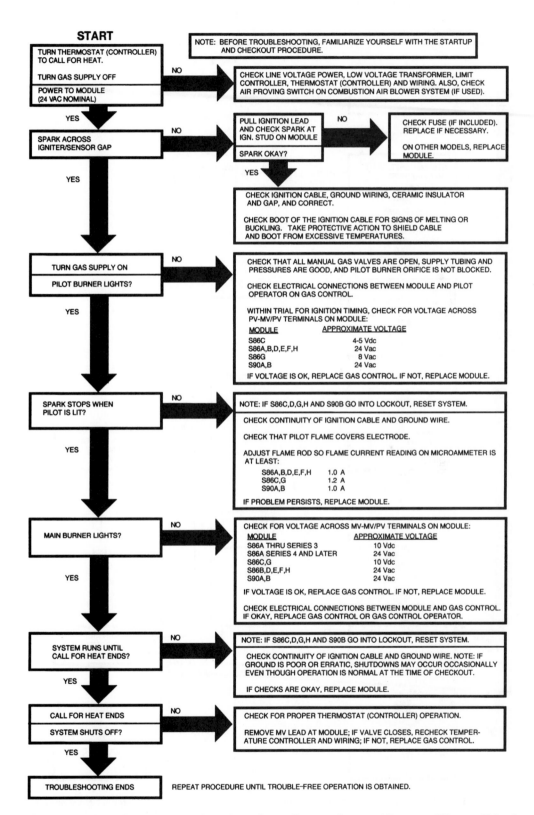

FIGURE 14.23 Troubleshooting chart for an intermittent pilot control system *(Courtesy of Honeywell, Inc.)*

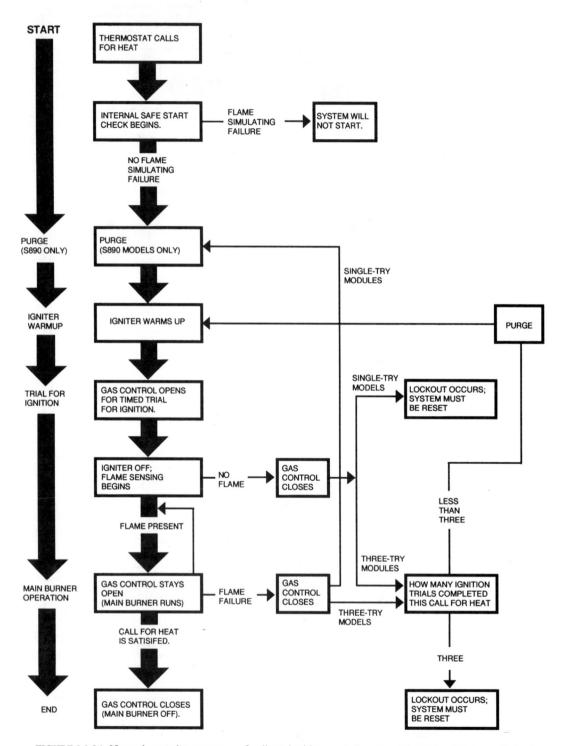

FIGURE 14.24 Normal operating sequence of a direct ignition control system *(Courtesy of Honeywell, Inc.)*

3. Reset the module by turning the thermostat to its lowest setting. Wait one minute and proceed to step 4.
4. Check the safety operation of the ignition module by closing the manual gas valve. Set the thermostat to a call for heat. The ignition control should create a spark or energize the hot surface ignition device immediately or following prepurge. Check the operation of the safety lockout. Open the gas supply and set the thermostat to its lowest setting.
5. Check the ignition module for normal operation by setting the thermostat to call for heat. Check for smooth ignition of the burners.

This procedure will ensure that the ignition control system is operating safely and properly.

A troubleshooting chart for a direct ignition control system is shown in Figure 14.25. Troubleshooting charts are available for most ignition modules in production.

14.9 OIL HEATING CONTROLS

Thermostats are used in oil furnaces just as they are in gas furnaces and basically for the same purpose. They are used as operating controls to stop and start the source of heat to the furnace, to stop and start fans according to the temperature of the furnace heat exchanger, and to de-energize components when unsafe operating conditions exist. The technician must know the function of these thermostats in the system in order to effectively troubleshoot the system. There are two types of primary controls that are used to supervise the combustion and operation of an oil burner: the stack switch and cad cell control. The stack switch senses the combustion by heat and the cad cell detects the combustion by sight. Other electrical devices in oil heating control systems have been discussed in other parts of this text and will not be covered again.

Troubleshooting Stack Switch Primary Controls

The stack switch is a heat-actuated control which uses the stack temperature to indicate the presence or absence of combustion. A bimetal is inserted into the stack that actuates a push rod on temperature rise to break the safety switch circuit. At the same time an alternate circuit is established to keep the burner operating. The stack switch should be located in the center of the stack where the element will be exposed to rapid temperature changes when combustion is established and extinguished. The temperature in the stack where the stack switch element is located should be less

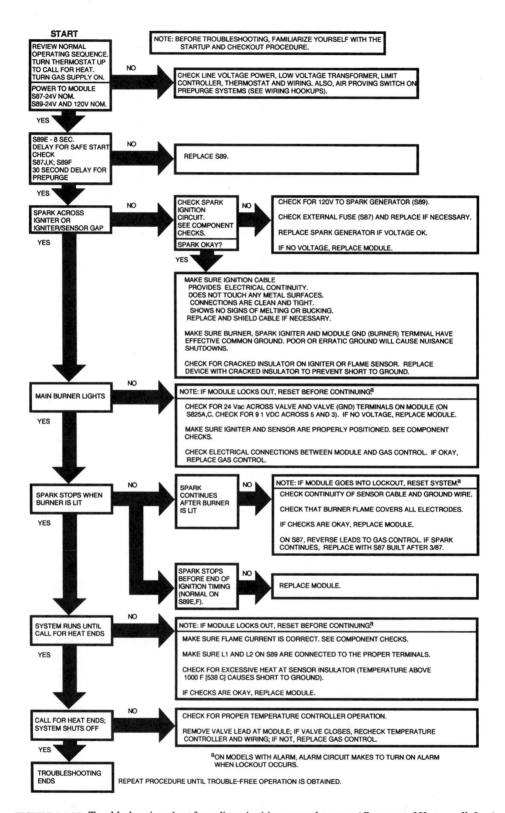

FIGURE 14.25 Troubleshooting chart for a direct ignition control system *(Courtesy of Honeywell, Inc.)*

than 1000°F. The stack switch must be mounted ahead of any draft regulator and, if installed in an elbow, should be in the outside curve of the elbow. With a call for heat the safety switch is energized until the bimetal senses the heat in the stack and opens, keeping the burner operating. If the bimetal does not sense that the temperature in the stack has increased, the safety switch will remain energized and open the contacts controlling the burner. If combustion has been established, the safety switch will open and the oil burner will continue to run. If at any time the stack temperature decreases sufficiently to cool the bimetal, the safety switch will energize and interrupt power to the burner. Regardless of the reason for lockout, a tripped safety switch must be manually reset before the stack switch will attempt to start the burner again.

There are certain safety checks that should be made on a stack switch primary control when the burner is initially started, during preseason start-up, or on other calls when checking the operation of the burner is necessary. Safety checks that should be made are flame failure, ignition or fuel failure, and power failure. It is essential that the oil burner primary control correctly shut down the burner if there is no combustion. The technician should shut the oil supply hand valve while the burner is operating. The control in some cases will attempt one restart before locking out. Some controls maintain ignition until the burner is locked out. The technician can check for ignition or fuel failure in the same manner as a flame failure. The power failure check can be made merely by removing power from the burner and, after waiting until the stack has cooled, restoring power and making sure the burner starts correctly. On a flame, ignition, or fuel failure, the safety switch will have to be manually reset.

When troubleshooting a stack switch primary control, the burner must be in satisfactory operating condition. If the problem does not seem to be in the burner and/or ignition system, then the primary control must be checked. If the burner does not start when a call for heat occurs, preliminary checks that should be made are the voltage to the control, checking limit switches, and resetting primary control. There are basically three conditions that can occur when a call for heat is made to the primary control: the burner starts, the burner does not start, or the burner starts then locks out on safety. If the burner does not start when the thermostat is set for heat, the first check is to see if the thermostat is actually closed. This can be done by jumpering the thermostat leads at the primary control. The operation of the oil burner would indicate that the thermostat is the problem. If the burner does not start, the contacts of the stack switch detector must be put in step by gently pulling the drive shaft lever out ¼ inch and releasing. The correct operation of the burner indicates that the primary control is good. If the burner will not start after this procedure, the detector contacts should be cleaned. Failure of the burner to operate at this point would indicate that the primary control must be replaced.

Another common problem with the oil burner controls is when combustion is indicated and it locks out on safety. The primary control should be reset. If the burner ignites and operates properly, the problem was only a temporary condition. After the primary control is reset, if the burner starts and locks out on safety again, there are two problems that could exist. The location of the bimetal element could be the problem if it is in a location under 300°F, or the entire bimetal element could need cleaning. After cleaning or moving the detector to a better location, reset the primary control. If the same results occur, the stack switch should be replaced. The technician should make certain that the installation is correct and no other control system components are faulty before replacing a stack switch.

Troubleshooting Cad Cell Primary Controls

The cad cell primary control consists of a primary control and a cad cell, which is light sensitive and can view the flame of the oil burner. When the cad cell views the light of an oil burner, its decreased resistance completes the flame detection circuit, preventing the primary control from locking out on safety. The fast response time of the cad cell to light eliminates the lag time found in the bimetal-controlled stack switch. The cad cell location is carefully determined by the manufacturer of the burner and is installed inside the air tube with a clear view of the oil burner flame. The cad cell should be able to view the flame directly, and care should be taken to prevent the cell from viewing external light sources because the cad cell responds to any light source. The location of the cad cell should not be changed. Normal service only involves cleaning accumulated dirt and soot from the cell or replacing the cell.

On a call for heat the burner is started and at the same time a bimetal-operated safety switch in the primary control starts to heat. If the cad cell does not view a flame, the safety switch will continue to heat and shutdown the burner. If flame is viewed by the cad cell, the resistance will drop, thus interrupting the power to the safety switch and allowing the burner to continue operating until the desired temperature is reached in the conditioned space. If at any time the flame goes out, the cad cell's resistance increases and the circuit containing the safety switch is energized, heating the bimetal and locking out the oil burner.

In order to effectively troubleshoot a cad cell primary control, the technician must make certain that the oil burner and its components are in good operating condition. If the burner does not start when the thermostat calls for heat, the first order of business is to check the power supply to the primary control and all limit switches. After the technician has determined that there is a call for heat to the primary control, then the technician must troubleshoot the cad cell and the primary control. A cad cell primary control tester is available, but it is not widely used in the industry because of its lack of compatibility with some types of cad cell primary controls.

The cad cell primary control is responsible for the safe operation of the oil burner. The primary control must stop the oil burner in the event that there is no combustion and keep the burner operating if combustion is proven. The technician should check the operation of the cad cell primary control, ensuring that it is operating correctly on initial installation and during any normal service call. A faulty cad cell could allow the burner to operate even without seeing a flame if it was shorted or viewing external light sources. This condition must be corrected by shielding the cad cell from the external light source or replacing the faulty cad cell. The cad cell can be checked with an ohmmeter while the cell is viewing a flame. When viewing a properly adjusted oil burner flame, the cad cell's resistance should be between 300 and 1000 ohms. If the cad cell's resistance is higher than 1600 ohms when viewing a properly adjusted oil burner flame, the cad cell must be replaced. If a 1500 ohm resistor is placed between the F terminals of the cad cell primary control, the system should operate correctly; care should be taken when testing in this manner because the technician is supervising the oil burner flame. The primary control is bad when it does not operate correctly with this resistor in place, and it must be replaced. Care should be taken when troubleshooting cad cell primary controls for the safety of the structure and occupants.

14.10 SERVICE CALLS

Service Call 1

Application: Residential conditioned air system

Type of Equipment: Packaged heat pump with electric resistance heaters used for supplementary heat

Complaint: Insufficient heat on cold days

Service Procedure:

1. The technician reviews the work order from the dispatcher for available information. The work order indicates that the heat pump does well on mild days but does not adequately heat the home on extremely cold days.

2. The technician informs the homeowner of his or her presence and obtains any additional information about the system problem. The homeowner states that at times, the heat pump is blowing extremely cold air into the structure.

3. Upon entering the residence, the technician makes certain that no dirt or foreign material is carried into the structure. The technician also takes care not to mar or damage interior walls.

4. The technician knows that most system items can be eliminated if the heat pump is heating the home to the desired temperature on most days. The technician determines that the heat pump refrigerant cycle is operating properly. With the system operating on the normal heating cycle the technician can eliminate the (1) transformer, (2) components that operate in the normal heating cycle, and (3) indoor blower circuits.

5. With these items eliminated the technician can assume that the supplementary heat is the problem because heat pumps need some form of supplementary heat in extremely cold weather and when the system is in defrost. The technician must check to see if this is the problem and if so, determine the corrections that are necessary.

6. One of the simplest methods that can be used to check the operation of the supplementary heat is to set the thermostat set point at least 3° above the room temperature; this setting should bring the supplementary heater on if the outdoor temperature is low enough. The technician will have to check the setting of the outdoor thermostats and make that determination.

7. If the supplementary heaters do not come on, the technician should check the control voltage being supplied to the supplementary heater control circuit. In this case, the technician measures no voltage to the heater control circuit, thus indicating that the second-stage heat section of the thermostats is inoperative.

8. The technician replaces the faulty thermostat with one of the correct design. Care is taken to make certain that the thermostats are compatible.

9. The technician informs the homeowner of the problem and that it has been corrected.

Service Call 2

Application: Domestic refrigeration

Type of Equipment: Frost-free refrigerator

Complaint: Temperature in fresh and frozen food sections is not cold enough

Service Procedure:

1. The technician reviews the work order from the dispather for available information. The work order reveals that the temperature in the fresh and frozen food sections of the refrigerator is not low enough. No one will be home, but the key will be in the mailbox. Upon leaving the residence, the technician is to lock door and place the key on top of the refrigerator.

2. Upon entering the residence, the technician makes certain that no dirt or foreign material is carried into the structure. The technician also takes care not to mar or damage interior walls.

3. The technician should makes a visual inspection of the refrigerator. While making this inspection, the technician notices a larger-than-normal amount of frost in the frozen food section of the refrigerator, which leads him to believe that the appliance could be experiencing a defrost problem.

4. The technician locates the defrost timer and rotates the timer until a click is heard, which stops the compressor and supplies power to the defrost heaters. The technician should check the voltage being supplied to the heater and the current of the heater, all of which can be done at the defrost timer. If voltage is available from the timer but no reading is observed on the ammeter, the reading indicates that the heater is not being energized.

5. Once the technician determines that there is no current being used in the heater circuit, the indication is that the circuit is open past the defrost timer or that the heater is bad. Most refrigerator defrost heater circuits have a defrost thermostat that breaks power going to the heater once the evaporator has reached a certain temperature.

6. Because they are in series, the only way to check these components is to isolate both components, which can only be done by removing the evaporator cover. Once the cover is removed, the technician has access to both the heater and the defrost thermostat.

7. The technician makes a resistance check of the heater and defrost thermostat and finds that the defrost thermostat is open.

8. The defrost thermostat should be replaced with one of the same temperature setting.
9. The technician replaces any cover that was removed and puts the refrigerator back in working order.
10. Because of the amount of time required for the refrigerator to drop the temperature, it is impractical for the technician to check the defrost cycle. The technician should call back in several days to check on the refrigerator's operation.
11. The technician makes certain that the homeowner's wishes are followed regarding the security of the home and the placement of the key.

Service Call 3

Application: Commercial refrigeration

Type of Equipment: Walk-in cooler

Complaint: No refrigeration

Service Procedure:
1. The technician reviews the work order from the dispatcher for available information. The work order indicates that the evaporator fan is not operating and the evaporator coil is completely covered with frost.
2. The technician informs the store owner of his presence and obtains any additional information about the problem.
3. The technician observes that the evaporator fan motor is not operating and checks the voltage being supplied. The voltage measured is the correct voltage for the fan motor. The technician notices that the fan motor is cold, indicating that it is not even trying to start. The technician measures infinite resistance at the motor windings, giving an indication that the motor windings are open.
4. The technician determines that the fan motor must be replaced.
5. The rotation, rpm, and horsepower of the replacement motor must be the same as those of the faulty motor.
6. The technician should cut off the condensing unit and start the fan motor to remove the frost from the evaporator.
7. The technician checks the operation of the new fan motor, including checking the current.
8. The store manager is informed that when the evaporator is defrosted, the disconnect to the condensing unit should be closed.

Service Call 4

Application: Residential conditioned air system

Type of Equipment: Electric furnace

Complaint: No heat

Service Procedure:

1. The technician reviews the work order from the dispatcher for available information. The work order reveals that the electric furnace is not heating, but the blower motor will operate when the fan switch on the thermostat is turned to the "on" position. The electric furnace is located in the attic with access through a disappearing stairway in the hall of the structure.
2. The technician informs the homeowner of his or her presence and obtains any additional information about the problem.
3. Upon entering the residence, the technician makes certain that no dirt or foreign material is carried into the structure. The technician also takes care not to mar or damage interior walls.
4. The technician sets the thermostat fan switch to the "on" position and the fan operates, indicating that control and line voltage are available to the unit.
5. The thermostat is set to the heating position and the technician proceeds to the electric furnace. The technician removes the access cover to the control panel of the electric furnace and measures the voltage available to the sequencer. It is 24 volts, indicating that there is a call for heat to the furnace.
6. The technician next checks the voltage across the contacts of the sequencer feeding the electric heaters and reads 240 volts, indicating that the contacts of the sequencer are not closing and with all three contacts not closing, the sequencer controlling element is bad.
7. The technician replaces the sequencer.
8. The operation of the electric furnace is checked to make certain that the furnace is operating properly.
9. The technician informs the homeowner of the problem and that it has been corrected.

Service Call 5

Application: Residential conditioned air system

Type of Equipment: Gas furnace with time-delay relay controlling fan

Complaint: No heat

Service Procedure:
1. The technician reviews the work order from the dispatcher for available information. The work order reveals that the burner is igniting and remaining on for only a short period of time. The blower is not operating. The furnace is located in hall closet of the home.
2. The technician informs the homeowner of his or her presence and obtains any additional information available about the problem. The homeowner states that a small amount of warm air is coming through the air vents.
3. Upon entering the residence, the technician makes certain that no dirt or foreign material is carried into the structure. The technician also takes care not to mar or damage interior walls.
4. The technician observes the operation of the gas furnace and notices that the burner ignites for approximately two minutes and is cut off by a limit switch.
5. The technician knows that the furnace is warm enough for the blower to be operating, but that the motor is off and is cool to the touch, indicating that it has not operated for some time.
6. The technician is unable to locate a temperature-controlled fan switch on the furnace and examines the diagram to determine what controls the blower motor. The technician finds that the blower motor is controlled by a time-delay relay.
7. The technician must now determine the condition of the time-delay relay. The relay controlling element is receiving 24 volts, which indicates a call for the relay to be energized. The relay contacts remain open, thus preventing the blower motor from starting. The indoor fan relay controlling the blower motor is faulty.
8. The technician replaces the time-delay relay with one of the same delay time period.
9. The homeowner is informed of the problem and the action taken by the technician.

Service Call 6

Application: Residential conditioned air system

Type of Equipment: Gas furnace with air-cooled condensing unit

Complaint: No cooling

Service Procedure:

1. The technician reviews the work from the dispatcher for available information. The work order information reveals that no part of the system is operating. The dispatcher requested that the homeowner place the fan switch in the "on" position when the call was received. The homeowner revealed that nothing happened. The gas furnace is in a utility room located off the carport.

2. The technician informs the housekeeper of his or her presence and obtains any additional information about the problem. The technician asks the housekeeper to set the thermostat to the cooling position.

3. Upon entering the residence, the technician makes certain that no dirt or foreign material is carried into the structure. The technician also takes care not to mar or damage interior walls.

4. The technician proceeds to the location of the furnace and finds that no voltage is available to the furnace.

5. The technician locates the electrical breaker panel for the structure and finds that no circuit breakers are in the tripped or overloaded position.

6. The thechnician removes the breaker panel cover in order to check voltage available from the breaker supplying power to the furnace. The technician measure 0 volts at the circuit breaker, thus determining that the circuit breaker is faulty.

7. The technician replaces the faulty circuit breaker with one of an equal ampacity rating.

8. The technician starts the system, reads the current being used by the furnace circuit, and determines that the furnace is operating properly.

9. The technician replaces the cover on the breaker panel.

10. The technician informs the housekeeper of the problem and the actions taken.

SUMMARY

Most heating, cooling, and refrigeration technicians are required to do some diagnosing of components. Thus, it is imperative for technicians to understand how components work, to know how to use electric meters to check components, and to know the proper procedures for checking electric components. In this chapter we have presented some guidelines that technicians may find helpful in troubleshooting electric systems and their components.

Contactors and relays can be effectively diagnosed by checking the contacts, the coil, or the mechanical linkage.

Several types of overloads are used in the industry. The fuse can be checked with an ohmmeter. The circuit breaker can often be checked with a voltmeter. The line voltage overload can be checked with an ohmmeter. The pilot duty overload must have its contacts and controlling line voltage elements checked. These components can usually be checked with an ohmmeter. The internal overload used in hermetic compressors can best be checked with an ohmmeter.

Two basic types of thermostats are used in the industry. The line voltage thermostat can be checked with an ohmmeter. The low-voltage thermostat can also be checked with an ohmmeter but care must be taken with this type of thermostat because of its many functions.

There are two types of common pressure switches: low pressure and high pressure. Both can be serviced with either a resistance check or a voltage check. Transformers can also be checked by using either an ohmmeter or a voltmeter.

There are three sources of energy used to supply heat to a structure: electricity, gas, and oil. Electrical energy is easy to control because the controls used do not have to supervise any type of combustion as do gas and oil heating controls. Electric heating control systems tend to use conventional controls such as contactors, sequencers, and time-delay relays. There are three types of ignition control systems for gas-fired equipment: standing pilot, intermittent ignition, and direct ignition systems. The standing pilot burns continuously and lights the main burner on a call for heat from the thermostat. A thermocouple is used to energize a pilot solenoid proving the pilot. The intermittent pilot is used only to light the main burner on a call for heat from the thermostat. The direct ignition control system lights the main burner from a spark igniter or a hot surface ignitor. A gas valve is used to control the flow of gas to the main burners in each of these ignition systems. The gas valves are usually 24 volts except in light commercial applications. An electronic ignition module controls both the intermittent and direct ignition control systems. The technician should check the power supply of these modules. Other checks are made using a manufacturer's troubleshooting chart(s). Oil-fired equipment uses two types of primary controls: stack switches and cad cell primary controls. The stack switch uses a bimetal element and the cad cell views the flame of the oil burner. The technician will have to determine if the primary control is receiving the signal that combustion has occurred. Once the correct signal is received, the technician must determine if the primary control is operating properly. If an unsafe condition occurs, the primary control must interrupt power to the burner. Safety is of the utmost importance when troubleshooting fossil fuel control systems that require some type of combustion.

 When installing safety components, make certain that they are installed to manufacturers' specifications.

REVIEW QUESTIONS

1. What is the first step in troubleshooting any system or component?

2. Open-type electric motors will fail in which of the following three areas?
   a. windings, bearings, and centrifugal switch
   b. windings, bearings, and contactor
   c. windings, centrifugal switch, and contactor
   d. centrifugal switch, contactor, and mobile module

3. Electric motor windings fail in which of the following conditions?
   a. open
   b. shorted
   c. grounded
   d. all of the above

4. How can bearing failure be determined in a hermetic compressor motor?

5. True or False: The best way to check the contacts of a contactor is by visual inspection.

6. In troubleshooting contactors and relays, three areas must be checked: the _____, _____, and _____.

7. How do you make a voltage check to diagnose the condition of a set of contacts in a relay or contactor?

8. What is the major cause of contact misalignment?
   a. coil failure
   b. pitted contacts
   c. faulty mechanical linkage
   d. none of the above

9. If the coil of a relay is shorted, the resistance reading will be _____ ohms.
   a. 0
   b. 500
   c. 10,000
   d. infinite

10. If the coil of a relay is open, the resistance reading will be _____ ohms.
    a. 0
    b. 500
    c. 10,000
    d. infinite

11. How do you check a fuse?

12. How do you check a circuit breaker?

13. When a line voltage overload is weak, an amperage check should be made to see what _____ is causing the overload to open.

14. True or False: The internal overload is difficult to check because it is in parallel with the common terminal of the compressor.

15. What precaution should a service technician take when diagnosing the condition of an internal overload in a hermetic compressor?

16. True or False: The low-voltage thermostat is easier to troubleshoot than the line voltage thermostat.

17. What procedure should be used to troubleshoot a low-voltage thermostat?

18. The line voltage thermostat usually has _____ function(s).
    a. 1
    b. 2
    c. 3
    d. multiple

19. Give some common applications for line and low-voltage thermostats.

20. What do the letters R, Y, G, and W represent on a thermostat subbase?

21. How do you diagnose a bad pressure switch?

22. The most important aspect of checking a pressure switch is to understand its _____ in the system.

23. What is the proper procedure for checking a transformer?

24. What thermostat is used in a forced-air furnace to start and stop the blower motor?
    a. limit switch
    b. fan switch
    c. combination switch
    d. none of the above

25. List three uses of limit switches in a forced-air furnace?

26. What is the millivolt reading of a good thermocouple?
    a. 5 mV
    b. 18 mV
    c. 50 mV
    d. 120 mV

27. What procedure is used to check a combination gas valve?

28. The combination gas valve has _____ valves.
    a. 1
    b. 2
    c. 3
    d. 4

29. What should be the first check a technician should make when troubleshooting ignition modules?
    a. voltage across gas valve terminals on module
    b. spark across igniter/sensor gap
    c. good ground connection
    d. power supply to module

30. True or False: The spark on an intermittent pilot ignition system sparks the entire call for heat cycle of a gas warm-air furnace.

31. What is the difference between spark ignition and hot surface ignition?

32. The resistance of the cad cell (increases or decreases) when the cell views a correctly adjusted oil burner flame.

33. The stack switch uses which of the following methods to determine combustion in an oil-fired furnace?

a. sight
b. pressure
c. heat
d. none of the above

34. How would you troubleshoot a stack switch and cad cell primary control?

35. What is the purpose of a rollout switch?

PRACTICE SERVICE CALLS

Determine the problem and recommend a solution for the following service calls. (Be specific; do not list components as good or bad.)

Practice Service Call 1

Application: Residential conditioned air system

Type of Equipment: Gas furnace with air-cooled condensing unit

Complaint: No cooling

Symptoms:
1. Correct line voltage to furnace and condensing unit.
2. Thermostat calling for cooling.
3. Condensing unit receiving 24 volts.
4. All safety controls in condensing circuitry are closed.
5. Contactor coil resistance reading is 0 ohms.

Practice Service Call 2

Application: Residential conditioned air system

Type of Equipment: Fan coil unit with air-cooled condensing unit

Complaint: No cooling

Symptoms:
1. Correct line voltage to furnace and condensing unit.
2. Thermostat calling for cooling.
3. Compressor contactor closed.
4. Compressor and condenser fan motor extremely hot.

5. Condenser fan motor resistance readings are C to R = 6 ohms, C to S = 28 ohms, and R to S = 34 ohms.
6. Compressor run capacitor in good condition.
7. Compressor motor and bearings in good condition.
8. Condenser fan blade turns with ease.
9. Condenser fan turning slower than normal until overload cuts out.

Practice Service Call 3

Application: Residential conditioned air system

Type of Equipment: Oil furnace with air-cooled condensing unit

Complaint: No cooling

Symptoms:

1. Air-cooled condensing unit operating correctly.
2. Indoor fan motor not operating.
3. Indoor fan motor cold to the touch.
4. Indoor fan relay receiving 24 volts.

Practice Service Call 4

Application: Commercial refrigeration

Type of Equipment: Commercial refrigerator (No contactor)

Complaint: No refrigeration

Symptoms:

1. Compressor not operating.
2. Condenser fan motor operating properly.
3. All starting components of compressor are good.
4. Compressor receiving correct line voltage.
5. Compressor body extremely hot.
6. Compressor overload open.
7. Compressor resistance readings are C to S = 18 ohms, C to R = 4 ohms, S to R = 22 ohms, and from case to S = 1000 ohms.

Practice Service Call 5

Application: Commercial refrigeration

Type of Equipment: Walk-in cooler with air-cooled condensing unit

Complaint: Not enough refrigeration

Symptoms:

1. Evaporator fan motor running.
2. Condensing unit receiving correct line voltage.
3. Contactor not closed.
4. Contactor coil not receiving 24 volts.
5. Safety components and operating controls in contactor coil control circuit are low-pressure switch, high-pressure switch, thermostat, and manual switch.
6. Hopscotching through the circuit, the technician places a voltmeter lead on $L2$ feeding the contactor coil.
7. Checking on the line side of the low-pressure switch, 240 volts is measured but on the load side of the low-pressure switch, 0 volts is measured.

Practice Service Call 6

Application: Commercial and industrial conditioned air system

Type of Equipment: Fan coil unit with large air-cooled condensing unit

Complaint: Not enough cooling

Symptoms:

1. Indoor blower operating properly.
2. Compressor contactor closed and compressor operating properly.
3. Condenser fan motors contactor closed (contactor controls both condenser fan motors).
4. One condenser fan motor operating.
5. Second condenser fan motor not operating.
6. Outdoor temperature is 100°F.
7. Second condenser fan motor good.
8. Thermostat is in series with second condenser fan motor, voltage across this thermostat is 240 volts.

Practice Service Call 7

Application: Residential conditioned air system

Type of Equipment: Gas furnace with air-cooled condensing unit

Complaint: No heat

Symptoms:
1. Line voltage is available to the furnace.
2. Indoor fan motor operates with thermostat fan switch in the "on" position.
3. Ignition module is receiving 24 volts.
4. Fuse good on ignition module.
5. No spark across igniter/sensor gap.
6. No spark at ignition stud on module.

Practice Service Call 8

Application: Residential conditioned air system

Type of Equipment: Oil furnace with air-cooled condensing unit

Complaint: No heat

Symptoms:
1. Line voltage available to furnace.
2. Cad cell primary control receiving correct line and control voltage.
3. Cad cell relay closing.
4. Spark across ignition points of oil burner.
5. Oil burner motor not operating.

Practice Service Call 9

Application: Commercial Refrigeration

Type of Equipment: Chest-type frozen food display case with CSR compressor motor

Complaint: Product thawing

Symptoms:
1. Compressor will not operate.
2. Compressor receiving correct line voltage.
3. Starting and running capacitor good.
4. Compressor motor and bearing good.

Practice Service Call 10

Application: Domestic refrigeration

Type of Equipment: Refrigerator

Complaint: No refrigeration

Symptoms:

1. Thermostat is closed, supplying 120 volts to compressor.
2. Compressor housing is cold to touch.
3. Compressor motor winding good.
4. Current-type starting relay coil good.

Practice Service Call 11

Application: Domestic refrigeration

Type of Equipment: Chest-type freezer with CS motor

Complaint: Food thawing

Symptoms:

1. Compressor hot to touch.
2. Compressor tries to start but overload opens.
3. Current type relay is good.
4. Compressor receiving correct voltage.

Practice Service Call 12

Application: Residential conditioned air system

Type of Equipment: Packaged heat pump

Complaint: Not enough heat

Symptoms:

1. Compressor and outdoor fan motor operating.
2. Indoor fan motor operating.
3. Supply of air is not warm enough.
4. System not calling for supplementary heat.
5. Outdoor coil frosted.
6. Heat pump is equipped with a differential pressure switch to initiate defrost cycle.

Practice Service Call 13

Application: Residential conditioned air system

Type of Equipment: Packaged air conditioner

Complaint: No cooling or fan operation

Symptoms:

1. Line voltage available to unit.
2. Closing of thermostat fan switch causes no action.
3. Voltage at terminals C and R on unit low-voltage terminal board is 0 volts.

Practice Service Call 14

Application: Residential conditioned air system

Type of Equipment: Gas furnace with air-cooled condensing unit

Complaint: System operates but home temperature goes from too hot to too cold

Symptoms:

1. Low-voltage thermostat closed.
2. Equipment operating properly except for large temperature swing.
3. Installation only 2 weeks old.
4. Fan switch good.
5. Fan operates longer than normal on "off" cycle before stopping.

Practice Service Call 15

Application: Commercial and industrial conditioned air system

Type of Equipment: Large condensing unit

Complaint: No cooling

Symptoms:

1. Indoor blower operating properly.
2. Control relay on condensing unit is closed, which should start condensing unit.
3. No voltage available to condensing unit.

LAB MANUAL REFERENCE

For experiments and activities dealing with material covered in the chapter, refer to Chapter 14 in the Lab Manual.

15

Air-Conditioning Control Systems

OBJECTIVES

After completing this chapter, you should be able to

- Understand the electrical circuitry of a residential condensing unit.
- Make all electrical connections to install a condensing unit in a residential application.
- Troubleshoot a residential condensing unit.
- Understand the basic control systems used in residential air-conditioning control systems.
- Understand the control systems used in light commercial air-cooled and water-cooled packaged units.
- Make all electrical connections for a complete residential installation.
- Understand the control systems used in a gas heat electric air-conditioning packaged unit.
- Troubleshoot residential air-conditioning systems.

KEY TERMS

Air-cooled packaged unit
Condensing unit
Factory-installed wiring
Field wiring
Gas pack

Indoor fan relay package
Internal pressure relief valve
Packaged air-conditioning unit
Short-cycling
Water-cooled packaged unit

INTRODUCTION

In the heating, cooling, and refrigeration industry there are many different types of control systems and they range from simple to very complex. The equipment used in residential air-conditioning units is usually simple and contains a limited number of components. Commercial and industrial air-conditioning equipment uses a more complex control system, with more emphasis on safety devices than is the case for residential control systems. The trend in the industry in the last few years has been to make residential air-conditioning control systems simpler without any reduction in safety. This has been done because of advances in control system design. At the present time the industry is using some solid-state control modules in control systems, but the control systems are by no means completely solid state.

The control systems used in the industry today are combinations of different control systems for separate pieces of equipment, such as a furnace with an air-conditioning unit containing some method of interlocking. The control system has the same function whether the system is a single component or a combination of components. An air-conditioning control system is designed to operate a system automatically, incorporating all the necessary safety equipment by the use of a thermostat. The safety devices of a control system ensure that the loads of the system operate without any chance of damage. However, these safety devices are not always effective in preventing damage to the components because of the economy factor that must always play a part in the design of equipment. Beyond the safety devices of a control system are devices that control the operation of equipment to maintain a temperature at a certain range.

The installation technician of air-conditioning or heating equipment must be able to connect the control system of the equipment and the thermostats—and in many cases the furnace. Most manufacturers furnish a wiring diagram with equipment that shows the proper hookup of the equipment being used. In many cases, however, the installation technician must install equipment without a diagram. Therefore, it is important for service technicians to know how to connect the controls correctly to ensure proper operation.

All heating, cooling, and refrigeration technicians are at one time or another required to be familiar with modern control systems. For example, salespeople may be called on to assist customers with control system design. Engineers are required to design control systems. Service technicians are required to maintain and repair control systems.

About 85% of the problems in heating, cooling, and refrigeration control systems can be traced to some electrical problem. Although it is impossible to study all the different control systems used in the industry today, familiarity with the more common control systems will enable you to understand special control systems because of the similarities of design. In this chapter we present the more common control systems in use in the industry today.

15.1 BASIC CONDENSING UNITS

A **condensing unit** is the portion of a split air-conditioning system that is mounted outside and contains the compressor, the condenser, the condenser fan motor, and the necessary devices to control these components. A split system is one that is divided into two parts, usually a condensing unit (outside) with a fan coil unit (inside). Figure 15.1 shows an air-cooled condensing unit.

In most cases, the condensing unit is used with some type of equipment that will produce the air flow, and a coil must be mounted in the air flow. Most condensing units used in the industry today are air-cooled condensing units, which means that the condenser is cooled by air. Water-cooled condensing units may still be found in the field, but these units are rapidly being replaced with the smaller and more economical air-cooled condensing units.

When a condensing unit is used in an air-conditioning system, it is separate from the rest of the equipment. The control system must be connected to allow for complete control of the entire system.

FIGURE 15.1 Equipment with air discharge out the top of the cabinet

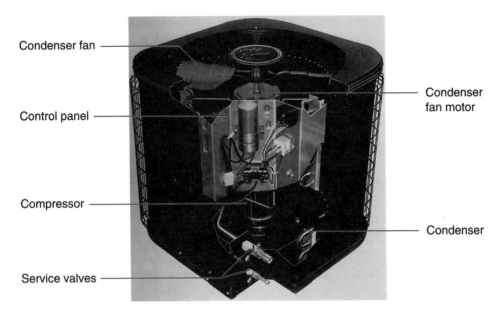

Condenser fan

Control panel

Compressor

Service valves

Condenser
fan motor

Condenser

FIGURE 15.2 Components of an air-cooled condensing unit *(Courtesy of Inter-City Products Corporation)*

The condensing unit takes the cool suction gas coming from the evaporator, compresses it, condenses it from a gas to a liquid, and forces it back to the evaporator. The standard components of an air-cooled condensing unit are the compressor, the condenser, the condenser fan motor, and all the devices used for control. The components of an air-cooled condensing unit are shown in Figure 15.2. Most of the controls used to operate the condensing unit are mounted somewhere in the unit itself with the exception of the interconnection between the thermostat and the condensing unit.

Condensing Unit Components for a Simple Control System

In early control systems, all components necessary for the operation of the total air-conditioning system were mounted in the condensing unit. This practice has all but disappeared in the industry because of the expense that is required and the realization that it was unnecessary. Most modern condensing units pick up their 24-volt power supply to operate the control system from the furnace transformer or from an indoor fan relay package. The evaporator fan motor usually is controlled by a relay when the system contains a furnace or fan coil unit and the condensing unit.

The compressor and condenser fan motor in a condensing unit are usually controlled by a contactor. The simplest control system used on condensing units today is a contactor that controls the operation of the entire condensing unit with the exception of the internal overloads in the compressor and condenser fan motor. With this type of control system, the contactor is energized whenever the thermostat is calling for cooling, even though one of the safety devices has the compressor or condenser fan motor cut out. This control system is the simplest and the least expensive of any condensing unit control system used today. A separate 24-volt source must be supplied from other components through the thermostat to the condensing unit.

Most condensing units without a high-pressure switch will have an **internal pressure relief valve** in the compressor that opens if the discharge pressure exceeds an unsafe level. This relief valve replaces the high-pressure switch on some units.

Figure 15.3 shows a control panel used on a condensing unit control system. Note the simplicity of the panel.

Figure 15.4 shows the schematic diagram of this type of control system. The control system requires a supply of 24 volts to the contactor through the thermostat to operate the condensing unit. When 24 volts are applied to the contactor through the thermostat, the contactor will close, allowing 240 volts to be applied to the compressor and condenser fan motor, and they will start. This control system is easily diagnosed for trouble, with the exception of the overloads in the compressor and the condenser fan motors.

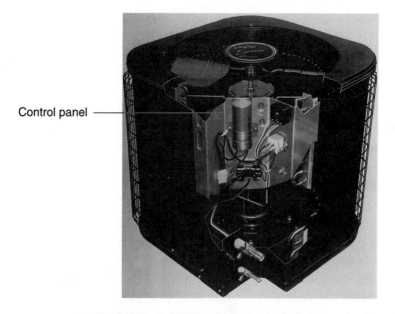

Control panel ———

FIGURE 15.3 Control panel of a simple air-cooled condensing unit
(Courtesy of Inter-City Products Corporation)

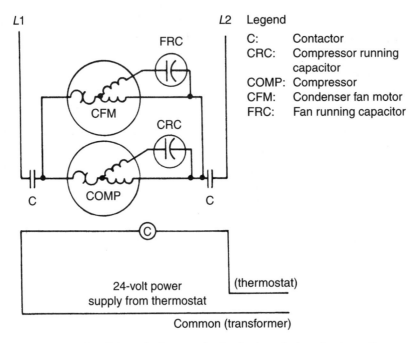

FIGURE 15.4 Schematic diagram of a simple air-cooled condensing unit

Condensing Unit Components for Complex Control Systems

Many manufacturers use a complex control system that contains certain safety devices to ensure the safe operation of the condensing unit. The control system discussed in the preceding section had only the minimum required safety devices, which are often inadequate for the completely safe operation of the condensing unit. The more complex systems incorporate a high-pressure and a low-pressure switch. Figure 15.5 shows the schematic diagram of a control system with the extra safety controls. If any of the safety controls open, the compressor and condenser fan motor stop.

Several manufacturers also use a device that protects the system from short-cycling. This device maintains a certain period of time between the cycles of the equipment so the system does not cut on and off in rapid succession (short-cycling). This control system adds a relay and a time clock mechanism for this purpose. Other systems use a special relay to lock out the entire condensing unit until it can be reset manually.

There are many other control system designs that use more components for specific purposes. For example, large condensing units use a control relay or compressor relay to control a 240-volt contactor coil and the condenser fan motor. This arrangement is used because for a large contactor,

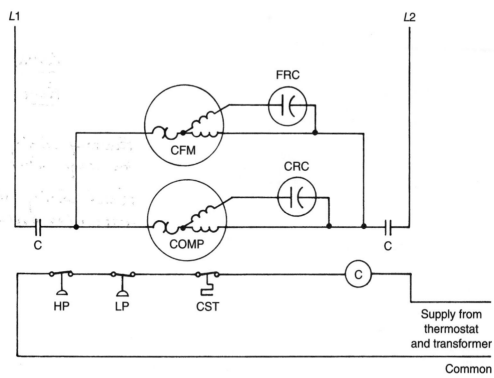

FIGURE 15.5 Schematic diagram of a condensing unit with safety controls

Legend

C:	Contactor	CRC:	Compressor running capacitor
COMP:	Compressor	HP:	High-pressure switch
CFM:	Condenser fan motor	LP:	Low-pressure switch
FRC:	Fan running capacitor	CST:	Compressor safety thermostat

24 volts are less effective in pulling the contactor in than are 240 volts. Many other designs, too numerous to discuss, are used in the industry for the control of condensing units. In most cases, though, control systems have several basic similarities that make them fairly easy to understand.

Wiring

All condensing units come from the manufacturers with a wiring diagram and, in some cases, with an installation wiring diagram. Usually it is not difficult to follow the installation instructions and wire the condensing unit correctly. Figure 15.6 shows an installation diagram used to install a condensing unit to a furnace for an add-on air-conditioning system. This diagram shows the wiring hookup of a condensing unit and a gas furnace. It is used when installing a condensing unit and furnace.

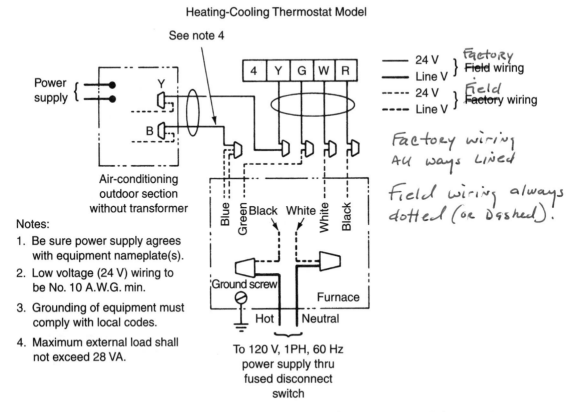

Heating-Cooling Thermostat Model

See note 4

Power supply

Y

B

Air-conditioning outdoor section without transformer

4 | Y | G | W | R

—— 24 V
—— Line V } Factory Field wiring

---- 24 V
---- Line V } Field Factory wiring

Factory wiring All ways Lined

Field wiring always dotted (or Dashed).

Blue Green Black White White Black

Ground screw

Furnace

Hot | Neutral

To 120 V, 1PH, 60 Hz power supply thru fused disconnect switch

Notes:

1. Be sure power supply agrees with equipment nameplate(s).

2. Low voltage (24 V) wiring to be No. 10 A.W.G. min.

3. Grounding of equipment must comply with local codes.

4. Maximum external load shall not exceed 28 VA.

FIGURE 15.6 Installation diagram for a condensing unit connected to a furnace *(Courtesy of The Trane Company)*

Troubleshooting

Any troubleshooting of the condensing unit can be done from the schematic because of the simplicity of most control systems. By using the schematic diagram, and having an understanding of the components, you should have no trouble with diagnosing a residential control system. However, you may have to study the more complex commercial and industrial control systems to diagnose problems in them.

15.2 PACKAGED UNITS

A **packaged air-conditioning unit** is built with all the components housed in one unit. In most cases, packaged units are complete except for the power connections and the control connections.

Air-cooled packaged units are the most widely used types of packaged units. The units are used on applications ranging from small residential

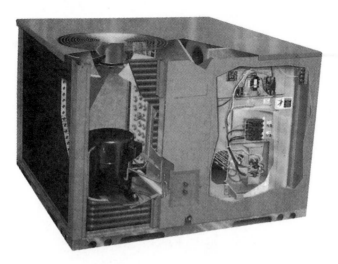

FIGURE 15.7 Air-cooled packaged unit *(Courtesy of Heil-Quacker Corporation)*

cooling units to large commercial and industrial air-cooled units. The smaller residential systems may be installed when the structure is built or at a later date.

An air-cooled packaged unit would usually be mounted outside the structure, as shown in Figure 15.7. In some cases, air-cooled packaged units can be installed inside with a remote condenser. The air-cooled packaged unit contains the compressor, evaporator fan motor, condenser fan motor if not of remote design, and all the necessary controls.

Water-cooled packaged units are usually used on commercial and industrial systems. A water-cooled packaged unit is shown in Figure 15.8. They are usually installed when the structure is built. The water-cooled unit requires a cooling tower or some means of supplying water to the condenser. If a water-cooled unit is used, there must be interlocks to ensure that the cooling tower pump is operating when the compressor is operating. Water-cooled packaged units come complete with the compressor, evaporator fan motor, all the necessary controls, and the condenser.

Control System for Packaged Air-Conditioning Systems

Packaged unit controls are usually simple on small residential equipment and become more complex on the larger commercial and industrial equipment. The smaller air-cooled packaged units use a fairly simple control system with an interlock for the condenser fan motors. The large air-cooled and water-cooled units have a fairly complex control system. The large air-cooled packaged units are complex because of the zoning requirements

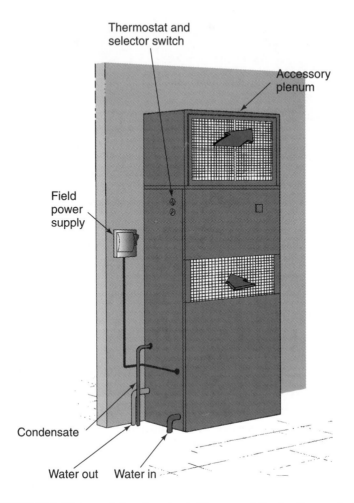

FIGURE 15.8 Water-cooled packaged unit *(Courtesy of Carrier Corporation)*

that these units must meet in the rooftop applications in which they are usually used. Water-cooled packaged units utilize a line voltage control system with some type of interlock to start the cooling tower pump or other accessories that are required.

Air-cooled packaged units can incorporate some form of heating as well as cooling. The smaller packaged units usually have some easy means of installing electric resistance heat into the equipment. If gas heating is desired, a **gas pack** should be used, which is an electric air-conditioning and gas-heating system mounted in one unit. The larger air-cooled packaged units make electric and gas heat available to the customer. Water-cooled packaged units and air-cooled packaged units usually use a hot-water coil, steam coil, or resistance heat for heating purposes.

Small Air-Cooled Packaged Units

The small air-cooled packaged units usually have a simple and easy-to-follow control system. The control system uses a contactor to operate the compressor and the condenser fan motor and an evaporator fan relay to control the evaporator fan motor. All necessary safety components are also included in the control system.

Not all manufacturers' control systems are alike. Many use only the necessary safety controls, while others use more safety controls to ensure complete and safe operation.

Figure 15.9 shows the schematic diagram of a small air-cooled packaged unit and its controls. This unit would be used in a residence or small commercial application. The thermostat at the bottom of the schematic controls the operation of the unit. When the fan switch is in the "on" position the indoor fan will operate, but its starting is delayed by the delay board. After the predetermined delay the indoor fan relay will energize, starting the indoor fan motor. When the thermostat calls for cooling, the compressor contactor is energized, closing the contacts and starting the compressor and outdoor fan motor. At the same time the indoor fan delay board is energized, and once the predetermined delay has been met the indoor fan relay will close, starting the indoor fan motor.

The installation of the air-cooled packaged unit is relatively simple because the only necessary connections are from the power source and the control source or thermostat. Most manufacturers include an installation diagram giving the correct connections and the control layout of the system.

Gas-Electric Air Conditioners

The combination of electric air conditioning and gas heating in a packaged unit gives a system that is somewhat more complex than a straight air-cooled packaged unit. The main difference between the two units is all the extra components needed to combine a heating system with an air-conditioning system. The basic control system for the air conditioning is the same as the one for most small residential units. However, several components are added to take care of the heating control system.

Figure 15.10 shows a schematic diagram of the control system used on a modern gas-heating and electric air-conditioning packaged unit. The installation instructions should be followed when installing this type of system.

Referring to Figure 15.10, the cooling mode of operation is energized through the thermostat to the contactor, which starts the compressor and outdoor fan motor. The indoor fan motor is started through the thermostat

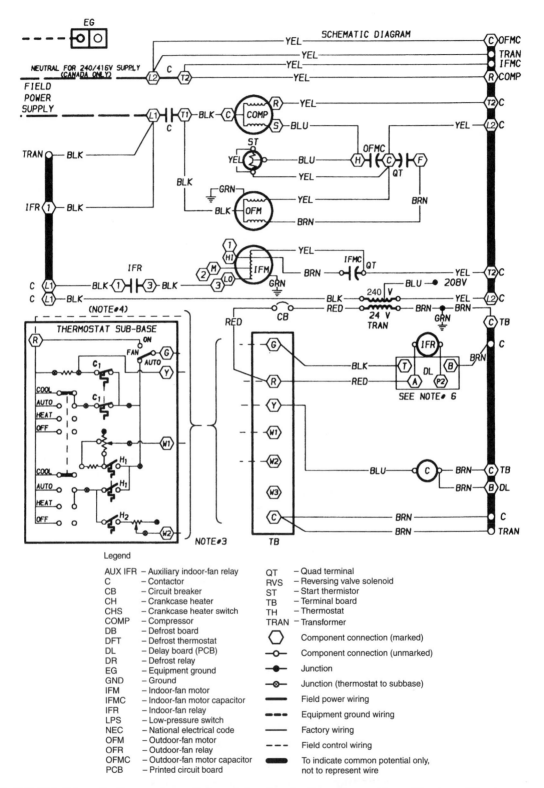

FIGURE 15.9 Schematic diagram of a small air-cooled packaged unit used in a residence *(Courtesy of Carrier Corporation)*

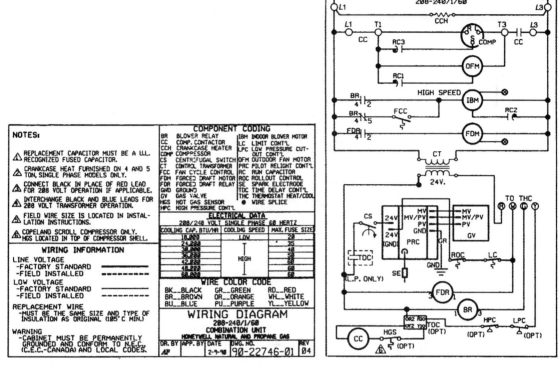

FIGURE 15.10 Schematic diagram for electric air-conditioning and gas-heating unit *(Courtesy of Rheen Air Conditioning Division)*

to the blower relay. The heating mode of operation is by the first stage of the thermostat (heating) energizing the gas valve.

Air-Cooled Packaged Units with Remote Condensers and Water-Cooled Packaged Units

The air-cooled packaged unit with the remote condenser and the water-cooled packaged unit use a line voltage control system and are similar in circuitry. These systems usually have line voltage control because they are shipped from the factory with all the controls mounted and wired, including the thermostat. These units are merely set in place when installed. The only necessary installation would be the power wiring connection, the condensate drain, and the piping.

The major difference between the water-cooled unit and the air-cooled unit is in the method the manufacturer uses to interlock the necessary components to cool the condenser. Figure 15.11 shows the schematic diagram

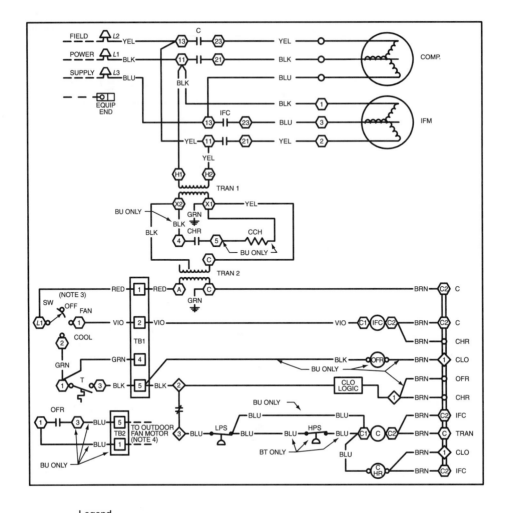

Legend

C   – Compressor contactor
CH or   – Crankcase heater
CCH
CHR   – Crankcase heater relay
CLO   – Compressor lockout
COMP – Compressor
CR   – Control relay
DU   – Dummy terminal
Equip   – Equipment ground
 Gnd
HPS   – High-pressure switch
IFC   – Indoor-fan contractor
IFM   – Indoor-fan motor
IP   – Internal protector
LLS   – Liquid line solenoid
LPS   – Low-pressure switch
OFR   – Outdoor-fan relay
OL   – Overload
QT   – Quadruple terminal
RC   – Run capacitor
S   – Compressor solenoid

SC   – Start capacitor
SR   – Start relay
SW   – Switch
T   – Thermostat
TB   – Terminal block (board)
TC   – Thermostat cooling
Tran – Transformer

⌁   Field splice

▢   Terminal block (board)

◇   Terminal compressor lockout (CLO)

○   Terminal (unmarked)

⬡   Terminal (marked)

——   Factory wiring

---   Field wiring

═══   Indicates common potential only;
       does not represent wire.

FIGURE 15.11 Schematic diagram of an air-cooled packaged unit with a remote condenser *(Courtesy of Carrier Corporation)*

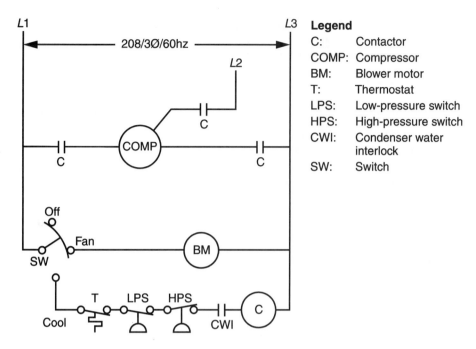

Legend

C:	Contactor
COMP:	Compressor
BM:	Blower motor
T:	Thermostat
LPS:	Low-pressure switch
HPS:	High-pressure switch
CWI:	Condenser water interlock
SW:	Switch

FIGURE 15.12 Schematic diagram of a water-cooled packaged unit

of an air-cooled packaged unit with a remote condenser interlocked in the system. Figure 15.12 shows the schematic of a water-cooled packaged unit with a cooling tower pump interlock.

The unit shown in the schematic diagram of Figure 15.11 has a line voltage control circuit. The switch marked "off," "fan," and "cool" operates the unit. With the switch set on "fan," the only portion of the unit that can operate is the indoor fan motor; it is energized through the manual switch. If cooling is desired, the switch must be set on "cool," which will not interrupt the operation of the indoor fan motor. The circuit energizes the outdoor fan motor through the outdoor fan relay. The compressor is energized through the contacts of the contactor, which is energized through the holding relay and the timer.

The unit shown in the schematic diagram of Figure 15.12, a small water-cooled packaged unit, also has a line voltage control circuit. The switch marked "fan" and "cool" actually stops and starts the unit. When the switch is in the "fan" position, the fan will run. When the switch is moved to "cool," the fan will continue to run and the cooling thermostat will operate the compressor through the contactor.

Installation of these air conditioners is simple because no connections are necessary, with the exception of the power wires. However, water connections are necessary on the water-cooled units, and refrigerant lines are needed between the outdoor remote condenser and the inside packaged unit.

Rooftop Units

The air-cooled packaged units used in rooftop applications are complex because they usually have some type of heating to go along with the air cooling and, in some cases, some type of zone control. The air-cooling and heating control system is usually simple. However, when it is connected to all the zone controls, the system becomes more difficult to service and install.

As their name implies, these units are usually installed on the roof of a structure. But they could just as easily be installed on ground level when the necessary changes are made in the ductwork connections. Often the electric connections of these systems are hard to install because of the number of wires required for the control circuit and the power wiring. The installation instructions furnished with these units are usually well written and give explicit instructions.

15.3 FIELD WIRING

In all heating, cooling, and refrigeration electric systems there is a certain amount of wiring that must be connected to the equipment after it has been set in place. This wiring must be installed by the installation technician and is called **field wiring**.

The **factory-installed wiring** is the wiring installed at the factory. It usually takes care of the connections between the components in the control panel and the system components. The factory wiring has been sized, color-coded, and installed in the control system to operate the equipment properly. The remainder of the wiring, whether it be the power wiring or control wiring, must be connected in the field.

Power Wiring

The power wiring of a cooling and heating system is usually simple and easy to install on residential systems. There are two power connections that must be made on a split air-conditioning and heating system: the connections to the condensing unit and the connections to the evaporator fan motor or furnace. Figure 15.13 shows the power wiring on a condensing unit and a furnace with the correct connections from the distribution panel in the residence.

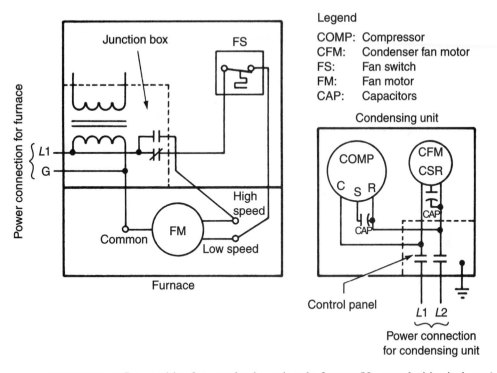

FIGURE 15.13 Power wiring for a condensing unit and a furnace (No control wiring is shown.)

The power connections for a residential air-cooled packaged unit are shown in Figure 15.14. This circuit is relatively simple, containing only the power connections and some type of bonding ground.

The electric supply wiring of a commercial and industrial system is somewhat more complex than the simple residential systems. The use of three-phase current in commercial and industrial structures does not add to the complexity of the system, only to the number of power wires that must be supplied to the system. The evaporator fans on any split commercial and industrial system would be supplied with a separate power source. The condensing unit would be supplied with its own power source. Figure 15.15 shows the supply wiring that must be installed with a fan coil unit that uses three-phase current.

Heating Systems. Commercial and industrial systems use various types of heating systems. In a system that uses electric resistance heat, a large power supply would be required to operate the heat. If some type of gas heating is used, the wiring would not have to be as large as that used

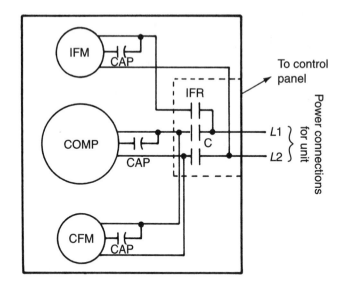

FIGURE 15.14 Power connections for a small air-cooled packaged unit (No control wiring is shown.)

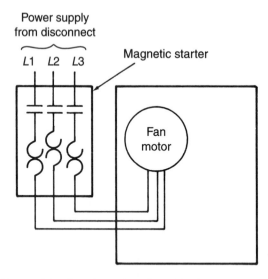

FIGURE 15.15 Power wiring for a three-phase fan coil unit (No control wiring is shown.)

with electric heat. Figure 15.16 shows the supply wiring hookup of a condensing unit, fan coil unit, and an electric duct heater. In many cases hot water or steam is used, but there are no additional power connections that must be made except for the pumps.

Large rooftop or packaged units equipped with gas heating usually require only one power source to the entire system. The rooftop units that

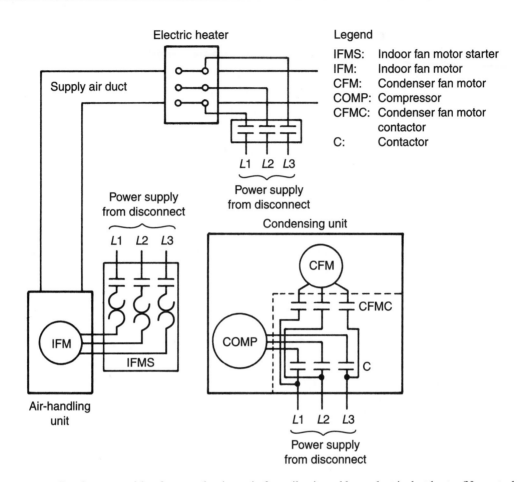

FIGURE 15.16 Supply power wiring for a condensing unit, fan coil unit, and large electric duct heater (No control wiring is shown.)

incorporate electric resistance heat are usually supplied by two or more power sources because of the large load created by the electric heat. One of the power sources feeds the cooling components and the evaporator fan motor; the other source feeds the electric heat.

There are many types and manufacturers of equipment in the industry today, and it is impossible to show all the supply power connections in this section. However, it is important that the installation technician study other methods of connecting the power wiring to modern equipment.

Sizing Wires and Fuses. When installing heating, cooling, and refrigeration systems, air-conditioning technicians in charge of the installation must make sure that all power wiring is of the correct size and type. The distance that the power wiring must be run is important because of the voltage drop that can occur on long circuit runs. The installation instructions usually give a wiring chart with the length that is allowable for each size of wire. Installation technicians should be capable of sizing the wire if it is not listed in the installation instructions (refer again to Chapter 8).

It is also important to follow the manufacturer's recommendations for the fuse or breaker size. Remember: Before any heating, cooling, or refrigeration system can be expected to operate properly, it must first be supplied with the correct size wire and fuse to deliver the proper voltage to the system.

Control Wiring

The control wiring of heating and air-conditioning systems is just as important as the supply power wiring. In most cases, low-voltage control systems are used on residential and small commercial applications. On large commercial and industrial applications, line voltage control systems are often used. Many large commercial and industrial systems are controlled by special control systems. These specialized control systems can be electric, pneumatic, or electronic. Specialized control systems are a complete subject in themselves and hence will not be covered in this section.

Residential. All residential packaged air conditioners use a low-voltage control system that connects the thermostat to the unit, as shown in Figure 15.17. These control systems are simple and easy to install by following the installation instructions.

Residential split systems are usually easy to install. They all incorporate a 24-volt control system, with the low-voltage power being supplied from an indoor fan relay package, the furnace, or the condensing units. Figure 15.18 shows one type of **indoor fan relay package** that is in common use in the industry today. The indoor fan relay package acts as a junction point between the thermostat, the furnace, and the condensing unit. The package has a terminal board labeled with easily identifiable letter identifications.

The control system of a residential split system can be connected several different ways, and the manufacturer's installation instructions should be followed. Some manufacturers use a transformer mounted in the condensing unit that supplies the low voltage to the system, but this practice is becoming unpopular because of the advancements made in the indoor fan relay packages. In some cases, the furnace transformer is used as the low-voltage power supply, but caution should be taken to ensure that the proper size is used. The more modern condensing units are usually designed without the low-voltage transformer but with two low-voltage connections

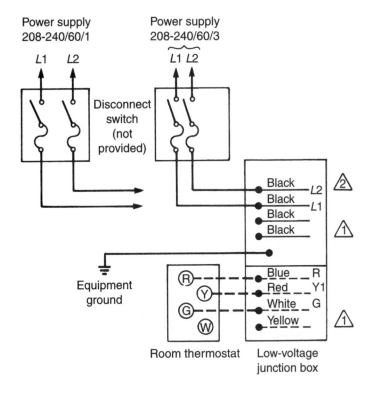

Power supply
208-240/60/1

Power supply
208-240/60/3

L1   L2

L1 L2

Disconnect
switch
(not
provided)

Black   L2   △2
Black   L1
Black   △1
Black

Equipment
ground

Blue   R
Red   Y1
White   G
Yellow   △1

Room thermostat   Low-voltage
junction box

△1 These leads are used only when an electric heater is installed.

△2 Do not cut excess length from these leads if an electric heater
will be installed at a later date.

FIGURE 15.17 Control connections for a residential packaged unit *(Courtesy of Lennox Industries, Inc.)*

FIGURE 15.18 A common indoor fan relay package

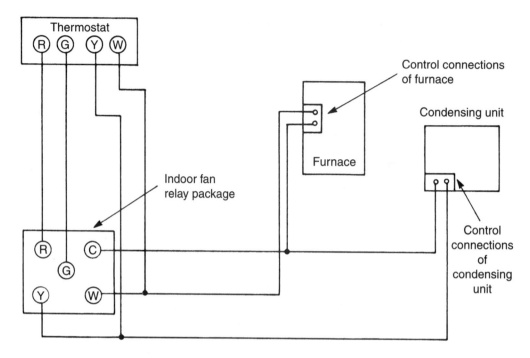

FIGURE 15.19 Control layout for a furnace, indoor fan relay package, thermostat, and condensing unit

instead. Figure 15.19 shows a control layout for a furnace, indoor fan relay package, thermostat, and the condensing unit. Figure 15.20 shows a control system of a furnace and a condensing unit with the thermostat and low-voltage power supply coming from the furnace transformer.

Industrial. Control systems for relatively small commercial units are similar in design to the residential low-voltage system. The only exception is that the system uses a control relay that supplies the condensing unit with 240 volts while using a normal 24-volt control system. Two-stage heating or two-stage cooling can also be used with a low-voltage control system.

Sizing Wire. Control wiring in most cases is a small wire (No. 18 to No. 20) that is usually covered by a rubber jacket when using a low-voltage control system. Thermostat wire can be purchased single-stranded or multi-stranded, with or without a rubber jacket protecting the small thermostat wires. The size of the control wiring should follow the manufacturer's specifications.

The wiring of air-conditioning and heating systems is one of the most important factors in the installation of equipment. The control system is

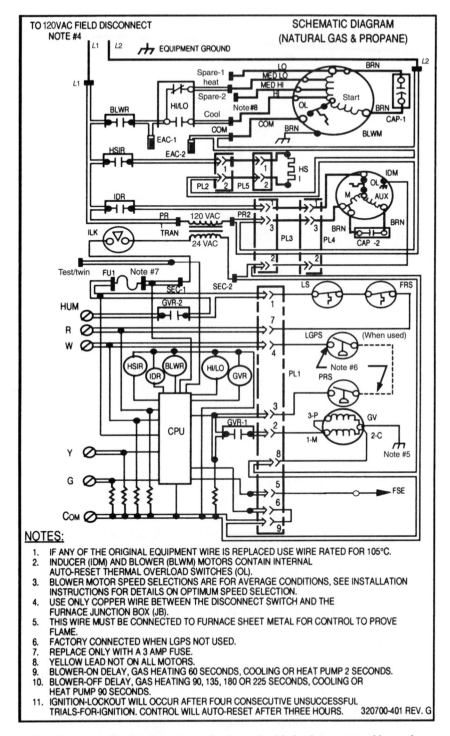

FIGURE 15.20 Control system of a furnace and a condensing unit with the thermostat and low-voltage power supplied from the furnace *(Courtesy of Carrier Corporation)*

actually the heart of the electric system because its function is to properly control the entire system. The life of the equipment can be drastically cut by undersized wire or by having loose connections. Therefore, wiring to all components must be sized correctly. The installation instructions are the best place to look for the proper method of wiring and the proper connection points. Overlooking the smallest detail in the electric connections of a system can give trouble and improper operation.

SUMMARY

Air-conditioning, heating, and refrigeration control systems range from simple control systems with a simple line voltage thermostat to very complex systems using many different devices to control the temperature of each zone in a large structure. It is understandable that a large air-conditioning unit with a very expensive compressor will require a larger and better control system than the smaller residential units.

The electric system of an air-conditioning unit covers the power wiring as well as the control wiring, and in most cases they should be treated as one.

The electric system of a small condensing unit used in the residential or small commercial area is designed with only the bare necessities in controls. Most of these smaller units are designed with a contactor and few safety controls in the condensing unit. The 24-volt supply must come from other equipment. The compressors of these units almost always contain an internal overload, and this should be taken into consideration when working with these units. Some of the smaller commercial systems will have various added features to prevent the equipment from short-cycling and to provide other important advantages.

Packaged air-conditioning equipment can take two different forms. The small residential and commercial packaged units are completely self-contained and must have an outside air source. The straight commercial packaged units are built in one piece and are used in many cases in a free-standing position within a structure; they may be water-cooled or air-cooled condensers that could be remote mounted.

The packaged units that are mounted outside come in various designs. They range from a two-ton air conditioner for a residence to a 50-ton or larger rooftop unit used in commercial buildings.

The straight packaged units usually use a low-voltage control system. Some of the larger units will have a line voltage control system within the unit. The free-standing packaged units almost always have a line voltage control system because they are prewired at the factory in this manner. All packaged units are prewired at the factory. In the straight packaged units, the control wiring must be field installed. The free-standing packaged units are completely wired at the factory.

All heating, cooling, and refrigeration equipment will require some kind of electric connection in the field, if nothing more than the power connections. Most equipment will require a control circuit and a power circuit before the equipment will operate. The power circuit consists of the electric source carried to the unit. In some cases, the evaporator fan motor will require a separate source. Control wiring is required between the thermostat and the equipment, or two sections of the equipment, to make one complete system. The manufacturer almost always supplies an installation diagram with the equipment. This diagram will give wire and fuse sizes that are recommended by the manufacturer and should be followed.

REVIEW QUESTIONS

1. What is the purpose of an air-conditioning control system?

2. Which of the following components is not in a typical condensing unit?
   a. compressor
   b. condenser
   c. condenser fan motor
   d. blower motor

3. Most modern residential condensing units require a control voltage of ___.
   a. 24 volts
   b. 120 volts
   c. 240 volts
   d. none of the above

4. Safety controls used for a condensing unit would be wired in _____ with the contactor coil.
   a. series
   b. parallel

5. The condenser fan motor and compressor in a residential condensing unit are generally wired in _____.
   a. series
   b. parallel

6. A packaged air-conditioning unit is built _____.
   a. with all components housed in one unit
   b. with the evaporator fan motor in a separate compartment
   c. with the compressor in a separate compartment
   d. none of the above

7. In which of the following applications would a water-cooled packaged unit most likely be used?
   a. residential
   b. commercial
   c. industrial
   d. commercial and industrial

8. What is a gas pack?

9. What is the major difference between the controls on a small residential unit and a large commercial condensing unit?

10. True or False: The simplest control system used on a condensing unit is a contactor.

11. An internal relief valve is sometimes used in a hermetic compressor in place of a(n) _____.
   a. low-pressure switch
   b. internal thermostat
   c. high-pressure switch
   d. none of the above

12. What is the purpose of the antishort-cycling device?

13. Briefly explain the operation of a condensing unit.

14. Draw a wiring diagram of a simple residential air-cooled condensing unit with a contactor, condenser fan motor, and compressor with starting components and the control voltage being supplied from the furnace.

15. What is a split air-conditioning system?

16. What would be the reason for using a 120-volt control system in a large condensing unit?

17. An air-cooled packaged unit with a remote condenser is usually mounted _____ with the condenser located _____.

   a. outside/outside
   b. outside/inside
   c. inside/outside
   d. inside/inside

18. Line voltage control systems are used with what type of packaged units?

19. Why is it necessary to interlock the cooling tower pump and fan motor into the control circuit of a water-cooled packaged unit?

20. What is the major difference between the water-cooled packaged unit and the air-cooled packaged unit?

21. Field wiring is installed by _____.
   a. assembly line personnel
   b. installation technicians
   c. service technicians
   d. all of the above

22. On an air-conditioning split system installation, the two parts of the system that require power connections are the _____ and _____.
   a. furnace and condensing unit
   b. indoor blower motor and condensing unit
   c. fan coil unit and condensing unit
   d. all of the above

23. True or False: Commercial and industrial systems usually use electric resistance heat.

24. What are two major considerations an installation technician must take into account when installing air-conditioning power wiring?

25. What type of control system is used on most residential and small commercial equipment?
    a. line voltage
    b. low voltage
    c. both a and b

26. What is the purpose of the indoor fan relay package?

27. The indoor fan relay package contains the _____.
    a. indoor fan relay and transformer
    b. compressor contactor and transformer
    c. power supply for the furnace
    d. none of the above

28. Approximately _____% of the problems in the residential air-conditioning industry can be traced to some electrical problem.
    a. 50
    b. 75
    c. 85
    d. a minimal

29. What is the purpose of control wiring?

30. How many power supplies would be required for a split system using electric resistance heat as the heating source?
    a. 1
    b. 2

LAB MANUAL REFERENCE

For experiments and activities dealing with material covered in this chapter, refer to Chapter 15 in the Lab Manual.

16

Control Systems Circuitry

OBJECTIVES

After completing this chapter, you will be able to

- Understand basic control circuits, including compressor, evaporator fan motor, condenser fan motor, and safety control circuits.
- Understand the control circuitry used in residential applications.
- Understand the basic circuitry of control systems used on light commercial and commercial and industrial applications.
- Identify the method of control for commercial and industrial systems.

KEY TERMS

Capacity control   Interlock
Electric control system   Pneumatic control system
Electronic control system   Pump-down control system
Head pressure   Water chiller

INTRODUCTION

Control systems used in the heating, cooling, and refrigeration industry are designed to control the electric loads of a system to maintain the desired temperature of a given area or medium. Control systems must contain the necessary safety components to ensure safe operation of the equipment.

Many different types of control systems are used in the industry today, but most systems have certain components and circuits in common, depending on the control voltage and the type of equipment. Low-voltage control circuits are different from line voltage control circuits because of the devices that can be used with each type of system. The basic control system contains several components common to all control systems, such as the compressor, the evaporator fan motor, and the condenser fan motor, along with their controlling devices.

It is essential that installation and service technicians understand the operation of the basic circuits of control systems to be able to install or troubleshoot the smaller air-conditioning and heating systems that are common to the industry. Not all equipment will have the same types of controls, and there are different methods of controlling certain devices. In this chapter we will be looking at basic control system circuits and procedures for troubleshooting control systems.

16.1 BASIC CONTROL CIRCUITS

All control systems have certain circuits and components in common. Most air-conditioning systems have a contactor to start and stop the compressor. The compressor is the largest electric load in an air-conditioning system. Therefore, it is necessary to use a contactor or starter to control its ampacity.

The control of the condenser fan motor is common on many systems, but it does vary due to the new high-efficiency equipment that is being produced. Circuits controlling the evaporator fan motor are common because of the use of an evaporator fan relay, but on heating systems different methods are used. The safety devices used on control systems vary greatly, but they are usually connected into the control circuit in series.

Now let us look at these control system components in detail.

Compressor Control Circuits

The first basic circuit of almost any residential or small commercial air-conditioning control system is the device that starts and stops the compressor. On a low-voltage control system, the thermostat opens or closes to energize or de-energize the contactor, which starts or stops the compressor, as shown in Figure 16.1. This circuit is common to almost all residential and small commercial air-conditioning systems in operation today.

The condenser fan motor is usually connected directly to the contactor to ensure that it is operating whenever the compressor is in operation, as shown in Figure 16.1.

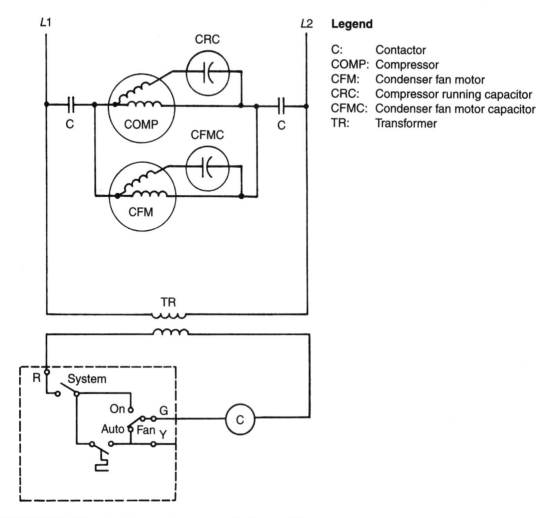

FIGURE 16.1 Schematic diagram of a contactor circuit controlling a compressor, showing the connection of the condenser fan motor

Evaporator Fan Motor Control Circuits

Evaporator fan motor circuits are almost alike in the smaller ranges of equipment. The evaporator fan motor operates from a relay controlled by the thermostat. If the thermostat fan switch is set on the "on" position, the indoor fan relay will be energized by the cooling function of the thermostat. Figure 16.2 shows the indoor fan relay with the connections of the fan motor and the thermostat. Notice the fan switch on the thermostat and note how the indoor fan is cut on or off by the use of the indoor fan relay. This control system is common only to air-conditioning systems. On heating installations, other controls are needed to start the fan. Figure 16.3 shows the evaporator fan circuits in a small packaged air conditioner.

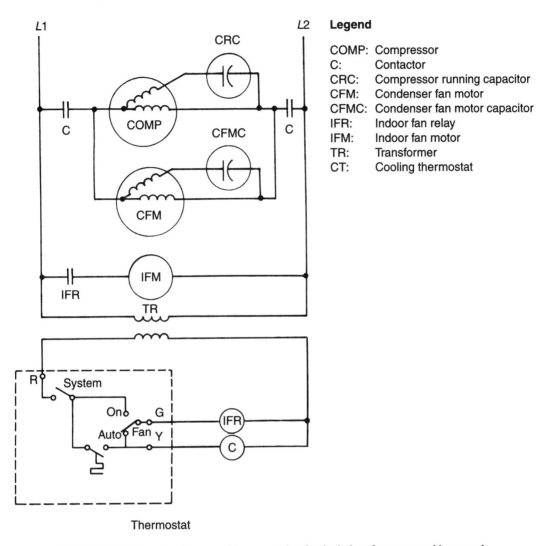

FIGURE 16.2 Schematic diagram of the connection for the indoor fan motor and its controls

Condenser Fan Motor Control Circuits

Most condenser fan motors on residential air-conditioning systems are controlled by the contactor because of the simplicity and the economy of using one control. Some condenser fan motors are cycled on and off through an outdoor thermostat set at a temperature that will maintain a constant head pressure. This allows better system efficiency. This type of control is accomplished on residential and small commercial systems by inserting the thermostat in series with the condenser fan motor. Other methods are used in the larger commercial and industrial applications.

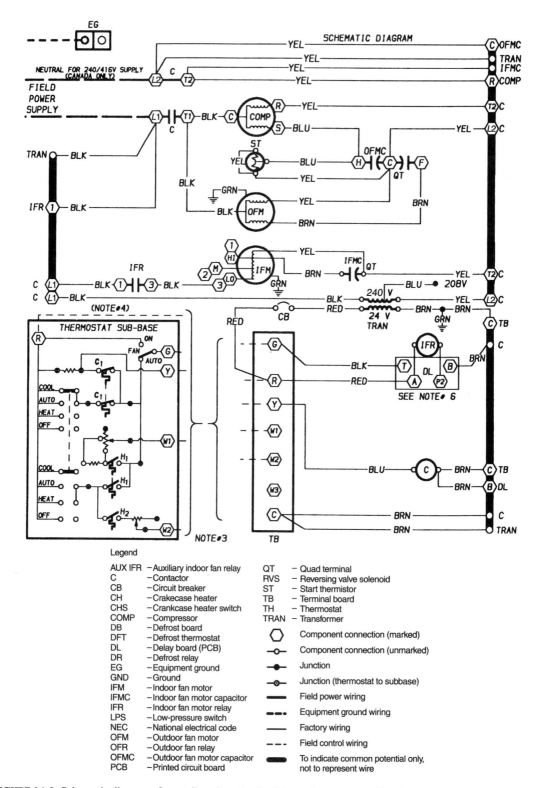

FIGURE 16.3 Schematic diagram of a small packaged unit with one fan motor used as the evaporator and the other as the condenser fan motor. *(Courtesy of Carrier Corporation)*

Safety Control Circuits

Most safety controls used in residential and small commercial systems are connected in series with the contactor coil. The only exception is the internal thermostat in a compressor, which breaks the power wires inside the compressor. Any other safety devices used in the control system, such as a high-pressure switch or a low-pressure switch, are connected in series with the contactor coil. The series connection ensures that in the case of an unsafe condition, the compressor will be de-energized.

Figure 16.4 shows a high-pressure and a low-pressure switch in the control circuit. Any additional safety devices would be connected in the same

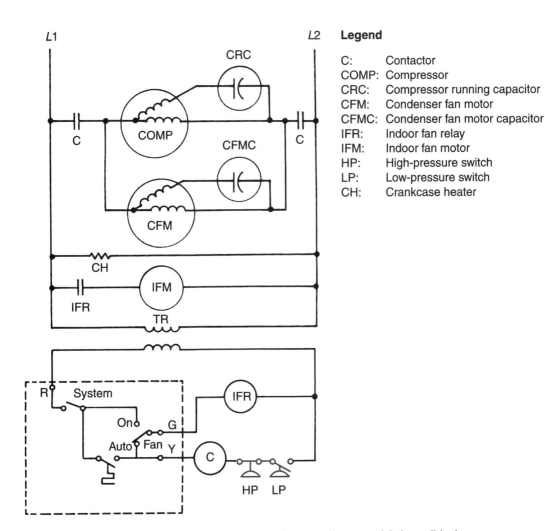

Legend	
C:	Contactor
COMP:	Compressor
CRC:	Compressor running capacitor
CFM:	Condenser fan motor
CFMC:	Condenser fan motor capacitor
IFR:	Indoor fan relay
IFM:	Indoor fan motor
HP:	High-pressure switch
LP:	Low-pressure switch
CH:	Crankcase heater

FIGURE 16.4 Total control circuit of a residential or small commercial air-conditioning system

manner. The internal overload in the compressor is connected in series with the common terminal of the compressor so the compressor will be de-energized when an overload occurs.

16.2 TOTAL CONTROL SYSTEM OF RESIDENTIAL UNITS

The total control circuit of a residential or small commercial air-conditioning system is shown in Figure 16.4. Each circuit can be discussed separately as long as each part in the circuit is complete in the total operation of the control system.

The thermostat is a switching point of the low-voltage supply to the necessary components. Figure 16.5 shows the schematic of the thermostat. From the schematic it can be seen that the indoor fan motor can operate continuously or cycle with the cooling thermostat, depending on the position of the fan function switch. The remainder of the thermostat operates the heating or cooling or both. The cooling thermostat will close or open to the correct temperature, as well as control the heating by the heating thermostat.

Figure 16.3 shows that the compressor and in most cases the condenser fan motor operate when the contactor is energized. The contactor is energized by the cooling thermostat until the temperature is decreased to the desired range. The evaporator fan motor is energized on the cooling cycle by the indoor fan relay and on the heating cycle by a fan switch contained in the furnace.

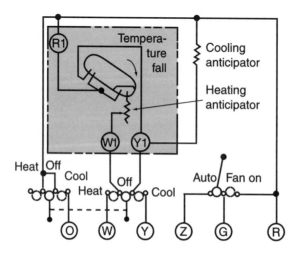

FIGURE 16.5 Schematic of a thermostat for a total control system (*Courtesy of Honeywell, Inc.*)

As can be seen in Figure 16.4, the safety devices are connected in series with the contactor coil, with the exception of the internal overload in the compressor, which is in series with the compressor common terminal. In some cases, manufacturers use a crankcase heater that is energized at all times to help keep refrigerant from migrating into the compressor crankcase during the off cycle where it could condense into a liquid and mix with oil, resulting in oil foaming and liquid slugging. This control system is commonly used on residential and small commercial air-conditioning and heating systems.

16.3 ADVANCED CONTROL SYSTEMS

All types and designs of control systems are used in commercial and industrial air-conditioning, heating, or refrigeration systems. The control systems used in equipment from 7.5 to 100 tons cannot afford to be as simple and use as few components as the smaller systems. This type of control system must ensure that the electric loads are operating in a safe manner because of their greater cost. The control of this type of equipment must also be more complex than the smaller systems because of its size and operation.

Most large systems use some method of capacity control because of the variations in the load in the conditioned areas. When **capacity control** is used on these larger air-conditioning systems, it usually involves a more complex control system. Also, most large air-conditioning and heating systems have several important control circuits that prevent short-cycling and operation of the equipment under conditions that would cause damage. Thus, we see that large equipment uses many more components than small equipment because of size, cost, capacity control, and general design of the larger systems.

Compressor Motor Controls

It would be impossible to cover all the control circuits used in the large equipment. However, the more common control circuits will be covered in detail.

There are two methods of energizing the compressor on the large systems, part winding and across the line. The contactor or magnetic starter is used to energize the compressor on all systems. An across-the-line type of motor starting requires only a single contactor, as shown in Figure 16.6(a). However, a part winding motor would require two contactors along with a time-delay relay. The part winding motor is used to allow the compressor

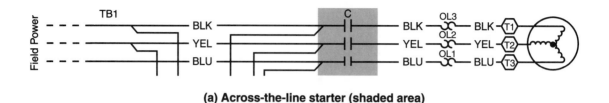

(a) Across-the-line starter (shaded area)

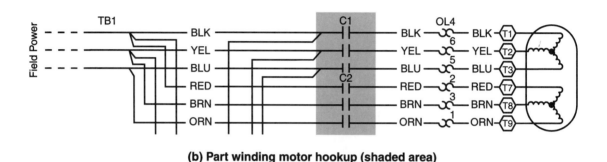

(b) Part winding motor hookup (shaded area)

FIGURE 16.6 Schematic diagram of an across-the-line motor hookup and part winding motor hookup (shaded area)

to start more easily, with less wear and tear. A part winding motor hookup is shown in Figure 16.6(b). The part winding motor is actually broken down into two separate windings, with the first winding energized 1.5 to 3 seconds before the second winding is energized. The purpose of a part winding motor is to reduce the initial starting ampacity of the motor.

Water Chiller Control Systems

Water chillers are refrigeration systems that cool the water pumped into other parts of the system to maintain the desired condition of a specific area. The control on a water chiller will usually come from the chilled water temperature, which will relate the information back to the refrigeration control system. This type of control can also include a pump-down system. The system maintains a certain chilled water temperature at all times. A typical control system for a water chiller is shown in Figure 16.7.

Evaporator Fan Motor Controls

The control of the evaporator fan motor in large commercial and industrial air-conditioning or heating systems comes from a source separate from the condensing unit. Fan motors are generally designed to operate at all times

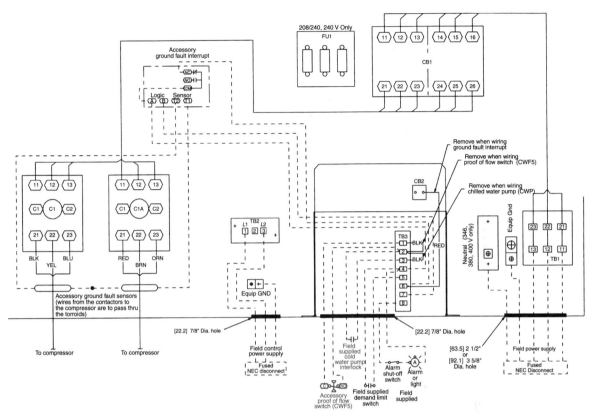

NOTES AND LEGEND TO TYPICAL WIRING SCHEMATIC

NOTES:
1. Factory wiring is in accordance with the NEC. Any field modifications or additions must be in compliance with all applicable codes.
2. Connect separate source of control power from field-supplied fused disconnect to terminal 1 of TB2. Neutral side must be connected to terminal 3 (50 Hz only). This provides power for the unit control circuit, cooler heater and compressor crankcase heater.
3. All field interlock contacts must have minimum rating of 360 va pilot duty plus capacity required for field installed equipment.
4. For internal unit wiring, reference wiring book or unit wiring label diagram.
5. Voltage requirements:

MAIN POWER	CONTROL POWER
208/230-3-60	115-1-60
460-3-60	115-1-60
575-3-60	115-1-60
380-3-60	220-1-60
230-3-50	230-1-50
346-3-50	200/230-1-50
400-3-50	230-1-50

LEGEND

C	—	Contactor
CB	—	Circuit Breaker
CWFS	—	Chilled Water Flow Switch
CWPS	—	Chilled Water Pump Starter Auxiliary Contact
EQUIP	—	Equipment
FU	—	Fuse
GND	—	Ground
NC	—	Normally Closed
NEC	—	National Electrical Code (U.S.A.)
NO	—	Normally Open
TB	—	Terminal Block

---- Field Power Wiring
---- Field Control Wiring
——— Factory Installed Wiring
—·— Field Installed Device

FIGURE 16.7 Typical control system for a water chiller *(Courtesy of Carrier Corporation)*

in the larger systems and are cut on and off by a clock or a system switch. These fan motors are interlocked into the control system of the condensing unit to ensure the operation of the fan motor whenever the condensing unit is operating.

The smaller commercial air-conditioning and heating units usually cut off their fan motors on the "off" cycle of the equipment. However, they are equipped with a fan switch that allows continuous operation.

Various other methods are used to control evaporator fan motors, but none is as common as the constant air flow and the fan cycling with equipment operation.

Condenser Fan Motor Controls

The condenser fan motor used on most commercial and industrial air-conditioning units is usually designed to maintain a constant head pressure by operating a certain number of condenser fans when the outside temperature drops below 75°F. This is accomplished by using thermostats to operate the condenser fan motors at certain outdoor temperatures or by using reverse-acting high-pressure switches to cycle the fans, as shown in Figure 16.8, only when they are needed. The condensing temperature of a condensing unit should be maintained at or around 90°F for proper operation of the air-conditioning equipment.

Water-Cooled Condenser Interlocks

Water-cooled condensing units require an interlock in the condensing unit control circuit. The interlock ensures that the air-conditioning unit does not start unless the cooling tower pump and the cooling tower fan motor are operating. If the air-conditioning equipment were to operate without the pump operating, there could be damage to the compressor because of high head pressure.

Water Chiller Interlocks

Water chillers require an interlock in the chiller control circuit to ensure that the chilled water pump is operating before the water chiller refrigeration unit will operate. This interlock ensures that the water chiller will not accidentally freeze the water in the evaporator, causing a rupture of the evaporator. Figure 16.7 shows two interlock circuits that ensure the operation of the chilled water pump before the chiller can operate. The interlocks

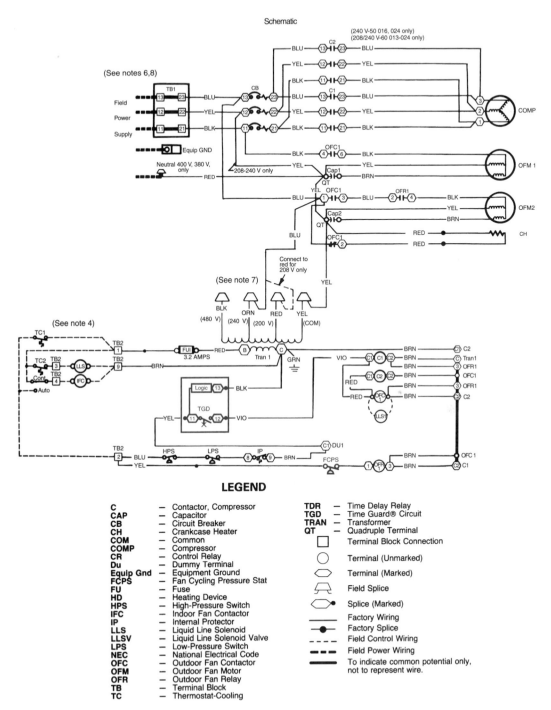

FIGURE 16.8 Schematic diagram of a large commercial and industrial condensing unit *(Courtesy of Carrier Corporation)*

are connected into TB3 (terminal board three) on the diagram and are the chilled water flow switch (CWFS) and the chilled water pump auxiliary contacts (CWP).

If the water chiller were to operate without the chilled water pump operating, there could be danger of the water freezing inside the chiller and causing tremendous damage to the system. With two interlocks in the chiller control circuit it is highly unlikely that the chiller could start without the chilled water pump operating.

Control System for Commercial Condensing Unit

A schematic diagram of a large condensing unit is discussed in this section. The unit is energized by applying 240 volts to the condensing unit. Unlike the small residential units, the control circuit of this unit is completely line voltage. The unit operates on 208 volts–three phase–60 hertz.

The schematic diagram for the unit discussed in this section is shown in Figure 16.8. The unit is started by the energization of the control relay. This relay could be energized by a thermostat, pressure switch, or any other controlling device. The diagram will be discussed in four sections: the timer and timer relay circuits, the compressor circuits, the condenser fan motor circuits, and the safety controls of the circuit.

Anti-Short-Cycling Device Circuits

The control circuit in Figure 16.8 is equipped with a time delay module (TCD). This module prevents the condensing unit from short cycling (rapid starting and stopping), which could occur from a faulty thermostat or possibly a safety control that is opening and closing rapidly. Short cycling of a compressor could damage or overload the compressor motor. The TCD module is an electronic timer that allows for a five-minute delay from the time the unit shuts down to when it starts again. Earlier control systems used a mechanical timer assembly like the Carrier Time Guard® to facilitate this task. The mechanical timers worked a timer relay to accomplish the same purpose, but they were more expensive and harder to troubleshoot. This timing method is one of several used as a protective circuit by the industry to prevent a heavy load from short-cycling.

Compressor Circuits

The compressor circuit is shown in Figure 16.9. The equipment is using a part winding motor to operate the compressor. A part winding motor has

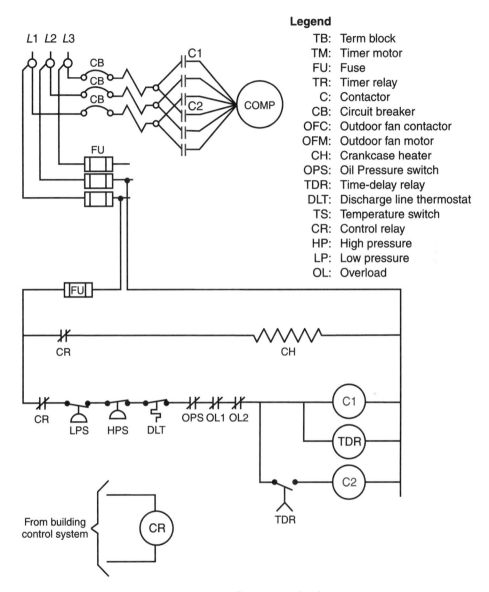

Legend

TB:	Term block
TM:	Timer motor
FU:	Fuse
TR:	Timer relay
C:	Contactor
CB:	Circuit breaker
OFC:	Outdoor fan contactor
OFM:	Outdoor fan motor
CH:	Crankcase heater
OPS:	Oil Pressure switch
TDR:	Time-delay relay
DLT:	Discharge line thermostat
TS:	Temperature switch
CR:	Control relay
HP:	High pressure
LP:	Low pressure
OL:	Overload

FIGURE 16.9 Compressor circuits

two separate motors, which are energized at different times (although close together), built in one housing. The second motor or winding is energized a few seconds after the first motor or winding by a time-delay relay. The motor is supplied by voltage through two contactors, C1 for the first motor or winding and C2 for the second.

The compressor motor circuits are equipped with a circuit breaker to protect the wiring and the motor. The power wiring is then split between the two contactors C1 and C2 that control the compressor. When the contacts of the contactor are closed, the compressor will operate.

The control circuit of the contactor is through the 208 volt–single phase–60 cycle section of the diagram. This includes the CR contacts, safety controls, the contactor coils, the time-delay relay, and crankcase heater. The crankcase heater operates when the normally closed contacts (CR contacts) are closed.

The compressor contactors control the operation of the compressor. When the CR contacts close, if all safety switches are closed, contactor coil C1 and the time-delay relay (TDR) are energized. The time-delay relay closes about one second after C1 has energized, and this energizes C2. The compressor is now operating.

Condenser Fan Motor Circuits

The condenser fan motors OFM1, OFM2, and OFM3 are shown in the circuits of Figure 16.10. The condenser fan motors are all controlled by the separate contactors OFC1, OFC2, and OFC3. The first condenser fan motor contactor OFC1 is energized by the closing of the control relay contacts in circuit 3. When the coil of contactor OFC1 is energized, the condenser fan motor OFM1 is started. The additional two condenser fan motors operate by the temperature surrounding the unit. If the temperature is high enough to close temperature switch TS1, then contactor OFC2 is energized and starts fan motor OFM2. Fan motor OFM3 follows the same operation as fan motor OFM2.

Circuit Safety Controls

The schematic diagram of the condensing unit in Figure 16.10 shows all the safety devices. The safety devices discussed here will not include fuses and circuit breakers.

All safety devices are added in one circuit and are connected between the CR contacts and the contactor coils. If any of the safety devices opens, the contactor coils are de-energized.

The high pressurestat (HP) and low pressurestat (LP) maintain the pressure in the refrigeration system within safe limits and open if the pressure exceeds the safe limit. The discharge line thermostat (DLT) opens to protect the compressor if the temperature of the discharge line exceeds a set temperature. The overload (OL) protects the compressor motor from overloading by opening if an overload occurs.

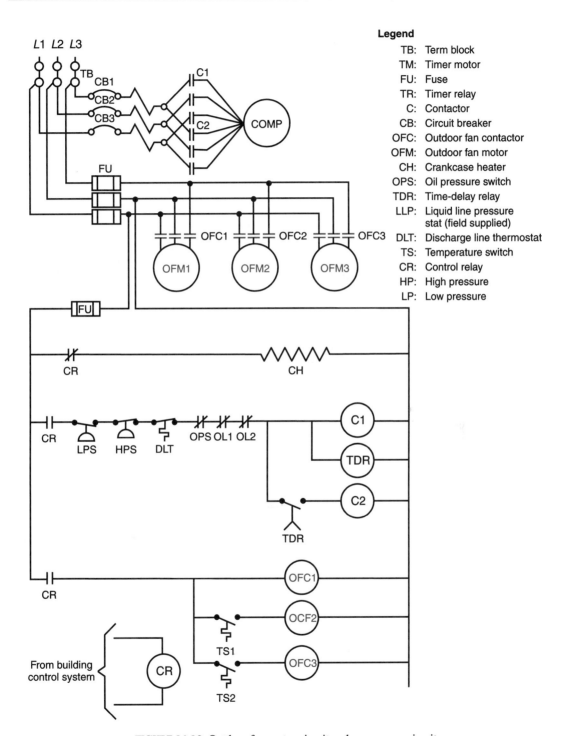

Legend

TB:	Term block
TM:	Timer motor
FU:	Fuse
TR:	Timer relay
C:	Contactor
CB:	Circuit breaker
OFC:	Outdoor fan contactor
OFM:	Outdoor fan motor
CH:	Crankcase heater
OPS:	Oil pressure switch
TDR:	Time-delay relay
LLP:	Liquid line pressure stat (field supplied)
DLT:	Discharge line thermostat
TS:	Temperature switch
CR:	Control relay
HP:	High pressure
LP:	Low pressure

FIGURE 16.10 Outdoor fan motor circuit and compressor circuit

The oil pressure switch (OPS) opens if the oil pressure of the compressor is too low for safe operation. The oil pressure switch that makes up the oil safety switch is controlled by a heater and a pressure switch shown in a circuit from A1 to *L2*. If the oil pressure switch remains closed, the heater heats up an element and opens the switch of the OPS thermostat. If the oil pressure is high enough, the OPS opens and cuts the heater out, allowing the unit to continue.

The schematic diagram covered in this chapter is not the most difficult diagram service technicians may encounter. However, it is a good example of what the service technician will come in contact with in the industry.

16.4 METHODS OF CONTROL ON ADVANCED SYSTEMS

There are many methods of controlling commercial and industrial systems. The total control system is often different from the control system for each piece of equipment, with the exception of the initiation and termination of the operation of the equipment.

The control initiation is different from that of a low-voltage control system because of the size of the contactors and the additional components. These systems sometimes use a low-voltage control system, but it would only be used to energize the main control system.

Most large commercial and industrial systems have a main control system that maintains the control of the total structure. It only initiates and terminates the operation of the equipment to deliver the heating and cooling medium at a desired temperature. The total control system takes care of the remaining conditions. Therefore, the air-conditioning control is accomplished by starting or stopping the equipment. The control system of the equipment takes care of the operation of the equipment.

Most systems are equipped with some method of capacity control, which allows the equipment to operate at from 100% to 33% of its capacity. The smaller commercial and industrial systems usually operate with a standard low-voltage staging thermostat that is capable of capacity control to some extent.

In this section we will discuss some basic control methods used on the larger commercial and industrial systems.

Small Commercial and Industrial Systems

The small commercial and industrial systems often use a simple low-voltage control system to maintain the desired temperature in the structure, with some staging accomplished by the thermostat. However, in many cases low voltage cannot be produced in sufficient strength to adequately close all the contactors or relays in a system. In this case, the low voltage would control a relay to supply 120 volts or 240 volts to the cooling equipment.

Figure 16.11 shows a common control system of a five-ton split air-conditioning system with a fan coil unit. A control relay starts and stops

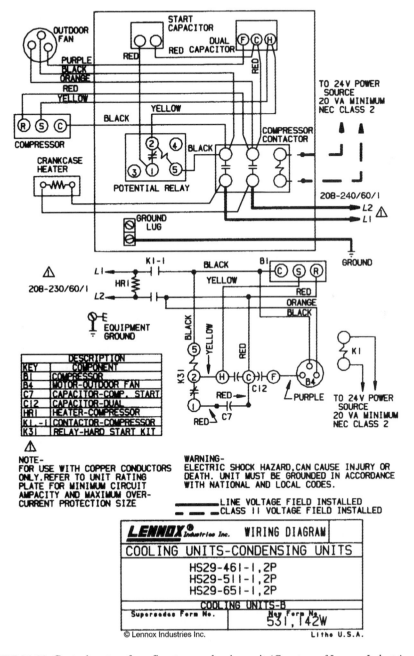

FIGURE 16.11 Control system for a five-ton condensing unit *(Courtesy of Lennox Industries, Inc.)*

the operation of the condensing unit by supplying line voltage to the unit. This control relay is operated by a thermostat. The thermostat also operates the evaporator fan motor by energizing a magnetic starter. After the initiation of the control of the entire system takes place, the condensing unit control system takes over and supervises the operation of the equipment. This control system has all the necessary safety devices for the major load and contains a timer with a relay to prevent rapid short-cycling. The timer unit allows for a five-minute delay between the cycles of the condensing unit.

It should be noted that many control systems will operate completely on low voltage and hence be very similar to the simple residential control systems.

Control Systems for Air-Cooled and Water-Cooled Packaged Units

Many air-cooled and water-cooled packaged units are designed strictly for commercial and industrial use. These packaged units are usually used in a freestanding installation put directly in the area that is to be conditioned, without any ductwork.

This type of unit is equipped with a line voltage control system that is totally housed in the equipment.

The unit has a manual switch that allows it to be operated by this switch and a common line voltage thermostat. The three-position switch has an "off" position, a "fan only" position, and a "cool" position. The "cool" position of the switch is connected in series with a line voltage thermostat. It cuts the compressor on and off as needed by energizing a contactor or magnetic starter. If the unit is large, the evaporator fan motor is also controlled by a magnetic starter. The smaller evaporator fan motors, being single phase, are controlled by the switch.

All the safety devices are connected in series with the coil of the contactor. In most cases, crankcase heaters are used on this type of system and are energized on the "off" cycle of the compressor. Figure 16.12 shows a schematic diagram of a common control system used on this type of equipment.

Pump-Down Control Systems

A large commercial and industrial air-conditioning condensing unit can be controlled by various types of initiation. The most common type of control is a **pump-down control system** that incorporates a solenoid valve in the liquid line of the equipment. The valve opens and closes on the direction of the overall control system of the structure. The action of the solenoid valve

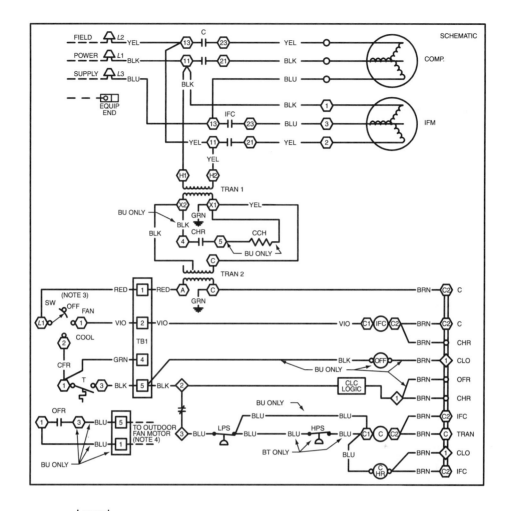

Legend

C	– Compressor contactor		SC	– Start capacitor
CH or CCH	– Crankcase heater		SR	– Start relay
			SW	– Switch
CHR	– Crankcase heater relay		T	– Thermostat
CLO	– Compressor lockout		TB	– Terminal block (board)
COMP	– Compressor		TC	– Thermostat cooling
CR	– Control relay		Tran	– Transformer
DU	– Dummy terminal			
Equip Gnd	– Equipment ground		⏚	Field splice
HPS	– High-pressure switch		☐	Terminal block (board)
IFC	– Indoor-fan contactor		◇	Terminal compressor lockout (CLO)
IFM	– Indoor-fan motor			
IP	– Internal protector		○	Terminal (unmarked)
LLS	– Liquid line solenoid			
LPS	– Low-pressure switch		⬡	Terminal (marked)
OFR	– Outdoor-fan relay		——	Factory wiring
OL	– Overload		- - -	Field wiring
QT	– Quadruple terminal		══	Indicates common potential only; does not represent wire.
RC	– Run capacitor			
S	– Compressor solenoid			

FIGURE 16.12 Schematic diagram of the control system of an air-cooled packaged unit with a remote condenser *(Courtesy of Carrier Corporation)*

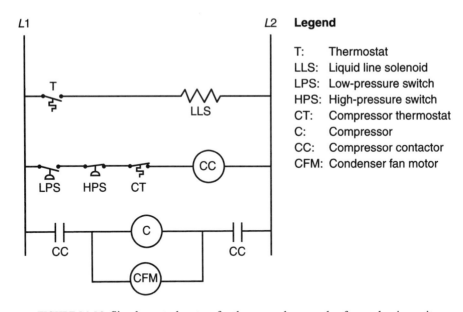

FIGURE 16.13 Simple control system for the pump-down cycle of a condensing unit

determines if the condensing unit is to operate. If the solenoid valve is open, the unit will operate; if the solenoid closes, the unit will continue to operate until the low-pressure switch cuts the compressor off. Figure 16.13 shows a simple pump-down hookup for a condensing unit. The figure shows the solenoid valve being controlled by the thermostat and the action of the condensing unit controlled by the low-pressure switch. Some condensing units use a solenoid relay mounted in the unit to control the solenoid valves.

The pump-down control method is in common use on commercial and industrial systems because it will not allow refrigerant to feed to the coil on the "off" cycle and migrate back to the compressor. The solenoid valve in this type of system can be controlled by almost any type of temperature-sensing device that can be mounted in the desired portion of the system. Or the valve can be controlled by the equipment to prevent it from operating in unsafe conditions.

16.5 TOTAL COMMERCIAL AND INDUSTRIAL CONTROL SYSTEMS

The commercial and industrial control system is complex because it controls virtually the entire structure. It controls the zones of the structure, the air-conditioning equipment, the heating equipment, the evaporator fan motor, and the temperature of the entire structure. These control systems are usually designed and installed by one of the major control manufacturers.

The controls of a large system can be electric, pneumatic, or electronic. The **electric control system** is not used as often as the pneumatic and electronic because the devices must be larger and they are less accurate than the others.

The **pneumatic control system** is operated by air pressure, which opens and closes switches and valves. An example of a pneumatic control system is shown in Figure 16.14.

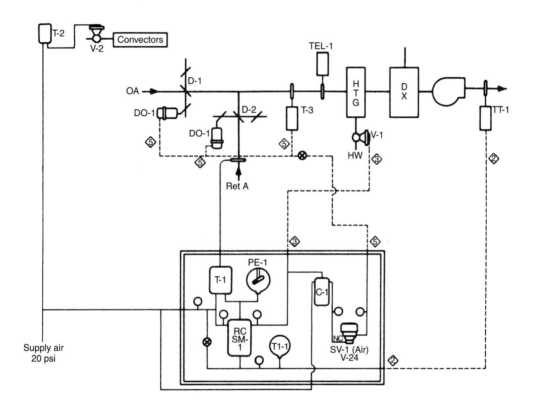

Legend

Ret A:	Return air	T-2:	Room thermostat
OA:	Outdoor air	T1-1:	Discharge air thermometer
C-1:	Minimum position controller	V-1:	Hot water valve
DO-1:	Outdoor air damper controller	V-2:	Room hot water valve
DO-2:	Return air damper controller		(convector)
D-1:	Outdoor air damper	TT-1:	Discharge air thermostat
D-2:	Return air damper	SV-1:	Air solenoid valve (outside
PE-1:	Cooling controller		air)
T-3:	Mixed air thermostat	TEL-1:	Freeze protector
RSCM-1:	Receiving controller	-----	Air lines field installed
T-1:	Return air thermostat	——	Supply air lines

FIGURE 16.14 Pneumatic control system *(Courtesy of Johnson Controls, Inc.)*

The **electronic control system** is shown in Figure 16.15. It controls the temperature in the zones by electronic devices.

These diagrams were shown here just as examples of some of the control systems in use today. Their circuitry will not be discussed.

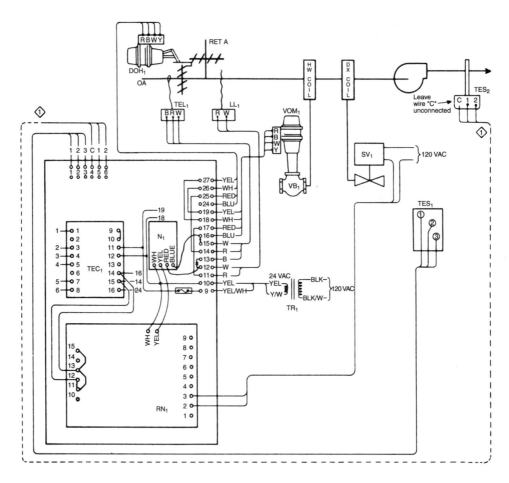

Legend

DOH:	Outdoor air damper	TR:	Transformer
TEL:	Outdoor air thermostat	TEC:	Main controller
LL:	Low-limit thermostat	N:	Outdoor air-adjustment relay
VB:	Hot-water valve	RN:	Staging relay
VOH:	Valve operator	TES$_1$:	Room sensor
SV:	Cooling solenoid valve	TES$_2$:	Discharge air sensor

FIGURE 16.15 Electronic control system *(Courtesy of Johnson Controls, Inc.)*

SUMMARY

There are certain similarities among control systems used on the smaller heating and air-conditioning equipment. The contactor in most cases will control the compressor and the condenser fan motor. In some cases, the condenser fan motors are controlled by a thermostat to maintain a constant head pressure in the system, but usually a contactor controls a two-speed fan motor. The evaporator fan motors are usually controlled by some temperature on the heating cycle and an indoor fan relay on the cooling cycle. The safety components are in series with the contactor controlling the compressor because it is the largest, most important, and most expensive load in the heating and air-conditioning system. From a familiarity with the common basic control circuits, an installation or service technician can easily adapt to the different control systems used in the industry. Usually some common type of control is used in all control circuits.

The advanced control system required on the larger commercial and industrial equipment is more complex and sophisticated than the smaller equipment controls because of its size and operation. In many large structures, the main control system signals for the cooling or heating equipment to operate. Once the equipment has been started, the control system of the equipment supervises the operation of the unit, including the capacity control.

There are basically two methods used to start condensing units and air-conditioning equipment. The pump-down method uses a solenoid valve to start and stop the flow of refrigerant to the evaporator. The other method used to start the equipment is by supplying a certain voltage to the equipment when it needs to be started. Large control systems have many basic circuits that are common to equipment throughout the industry.

REVIEW QUESTIONS

1. Briefly explain the operation of the unit in Figure 16.16.

2. Briefly explain the operation of the unit in Figure 16.17.

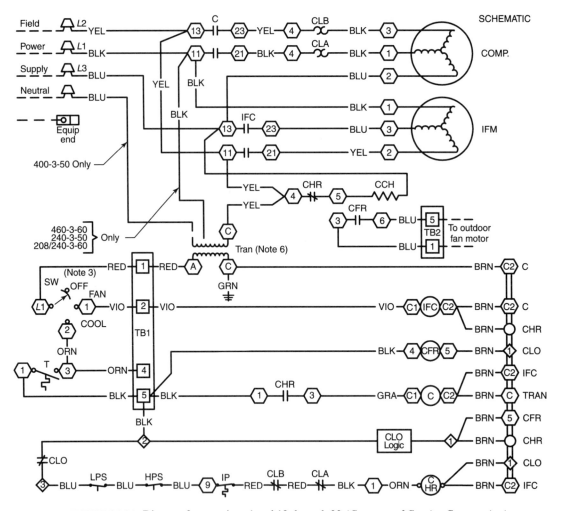

FIGURE 16.16 Diagram for questions 1 and 19 through 22 *(Courtesy of Carrier Corporation)*

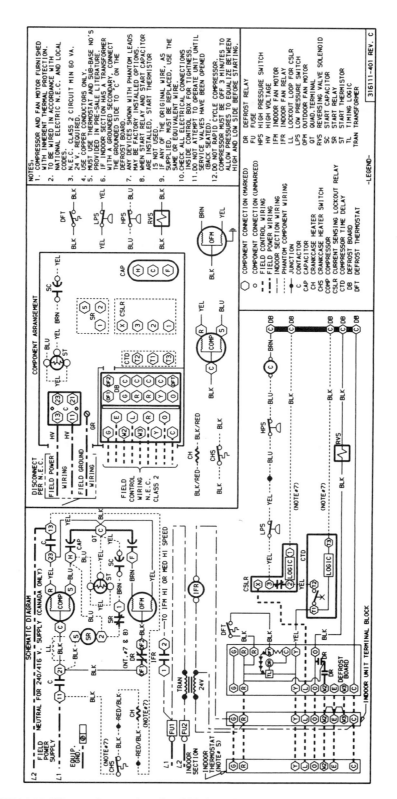

FIGURE 16.17 Diagram for questions 2 and 23 through 26 *(Courtesy of Carrier Corporation)*

3. Which of the following components are common to most automatic control systems?

   a. thermostat, contactor, transformer
   b. thermostat, electric motor, contactor
   c. thermostat, oil safety switch, transformer
   d. thermostat, pressure switch, contactor

4. True or False: Evaporator fan motors used in small residential units are alike.

5. On most small residential air-conditioning condensing units the condenser fan motor is controlled by the

   _____.

   a. transformer
   b. thermostat
   c. compressor contactor
   d. indoor fan relay

6. True or False: Safety controls used in control systems under 10 tons are connected in parallel with the contactor coil.

7. Which of the following safety components would not be connected in series with the contactor coil?

   a. compressor internal overload
   b. low-pressure switch
   c. high-pressure switch
   d. oil safety switch

8. Which of the following is an advantage of using a pilot duty instead of a line break control?

   a. control cost is less
   b. wiring costs are less

   c. sized for light load
   d. all of the above

9. Why would the control system for a 25-ton light commercial air-cooled condensing unit be simpler that the control system for a 3-ton residential air-cooled condensing unit?

10. What are the two methods used to energize a compressor motor on a large condensing unit?

    a. part winding and full winding
    b. full winding and load winding
    c. part winding and across-the-line winding
    d. across-the-line winding and full winding

11. True or False: In the larger systems, many fan motors are designed to operate at all times.

12. What is the approximate condensing temperature for proper operation of a condensing unit?

    a. 50°F
    b. 70°F
    c. 90°F
    d. 110°F

13. What are some methods used to maintain a constant discharge pressure in a light commercial air-cooled condensing unit?

14. What is the purpose of an interlock between the compressor and cooling tower pump in a unit with a water-cooled condensing unit?

15. What is a pump-down control system? How does it operate?

16. The control system used in a commercial and industrial structure can be

    _____,

    _____, and

    _____.

17. True or False: The compressor is the most expensive component of an air-cooled condensing unit.

18. Which of the following should be the first check made by a technician when troubleshooting a small residential condensing unit?

    a. line voltage power supply
    b. open internal overload
    c. open low-pressure switch
    d. contactor coil

Answer the following questions from Figure 16.16.

19. The control system of the unit in the diagram is _____.

    a. line voltage
    b. low voltage

20. The indoor fan operates when _____.

    a. the thermostat closes
    b. the unit is turned on
    c. the disconnect is closed
    d. OFR is closed

21. The compressor motor is _____.

    a. single-phase
    b. three-phase

22. How many safety components are in series with the contactor coil?

    a. 2
    b. 3
    c. 4
    d. 5

Answer the following questions from Figure 16.17.

23. The defrost cycle of the heat pump can be terminated by which of the following components?

    a. timer logic
    b. defrost thermostat
    c. both a and b
    d. contactor

24. The control system of the unit in the diagram is _____.

    a. line voltage
    b. low voltage

25. What is the purpose of ST?

26. What controls the crankcase heater?

LAB MANUAL REFERENCE

For experiments and activities dealing with material covered in this chapter, refer to Chapter 16 in the Lab Manual.

17

Troubleshooting Modern Refrigeration, Heating, and Air-Conditioning Control Circuitry and Systems

OBJECTIVES

After completing this chapter, you should be able to

- Use the proper safety procedures when troubleshooting HVAC control systems.
- Determine and use the correct electrical instrument to check the electrical characteristics (potential, current, and resistance) in an HVAC electrical system.
- Troubleshoot any electrical component in an HVAC electrical system.
- Isolate electrical circuits that are operating incorrectly by reading electrical wiring diagrams and using electrical meters.
- Troubleshoot a line voltage control system.
- Troubleshoot a residential packaged unit.
- Troubleshoot a residential gas heating and electric air conditioning split system.
- Troubleshoot a heat pump.

KEY TERMS

Current-sensing lockout relay
Electronic self-diagnostic feature
Fault isolation diagram
Hopscotching

Installation and service instructions
Troubleshooting chart
Troubleshooting tree

INTRODUCTION

A large percentage of the problems that occur in refrigeration, heating, and air-conditioning equipment and systems are electrical problems. The electrical components in most control systems number from a few to many, depending upon the size and complexity of the equipment and control system. It is the job of the service technician to diagnose and repair the problem with the equipment in a timely manner no matter how simple or complex the electrical system may be.

The service technician will be required to properly diagnose problems in electrical components and electrical circuits that make up the electrical system. Troubleshooting procedures include diagnosing the condition of electrical components. Troubleshooting electrical components individually has been covered in earlier chapters of this text. Common electrical loads that are found in most refrigeration, heating, or air-conditioning systems include motors, heaters, solenoids, signal lights, and others. Common electrical switches used in the industry include thermostats, pressure switches, manual switches, and other miscellaneous switches. Contactors and relays are often used to control loads; these contain switches or contacts that close when a solenoid coil is energized. Transformers are used to transform the incoming voltage to the desired control voltage, 24 volts in most cases. It is the technician's responsibility to locate the suspected faulty component and determine its condition.

The two most important tools available to technicians are electrical meters and schematic wiring diagrams. A technician must know how to interpret schematics in order to know how the unit should operate. Schematic diagrams make the technician's job easier because the electrical system is broken down into a circuit-by-circuit arrangement. The technician must know how to correctly read electrical test instruments in order to measure the voltage being supplied to a unit, a circuit, or an electrical component. The current draw of an electrical load can determine if an electrical load is operating properly; the current draw is measured in amperes. When troubleshooting electrical components, technicians will often have to check the resistance to determine its condition; the resistance is measured in ohms.

Technicians will in most cases develop their own troubleshooting procedures after several years' experience. The more experience technicians have in troubleshooting circuits and equipment, the better and faster they become. As simple as it sounds, the process of elimination is used on many service calls. Many devices can be eliminated from consideration as the problem by observing how the system or component operates.

Manufacturers' troubleshooting charts and tables are available for most equipment used in the industry. Many of these troubleshooting charts or tables come packed with the equipment installation instructions. The troubleshooting chart or table breaks down the troubleshooting procedure step-by-step. Many motor manufacturers have developed troubleshooting tables for motors.

Technicians must develop a troubleshooting method that leads them to a particular circuit and to the component that is at fault. Hopscotching is one such method; it is used to determine which component in a circuit is open and preventing the load from operating properly. This method is especially helpful when circuits include many safety components. Hopscotching is discussed in more detail later in this chapter.

To sum up, technicians must diagnose what the unit is doing, isolate the problem circuit, and repair the problem. To do this they must know what the equipment should do and what the equipment is doing. Along with diagnosing problems and repairing the equipment, technicians must develop safe working habits and good customer skills.

17.1 DIAGNOSIS OF ELECTRICAL COMPONENTS

The diagnosis of electrical components is a fairly easy task for the technician who understands the device. Service technicians often assume the role of parts changers, instead of actually diagnosing each component that might be faulty. A good service technician will change some electrical components that are perfectly good, but this should happen only rarely.

Electrical problems are very common throughout the industry and are said to cause 80% to 90% of the reported system failures. Thus, it is important for service technicians to be able to diagnose and correct electrical problems in a system.

Motors

The electric motor is the largest load used in most heating, cooling, and refrigeration systems. The compressor motor is the most expensive component of the system, and care should be taken to correctly diagnose this component because of the expense that can be needlessly incurred with a faulty diagnosis. In fact, all electric motors are expensive and should be accurately diagnosed. For details about electric motors, refer again to Chapters 9 and 10.

Switches

Electrical switches are used in every piece of equipment and control system that is installed in the industry today. Many types of automatic switches are used in the industry, from simple manual switches to thermostats that have several sets of contacts. It would be impossible to provide adequate safety and temperature or humidity control without some type of switch to determine when the equipment needs to operate and cut off.

Manual switches are basic devices that are used to turn the power on or off to an appliances. Examples include disconnect switches and on/off switches on some appliances.

Thermostats and pressure switches are controlled by the temperature of the medium surrounding them or by the pressure at the point where they are connected to the system. Simple thermostats or pressure switches can easily be diagnosed because they usually have only one set of contacts. Before blaming a thermostat or pressure switch for a problem, the technician must be absolutely certain that the problem is not due to some other device malfunctioning. In other words, the technician must know the function of the switch in the system.

The low-voltage thermostat used on residential and light commercial conditioned air systems is more difficult to diagnose because of its complexity and the fact that it is used for several purposes in a heating and cooling system. The service technician should be careful not to condemn as faulty a thermostat that is merely out of calibration by a few degrees. For details about thermostats and pressure switches, refer again to Chapter 12 and sections 14.4 and 14.5 in Chapter 14.

Contactors and Relays

Electric switching devices such as contactors and relays can cause many electrical problems. These devices are used to control some load in the system. The relay and contactor are basically the same device except for their size and ampacity. These devices must be checked for three possible malfunctions. The contacts, the coil, and the mechanical linkage can all malfunction, rendering the contactor or relay bad. These devices are easily diagnosed for problems if the technician understands their operation and construction. For details about contacts and relays, refer again to Chapter 11 and section 14.2 in Chapter 14.

Other Electrical Devices

Many other electrical components used in control systems, such as transformers, overloads, and others, will not be covered in this review but are covered in previous chapters of this book.

 CAUTION A technician should use extreme caution when working around live circuits.

17.2 TROUBLESHOOTING TOOLS

Service technicians are expected to correctly diagnose and repair most problems to which they are assigned. No matter how good the technician or how hard he or she tries, there will be occasional callbacks because of a misdiagnosis. You can minimize the risk of such embarrassing situations by learning to use all the troubleshooting tools at your disposal.

Electrical Meters

The most important tools that the technician has for troubleshooting are electrical meters. Meters commonly used in the industry are the volt-ohm meter and the clamp-on ammeter. The volt-ohm meter is used to measure the voltage and resistance of an electrical circuit and the milliamperes in certain diagnostic testing of electronic controls. It can be used to read the voltage being supplied to a piece of equipment or load or to check the resistance of a circuit or component to determine its condition. Clamp-on ammeters are used to determine if an electrical load is operating within an acceptable current flow. The current draw of a load can determine if it is operating in an unsafe condition. This information is especially useful when troubleshooting motors. The technician should never attempt to check voltage when the meter scale is set on resistance.

Electrical Wiring Diagrams

The electrical wiring diagram will be the technician's primary tool in electrical troubleshooting. The schematic diagram, sometimes called a ladder diagram, will be most useful in troubleshooting because it breaks down the equipment wiring or control system into a circuit-by-circuit arrangement. The technician can easily determine what the unit should be doing by read-

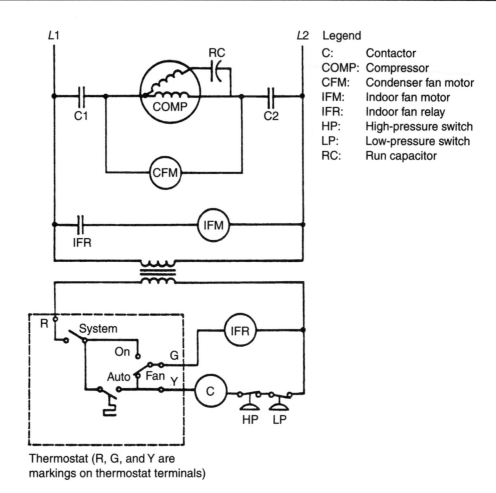

L1   L2

Legend

C:	Contactor
COMP:	Compressor
CFM:	Condenser fan motor
IFM:	Indoor fan motor
IFR:	Indoor fan relay
HP:	High-pressure switch
LP:	Low-pressure switch
RC:	Run capacitor

Thermostat (R, G, and Y are
markings on thermostat terminals)

FIGURE 17.1 Small residential packaged unit

ing the schematic diagram. A schematic diagram for a small residential packaged unit is shown in Figure 17.1.

The pictorial wiring diagram is a picture of the control panel and is sometimes called a component arrangement diagram. This type of diagram helps the service technician to locate and identify an electrical component in a control panel. A pictorial diagram is shown highlighted in Figure 17.2.

Most diagrams have a legend that helps the technician to interpret the abbreviations used in the diagram. A legend for a wiring diagram is shown highlighted in Figure 17.3. Along with the legend, many diagrams have notes that are extremely important to the installation and service of the equipment. The notes of a diagram are shown highlighted in Figure 17.4.

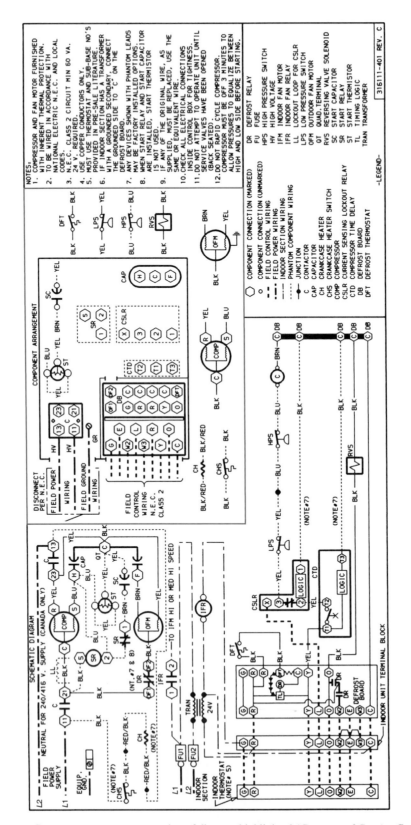

554 FIGURE 17.2 Component arrangement section of diagram highlighted *(Courtesy of Carrier Corporation)*

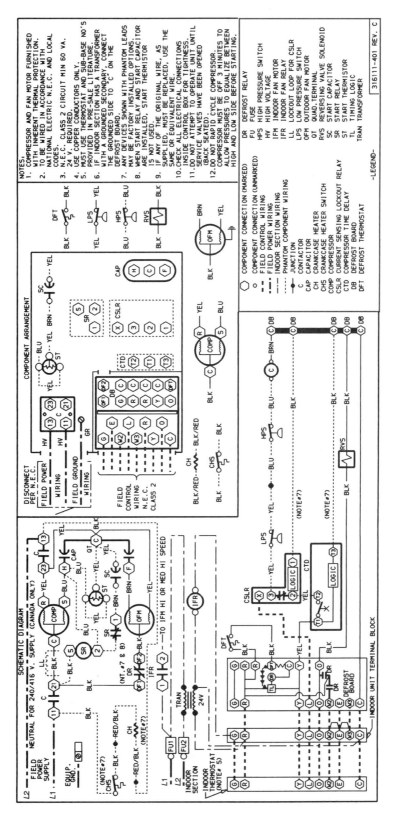

FIGURE 17.3 Legend of diagram highlighted (*Courtesy of Carrier Corporation*)

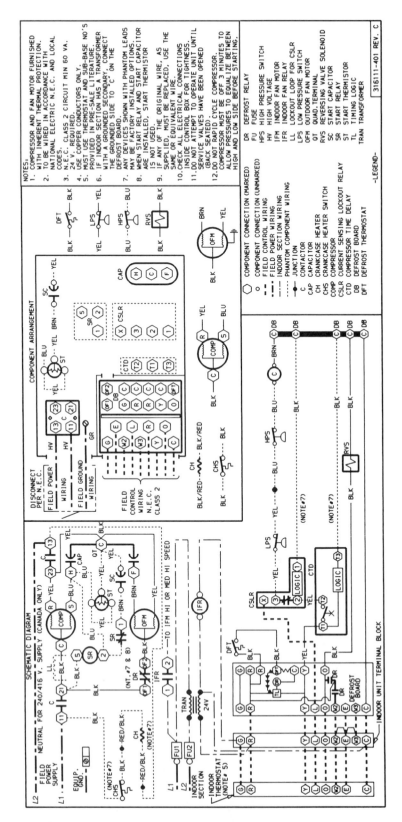

FIGURE 17.4 Notes of a diagram highlighted *(Courtesy of Carrier Corporation)*

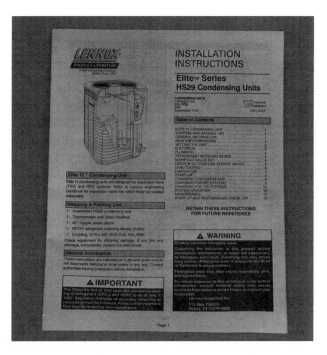

FIGURE 17.5 Instructions shipped with equipment

Installation and Service Instructions

A tool that is often overlooked by many technicians is the set of **installation and service instructions** packed with the equipment at the factory. This package of material is very important because it gives the installation technician the information needed to correctly install the equipment electrically and mechanically. This information may include troubleshooting charts and specific information about using and interpreting the built-in diagnostic test. An assortment of material that is shipped with equipment is shown in Figure 17.5.

Troubleshooting Charts

A **troubleshooting chart** is a graphical representation of a troubleshooting procedure for a particular unit. Some manufacturers make troubleshooting charts available to air-conditioning dealers that handle their line of equipment. Many manufacturers include troubleshooting charts with the installation and service instructions that are included with equipment at the time of shipment. Troubleshooting charts are intended to guide the service technician through a systematic approach to a correct diagnosis and a corrective action that will repair the problem in the system. There are many different types of troubleshooting charts. Figure 17.6

COOLING TROUBLESHOOTING CHART

SYMPTOM	CAUSE	REMEDY
Compressor and condenser fan will not start.	Power failure	Call power company.
	Fuse blown or circuit breaker tripped	Replace fuse or reset circuit breaker.
	Defective thermostat, contactor, transformer, or control relay	Replace component.
	Insufficient line voltage	Determine cause and correct.
	Incorrect or faulty wiring	Check wiring diagram and rewire correctly.
	Thermostat setting too high	Lower thermostat setting below room temperature.
	Single phase units with scroll compressor (555A048,060 and 557A) have a 5-minute time delay	DO NOT bypass this compressor time delay — wait for 5 minutes until time-delay relay is deenergized.
Compressor will not start but condenser fan runs.	Faulty wiring or loose connections in compressor circuit	Check wiring and repair or replace.
	Compressor motor burned out, seized, or internal overload open	Determine cause. Replace compressor.
	Defective run/start capacitor, overload, or start relay	Determine cause and replace.
	One leg of 3-phase power dead	Replace fuse or reset circuit breaker. Determine cause.
Three-phase scroll compressor (555A048,060, and 557A036-060 units only) makes excessive noise, and there may be a low pressure differential	Scroll compressor is rotating in the wrong direction	Correct the direction of rotation by reversing the 3-phase power leads to the unit.
Compressor cycles (other than normally satisfying thermostat).	Refrigerant overcharge or undercharge	Recover refrigerant, evacuate system, and recharge to capacities shown on nameplate.
	Defective compressor	Replace and determine cause.
	Insufficient line voltage	Determine cause and correct.
	Blocked condenser	Determine cause and correct.
	Defective run/start capacitor, overload, or start relay	Determine cause and replace.
	Defective thermostat	Replace thermostat.
	Faulty condenser-fan motor or capacitor	Replace.
	Restriction in refrigerant system	Locate restriction and remove.
Compressor operates continuously.	Dirty air filter	Replace filter.
	Unit undersized for load	Decrease load or increase unit size.
	Thermostat set too low	Reset thermostat.
	Low refrigerant charge	Locate leak, repair, and recharge.
	Leaking valves in compressor	Replace compressor.
	Air in system	Recover refrigerant, evacuate system, and recharge.
	Condenser coil dirty or restricted	Clean coil or remove restriction.
Excessive head pressure.	Dirty air filter	Replace filter.
	Dirty condenser coil	Clean coil.
	Refrigerant overcharged	Recover excess refrigerant.
	Air in system	Recover refrigerant, evacuate system, and recharge.
	Condenser air restricted or air short-cycling	Determine cause and correct.
Head pressure too low.	Low refrigerant charge	Check for leaks, repair and recharge.
	Compressor valves leaking	Replace compressor.
	Restriction in liquid tube	Remove restriction.
Excessive suction pressure.	High heat load	Check for source and eliminate.
	Compressor valves leaking	Replace compressor.
	Refrigerant overcharged	Recover excess refrigerant.
Suction pressure too low.	Dirty air filter	Replace filter.
	Low refrigerant charge	Check for leaks, repair, and recharge.
	Metering device or low side restricted	Remove source of restriction.
	Insufficient evaporator airflow	Increase air quantity. Check filter, and replace if necessary.
	Temperature too low in conditioned area	Reset thermostat.
	Outdoor ambient below 40 F	Install low-ambient kit.
	Field-installed filter-drier restricted	Replace.

FIGURE 17.6 Manufacturer's troubleshooting chart *(Courtesy of Carrier Corporation)*

COOLING TROUBLESHOOTING CHART (cont)

SYMPTOM	CAUSE	REMEDY
Integrated control motor (units 557A048,060 208/240 V), IFM does not run.	Blower wheel not secured to shaft	Properly tighten blower wheel to shaft.
	Insufficient voltage at motor	Determine cause and correct.
	Power connectors not properly seated	Connectors should snap easily; do not force.
Integrated control motor (units 557A048,060 208/240 V) IFM runs when it should be off.	Motor programmed with a delay profile	Allow a few minutes for motor to shut off.
	With thermostat in OFF state, the voltage on G,Y1,Y/Y2,W with respect to common, should be less than ½ of actual low voltage supply	If measured voltage is more than ½, the thermostat is incompatible with motor. If voltage is less than ½, the motor has failed.
Integrated control motor (units 557A048,060 208/240 V) IFM operation is intermittent.	Water dripping into motor	Verify proper drip loops in connector wires.
	Connectors not firmly seated	Gently pull wires individually to be sure they are crimped into the housing.

IFM — Indoor (Evaporator) Fan Motor

FIGURE 17.6 Continued

shows a troubleshooting chart of one manufacturer for a total cooling system. Figure 17.7 shows another type of troubleshooting chart that is for the heating and cooling cycles of a heat pump. Figure 17.8 shows a different type of troubleshooting chart used to troubleshoot an ignition module on a gas furnace. Manufacturers develop troubleshooting charts for almost all models of equipment they produce. If troubleshooting charts are available, it is advisable for service technicians to use them, especially if they are just beginning their careers.

Electronic Self-Diagnostic Feature

Many newer-model appliances contain an **electronic self-diagnostic feature** that helps the technician to isolate malfunctions in system operation. This type of diagnostic tool comes in the form of a light that might blink in a specific sequence or a specific number of times to indicate the possible malfunction of a component. The chart shown in Figure 17.9 is representative of a reference chart for the self-diagnostic check of a unit. This type of troubleshooting tool is becoming increasingly popular as electronic circuitry becomes more cost effective.

HEAT PUMP
TROUBLESHOOTING–HEATING CYCLE

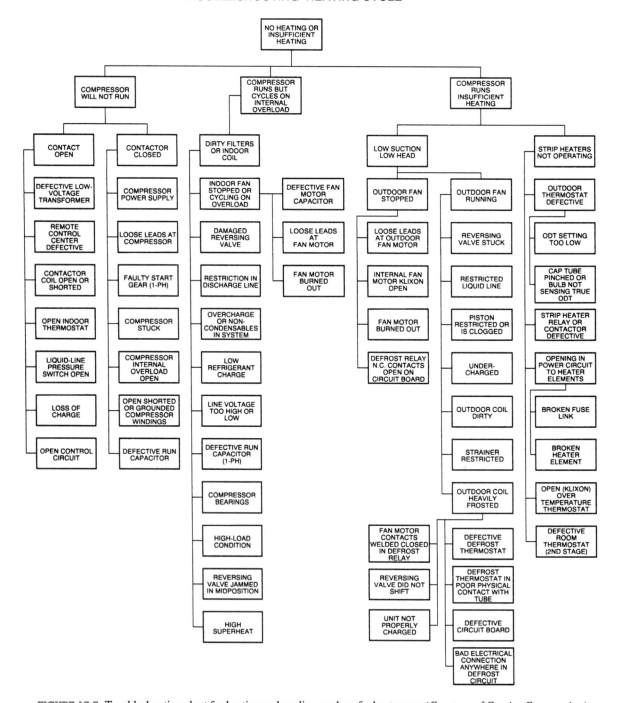

FIGURE 17.7 Troubleshooting chart for heating and cooling cycles of a heat pump *(Courtesy of Carrier Corporation)*

HEAT PUMP
TROUBLESHOOTING—COOLING CYCLE

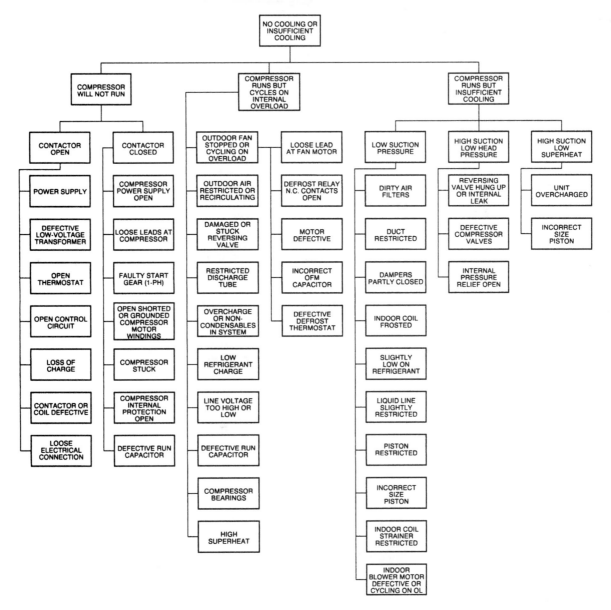

FIGURE 17.7 Continued

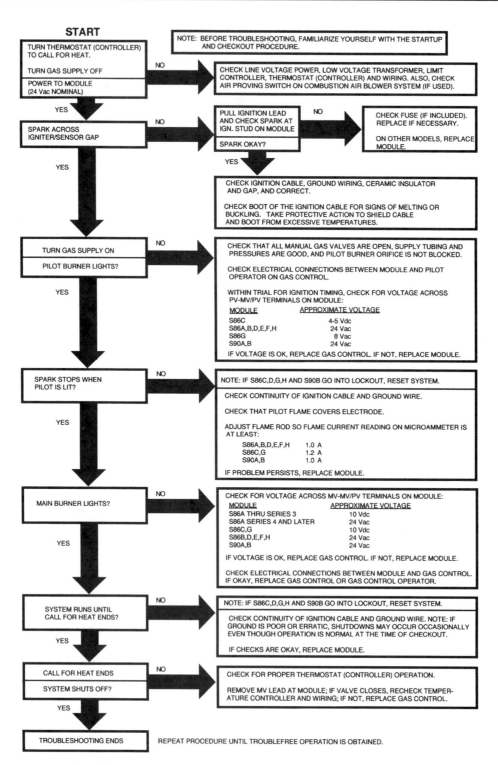

FIGURE 17.8 Troubleshooting chart of an intermittent pilot control system *(Courtesy of Honeywell, Inc.)*

ELECTRIC FAULT CODE	
Fault Light Flashes	**Fault**
2	Indoor Blower
3	Indoor Thermostat
4	Induced Draft Blower
5	Insufficient Ground
6	Fault Igniter
7	No Pilot Established
8	Ignition Module

FIGURE 17.9 Diagnostic codes for gas furnace

17.3 TROUBLESHOOTING WITH ELECTRICAL METERS

Electrical meters are a technician's most important troubleshooting tools. With the price of electrical test equipment at an all-time low, no technician should be without a good volt-ohm meter and a good clamp-on ammeter. Many volt-ohm meters are multifunction and can measure a multitude of electrical characteristics such as volts, ohms, milliamps, microfarads, and amperage with an adapter and checking diodes. The volt-ohm meter, the clamp-on ammeter, and the knowledge to use them are essential to the success of the technician when troubleshooting refrigeration, heating, and air-conditioning equipment and circuits.

 CAUTION Use the proper electrical test equipment for the job being performed.

One of the most important elements to consider when troubleshooting is whether the correct voltage is available to the equipment and control system. This is the first check that a technician should make on a service call. In some cases this can be accomplished by merely turning on an electrical device through a switch, but often the technician will have to use a volt-ohm meter to determine the voltage being supplied to the equipment as shown in Figure 17.10. If electrical energy is available to the equipment, the technician must proceed to a suspicious circuit or electrical component. If the correct voltage is not being supplied to the equipment, the technician must find out why not. If supply voltage is available, the next check in this procedure on a system that has a 24-volt control circuit would be to check

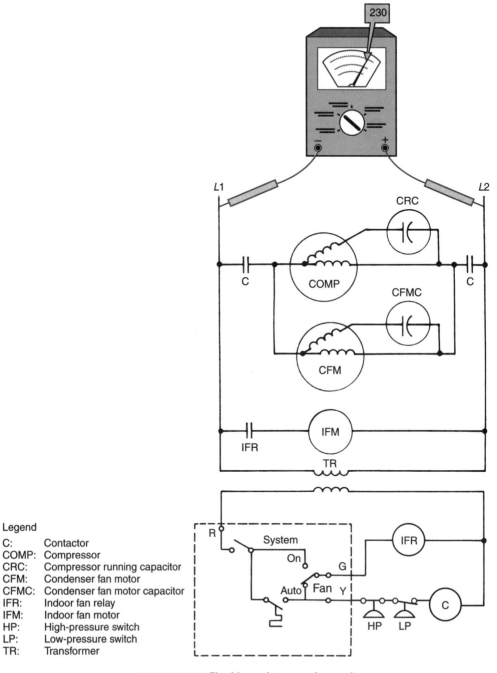

Legend

C: Contactor
COMP: Compressor
CRC: Compressor running capacitor
CFM: Condenser fan motor
CFMC: Condenser fan motor capacitor
IFR: Indoor fan relay
IFM: Indoor fan motor
HP: High-pressure switch
LP: Low-pressure switch
TR: Transformer

FIGURE 17.10 Checking voltage supply to unit

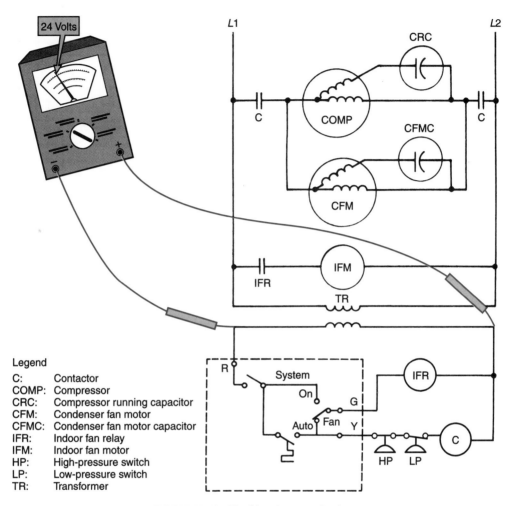

Legend

C:	Contactor
COMP:	Compressor
CRC:	Compressor running capacitor
CFM:	Condenser fan motor
CFMC:	Condenser fan motor capacitor
IFR:	Indoor fan relay
IFM:	Indoor fan motor
HP:	High-pressure switch
LP:	Low-pressure switch
TR:	Transformer

FIGURE 17.11 Checking the control voltage

the control voltage as shown in Figure 17.11. If no control voltage is available and electrical energy is being supplied to the transformer, then the transformer should be checked. If control voltage is available, the technician must then isolate the circuit that he or she thinks is at fault.

For example, upon examination of the equipment the technician finds that the compressor is operating and the condenser fan motor is not. The schematic for this unit is shown in Figure 17.12. Upon examination of the wiring diagram, the technician can assume that voltage is being supplied to the condenser fan if the compressor is operating. Up to this point the technician has determined that (1) the equipment is being supplied with the correct voltage, (2) the control system is being supplied with 24 volts,

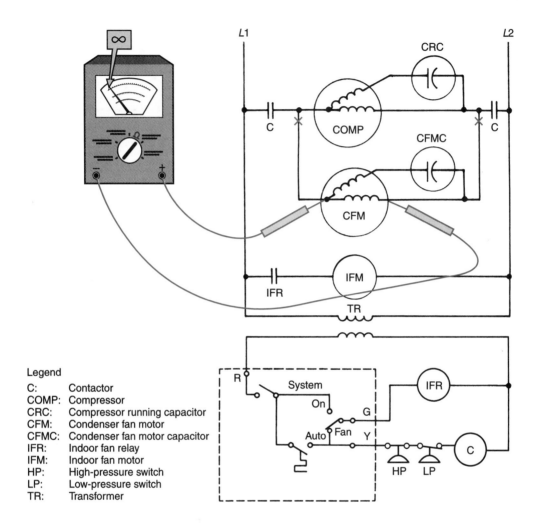

FIGURE 17.12 Checking the resistance of condenser fan motor

(3) the compressor is being supplied with 240 volts because it is operating, and (4) unless there is a broken wire, the condenser fan motor is being supplied with the correct voltage.

Once the technician has determined that the correct voltage is being supplied to the circuit in question and the loads are not operating, he or she must continue to search for the problem. The next step would be to check the condition of the condenser fan motor and condenser fan motor capacitor. The technician should, if possible, touch the condenser fan motor to determine how hot the motor is. If the motor is cool to the touch, then the windings of the condenser fan motor could be open. Chances are if the motor is cool to the touch the motor windings are open. Even though the technician suspects that the motor is bad, he or she should check resistance of the condenser fan motor windings with an ohmmeter to be sure. The resistance of the condenser fan motor can be easily checked by disconnecting the leads of the motor from the contactor as shown in Figure 17.12. If the resistance reading of the fan motor is infinite ohms, the motor windings are open and the motor should be replaced. The resistance check of the condenser fan motor could also turn up a shorted motor winding, but this is not likely because a shorted motor would cause a circuit overload.

If the motor is warm to the touch, the technician must approach the problem with another procedure. The technician will have to determine whether the motor or the capacitor is at fault. The technician first should see if the condenser fan blade turns freely; if not, the motor has bad bearings and should be replaced. If the blade turns freely and the fan motor runs, the technician should turn the condensing unit on and check the current draw of the condenser fan motor as shown in Figure 17.13. If the current draw is

FIGURE 17.13 Technician reading the current draw of a fan motor

higher than the running amps of the motor, found on the motor nameplate, there is a problem with the motor or the load on the motor is higher than normal. If the motor checks out to be good, then the technician must check the condenser fan motor capacitor with an ohmmeter, a capacitor tester, or a volt-ohm meter with a capacitor tester function. If the capacitor checks out faulty, then the capacitor should be replaced.

This procedure is only an example of the method that a technician would use to check a problem with an air-conditioning system. The importance of technicians' being able to correctly use the test instruments that are available cannot be overemphasized.

17.4 USING TROUBLESHOOTING CHARTS

Troubleshooting charts are designed to guide the technician to the correct diagnosis and corrective action, but in some cases, can be misleading. There are several different types of troubleshooting tables or charts that are used in the industry. Figure 17.6 shows a troubleshooting table or chart that gives the symptoms in the left-hand column, the cause in the center column, and the remedy in the right-hand column. For example, follow the highlighted area of the troubleshooting chart in Figure 17.14. The technician has determined that the compressor cycles but not by the thermostat. One item that could cause the problem is insufficient line voltage; the technician determines that the line voltage supplying the unit is low. The remedy for the problem is found in the right-hand column: determine cause and correct. Figure 17.7 shows a different type of troubleshooting chart. A problem with a heat pump system set to operate in the cooling cycle is not cooling. In Figure 17.15, the highlighted line would be the procedure used to diagnose the problem. The call is a no cool or insufficient cool situation; the technician checks to determine if the compressor is operating. The compressor is not operating, and the contactor is open. The technician would then follow the "contactor open" column and determine that the contactor coil is receiving 24 volts but not closing. The technician checks the contactor coil with an ohmmeter and determines that it is open. The technician changes the contactor and starts the heat pump. The service call would not be complete until the operation of the system is examined by the technician.

COOLING TROUBLESHOOTING CHART

SYMPTOM	CAUSE	REMEDY
Compressor and condenser fan will not start.	Power failure	Call power company.
	Fuse blown or circuit breaker tripped	Replace fuse or reset circuit breaker.
	Defective thermostat, contactor, transformer, or control relay	Replace component.
	Insufficient line voltage	Determine cause and correct.
	Incorrect or faulty wiring	Check wiring diagram and rewire correctly.
	Thermostat setting too high	Lower thermostat setting below room temperature.
	Single phase units with scroll compressor (555A048,060 and 557A) have a 5-minute time delay	DO NOT bypass this compressor time delay — wait for 5 minutes until time-delay relay is deenergized.
Compressor will not start but condenser fan runs.	Faulty wiring or loose connections in compressor circuit	Check wiring and repair or replace.
	Compressor motor burned out, seized, or internal overload open	Determine cause. Replace compressor.
	Defective run/start capacitor, overload, or start relay	Determine cause and replace.
	One leg of 3-phase power dead	Replace fuse or reset circuit breaker. Determine cause.
Three-phase scroll compressor (555A048,060, and 557A036-060 units only) makes excessive noise, and there may be a low pressure differential	Scroll compressor is rotating in the wrong direction	Correct the direction of rotation by reversing the 3-phase power leads to the unit.
Compressor cycles (other than normally satisfying thermostat).	Refrigerant overcharge or undercharge	Recover refrigerant, evacuate system, and recharge to capacities shown on nameplate.
	Defective compressor	Replace and determine cause.
	Insuffecint line voltage	Determine cause and correct.
	Blocked condenser	Determine cause and correct.
	Defective run/start capacitor, overload, or start relay	Determine cause and replace.
	Defective thermostat	Replace thermostat.
	Faulty condenser-fan motor or capacitor	Replace.
	Restriction in refrigerant system	Locate restriction and remove.
Compressor operates continuously.	Dirty air filter	Replace filter.
	Unit undersized for load	Decrease load or increase unit size.
	Thermostat set too low	Reset thermostat.
	Low refrigerant charge	Locate leak, repair, and recharge.
	Leaking valves in compressor	Replace compressor.
	Air in system	Recover refrigerant, evacuate system, and recharge.
	Condenser coil dirty or restricted	Clean coil or remove restriction.
Excessive head pressure.	Dirty air filter	Replace filter.
	Dirty condenser coil	Clean coil.
	Refrigerant overcharged	Recover excess refrigerant.
	Air in system	Recover refrigerant, evacuate system, and recharge.
	Condenser air restricted or air short-cycling	Determine cause and correct.
Head pressure too low.	Low refrigerant charge	Check for leaks, repair and recharge.
	Compressor valves leaking	Replace compressor.
	Restriction in liquid tube	Remove restriction.
Excessive suction pressure.	High heat load	Check for source and eliminate.
	Compressor valves leaking	Replace compressor.
	Refrigerant overcharged	Recover excess refrigerant.
Suction pressure too low.	Dirty air filter	Replace filter.
	Low refrigerant charge	Check for leaks, repair, and recharge.
	Metering device or low side restricted	Remove source of restriction.
	Insufficient evaporator airflow	Increase air quantity. Check filter, and replace if necessary.
	Temperature too low in conditioned area	Reset thermostat.
	Outdoor ambient below 40 F	Install low-ambient kit.
	Field-installed filter-drier restricted	Replace.

FIGURE 17.14 Using a manufacturer's troubleshooting chart *(Courtesy of Carrier Corporation)*

COOLING TROUBLESHOOTING CHART (cont)

SYMPTOM	CAUSE	REMEDY
Integrated control motor (units 557A048,060 208/240 V), IFM does not run.	Blower wheel not secured to shaft	Properly tighten blower wheel to shaft.
	Insufficient voltage at motor	Determine cause and correct.
	Power connectors not properly seated	Connectors should snap easily; do not force.
Integrated control motor (units 557A048,060 208/240 V) IFM runs when it should be off.	Motor programmed with a delay profile	Allow a few minutes for motor to shut off.
	With thermostat in OFF state, the voltage on G,Y1,Y/Y2,W with respect to common, should be less than ½ of actual low voltage supply	If measured voltage is more than ½, the thermostat is incompatible with motor. If voltage is less than ½, the motor has failed.
Integrated control motor (units 557A048,060 208/240 V) IFM operation is intermittent.	Water dripping into motor	Verify proper drip loops in connector wires.
	Connectors not firmly seated	Gently pull wires individually to be sure they are crimped into the housing.

IFM — Indoor (Evaporator) Fan Motor

FIGURE 17.14. Continued

Another type of troubleshooting chart that is used by the industry is a **troubleshooting tree** or a **fault isolation diagram** such as that shown in Figure 17.8. This type of diagram asks questions that require a response before the technician proceeds to the next step.

The technician closes the thermostat and follows the highlighted procedure on the diagram as shown in Figure 17.16. The technician checks and finds 24 volts to the ignition module and discovers that there is no spark. He or she proceeds to the next step by following the arrow and pulls the ignition lead to check the spark at the ignition stud on the module. There is no spark, so the technician proceeds to the next step and finds that the module has no fuse, so the ignition module must be replaced. This type of troubleshooting chart is available on many of the more complex electrical components and on some pieces of equipment.

HEAT PUMP
TROUBLESHOOTING–HEATING CYCLE

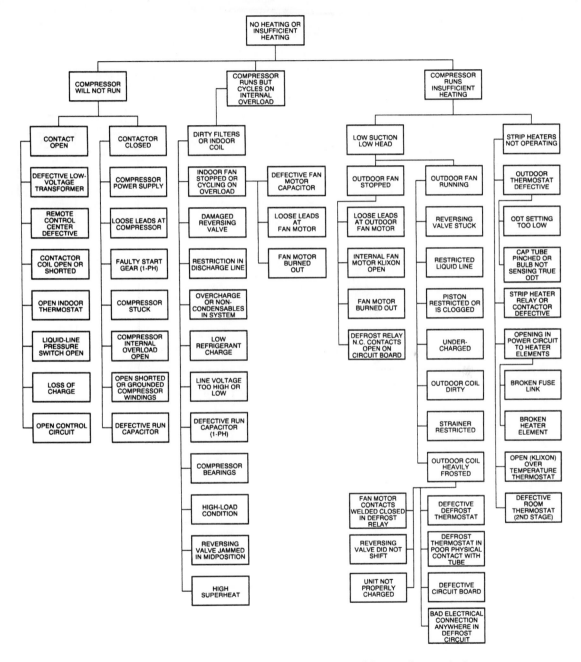

FIGURE 17.15 Using a manufacturer's troubleshooting chart *(Courtesy of Carrier Corporation)*

HEAT PUMP
TROUBLESHOOTING—COOLING CYCLE

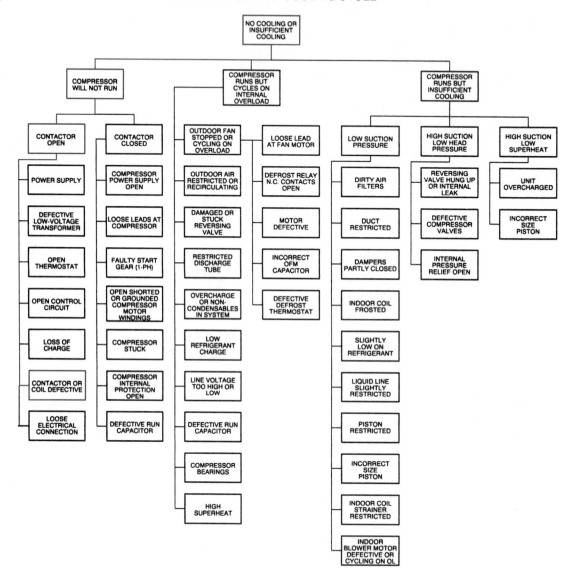

FIGURE 17.15. Continued

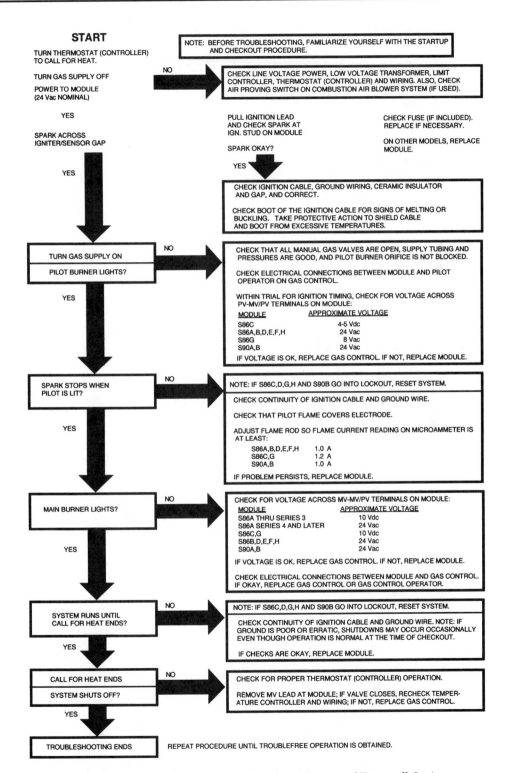

FIGURE 17.16 Using a troubleshooting chart *(Courtesy of Honeywell, Inc.)*

17.5 HOPSCOTCHING: A USEFUL TOOL FOR TROUBLESHOOTING

One frequently used procedure for troubleshooting an electric circuit is called **hopscotching**. Figure 17.17 shows an electric circuit that will be used as an example to follow the procedure. The contactor coil in the circuit closes, starting the compressor. Let's assume that we are getting voltage to $L1$ and $L2$ but not to the contactor coil. If the correct voltage is not available to the coil, one of the switches in the contactor coil circuit must be open. The question is which one. First connect one lead of the voltmeter to $L2$. The other lead of the voltmeter should be placed at point A as shown in Figure 17.18. If line voltage is read, the switch at A is closed and not at fault. Then hop to points to B, C, and D with the meter leads. At the point where voltage is not read, that switch is open, as shown in Figure 17.19. The discharge line thermostat is open.

Using Figure 17.20, follow another example of hopscotching. The customer has complained that a heat pump is not cooling. The technician arrives on the job and communicates with the homeowner that he or she is on the premises and will be checking their system. The technician examines the system and determines that (1) the indoor fan is operating, (2) the thermostat is correctly set for cooling, (3) 240 volts are available to the outdoor unit, (4) the reversing valve solenoid is being supplied with 24 volts, and (5) 24 volts is being supplied to the Y terminal of the defrost board. The technician determines from this examination that the outdoor unit contactor should be closed because the circuit is receiving 24 volts at Y on the defrost board. However, the contactor is not closed. The technician determines that there is no voltage to the contactor coil. The circuit that is in question and must be checked is highlighted in Figure 17.21. The

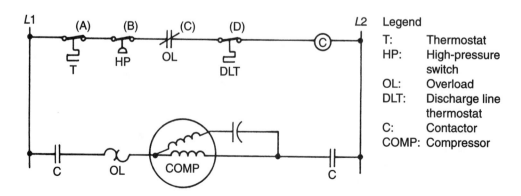

FIGURE 17.17 Simple electric circuit for hopscotching

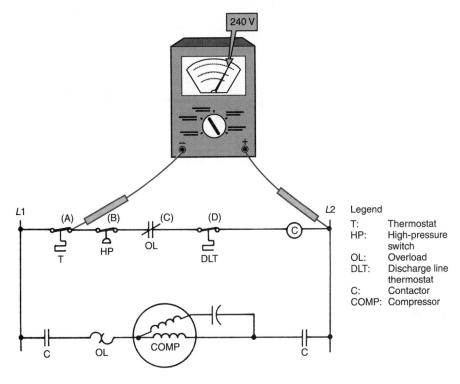

FIGURE 17.18 Hopscotching an electrical circuit

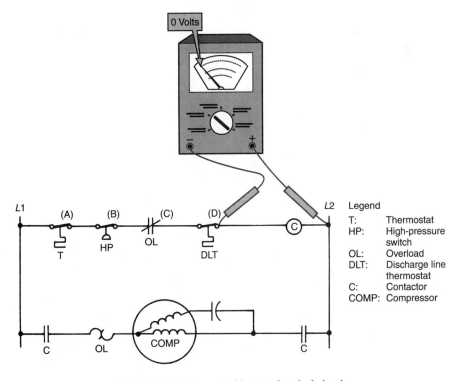

FIGURE 17.19 Hopscotching an electrical circuit

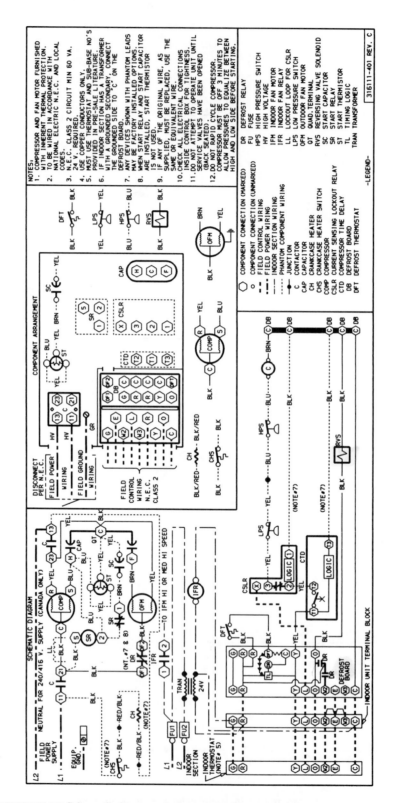

FIGURE 17.20 Schematic diagram of a heat pump *(Courtesy of Carrier Corporation)*

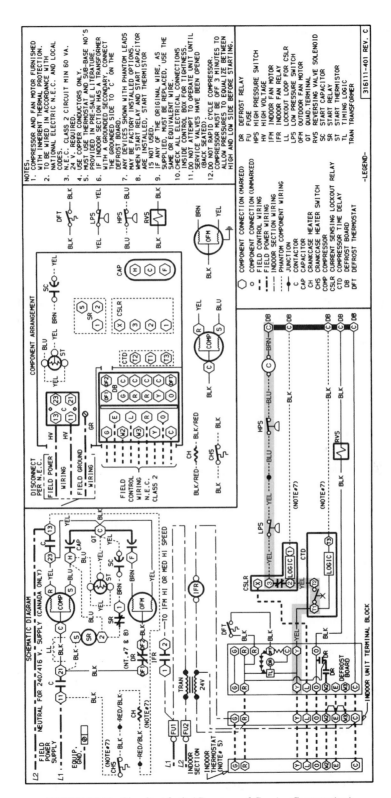

FIGURE 17.21 Circuit at fault *(Courtesy of Carrier Corporation)*

easiest way to check the circuit is by hopscotching. The technician places one lead of the voltmeter on the contactor coil where the brown wire is connected, as shown in Figure 17.22. The technician then proceeds to check the circuit, beginning with the Y terminal of the defrost board, and finds that 24 volts is available as shown in Figure 17.23. The technician continues in the circuit to T2 of the compressor time delay and finds that 24 volts is available, as shown in Figure 17.24. Continuing to terminal 3 of the current-sensing lockout relay, the technician finds that no voltage is available, as shown in Figure 17.25. A **current sensing lockout relay** is an overload device that senses the current draw of the compressor through the lockout loop (LL). The current-sensing lockout relay is open between terminals 2 and 3. The technician has determined that the CSLR is the reason that the contactor coil is not receiving 24 volts. Before the technician can put the blame on the CSLR, however, he or she must make sure that there is no other reason for the compressor to be operating in a high-amperage condition.

This method of troubleshooting is one of many that are used in the industry. Good service technicians will usually develop their own procedures as they gain experience with various types of equipment.

17.6 TROUBLESHOOTING CONTROL SYSTEMS

As you develop your own troubleshooting procedures, you must take a systematic approach. It is no easy task to diagnose problems in electric systems, especially the larger commercial and industrial equipment. The smaller residential and commercial equipment, fortunately, is built and designed with a simpler control system and fewer components than the larger equipment.

Service technicians must understand how, when, and why a certain piece of equipment operates as it does. Wiring diagrams are extremely important because they tell the technician how the equipment operates, what components are used, and where they are located in the system. The schematic diagram is usually the easiest tool to use for diagnosing problems. However, in many cases manufacturers will provide a pictorial diagram or a combination of a pictorial and schematic with the equipment. No matter what type of diagram is furnished with the equipment, the service technician must find and repair the malfunction.

At all times the technician must be careful not to overlook the small things that homeowners or inexperienced service mechanics could have done to an electric system, such as changing wires, jumping controls out, and removing components they felt were unnecessary.

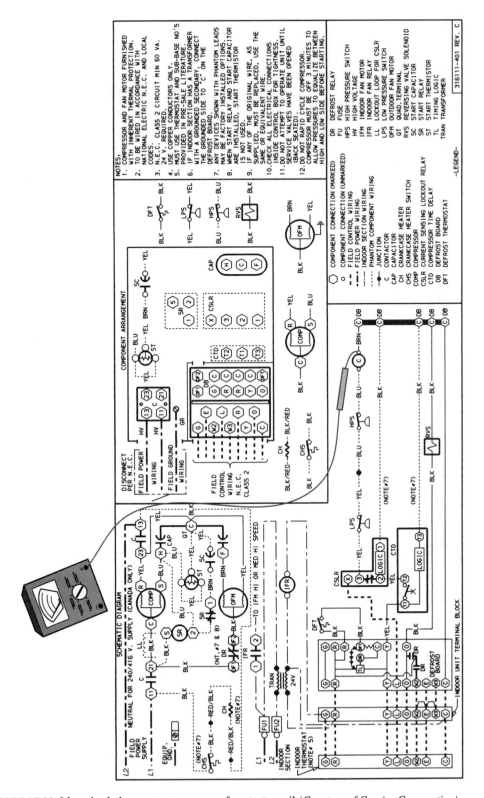

FIGURE 17.22 Meter lead placement at common of contactor coil *(Courtesy of Carrier Corporation)*

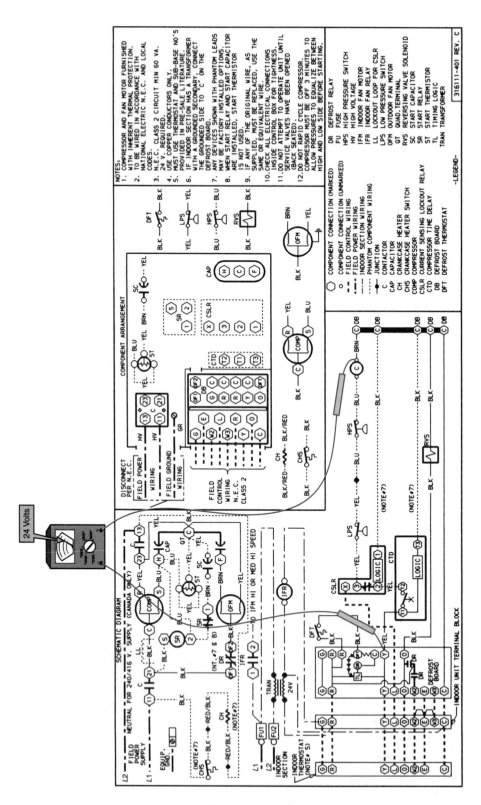

FIGURE 17.23 Hopscotching circuit *(Courtesy of Carrier Corporation)*

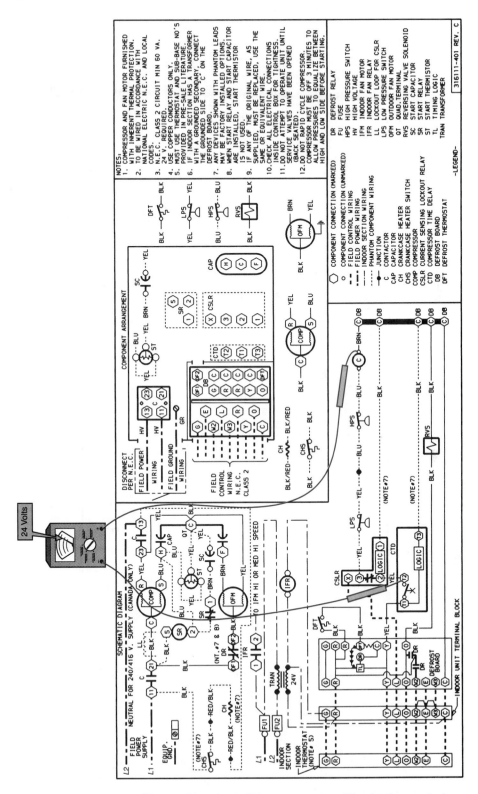

FIGURE 17.24 Hopscotching circuit *(Diagram courtesy of Carrier Corporation)*

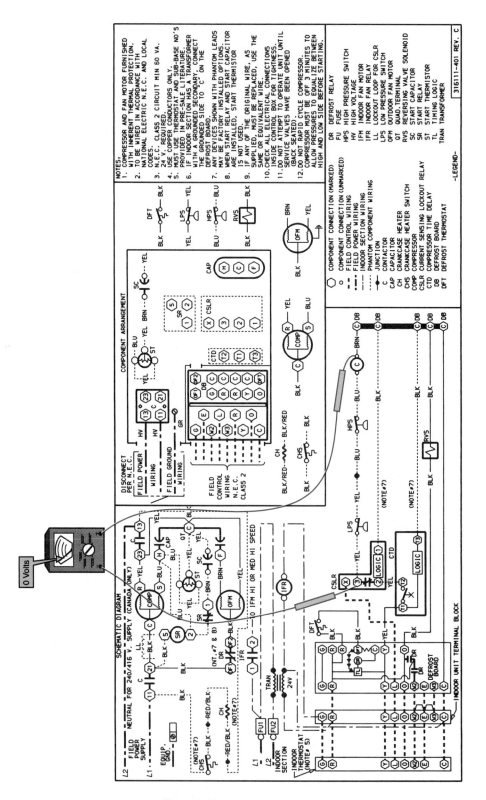

FIGURE 17.25 Hopscotching circuit *(Courtesy of Carrier Corporation)*

Diagnosis of Control Systems

In many cases, the components are at fault when the system is not operating properly. But it is sometimes difficult to identify the faulty component unless the technician uses the wiring diagram to identify the circuit that is not operating properly and then goes to the load or to the device that controls the load. A technician can usually identify the load that is not operating properly. Therefore, the circuit should be checked for any malfunctions or faulty components.

Troubleshooting the electrical control circuits in a cooling, heating, or refrigeration system is basically common sense. It is essential that the service technician understand how the unit functions electrically. With a good knowledge of how the unit functions electrically and a good troubleshooting approach, the service technician can eliminate guesswork and parts changing when servicing a unit. Observing the unit in operation will lead the technician to the electrical circuits that can be eliminated as trouble sources. This process of elimination is based on a systematic analysis of the unit's operation. Most service mechanics quickly perform this task mentally once the operation of the unit has been observed. For example, if an air-conditioning system has three major loads, as shown in Figure 17.26, and two of the three loads are operating correctly, then the technician only has to check the circuit that is controlling the load that is not operating properly, along with checking the load itself. In Figure 17.26, if the indoor fan motor is not operating, then the technician should concentrate on the indoor fan motor and its controlling circuits. The service technician must be able to recognize and separate the load circuits.

Residential Systems. Figure 17.27 shows a basic circuit in a residential air conditioner condensing unit. The circuit contains the contactor and the compressor, with the contactor controlling the load. If the compressor is not operating, it is logical to assume that the problem is with the power supply to the contactor or the unit. The first check should be the incoming power to the contactor. If it is available, continue the voltage check to the compressor. If voltage is not available to the compressor, check the contacts and make certain of their position. If the contactor is closed and the compressor is being supplied with the correct power, diagnose the compressor, making certain to consider the overload, if it is internal. Check the contacts of the contactor if the contactor is closed but no voltage is going through its contacts. However, if the contacts are not closed, you will have to look at the circuit that controls the contactor coil.

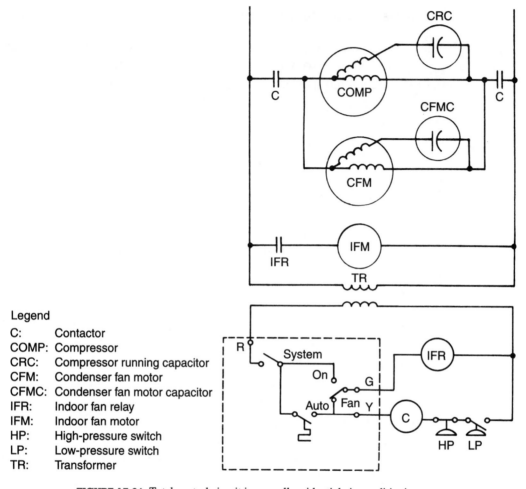

Legend

C:	Contactor
COMP:	Compressor
CRC:	Compressor running capacitor
CFM:	Condenser fan motor
CFMC:	Condenser fan motor capacitor
IFR:	Indoor fan relay
IFM:	Indoor fan motor
HP:	High-pressure switch
LP:	Low-pressure switch
TR:	Transformer

FIGURE 17.26 Total control circuit in a small residential air-conditioning system

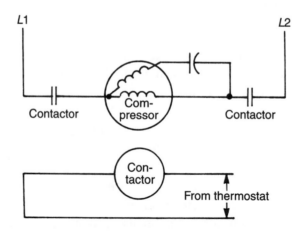

FIGURE 17.27 Basic circuit in a residential air-conditioning condensing unit

Total Circuit. The total circuit schematic of a residential air-conditioning system is shown in Figure 17.26, including the total wiring diagram. The methods used in the previous paragraphs can be followed here. The low-voltage control circuit can be checked out if the contactor is not closing, assuming the compressor is good and the contactor is not closed. The first check that should be made is the low-voltage power supply to ensure that voltage is available to the control circuit. The line voltage power supply has been checked and is correct.

Once it is determined that voltage is available to the control circuit, the only items left to check are the devices in the circuit. In the control circuit there are four devices that should be checked: the low-pressure and high-pressure switches, the thermostat, and the contactor coil. The contactor coil can be checked by reading the voltage available to it. If 24 volts are available but the contactor is not closed, then the coil or mechanical linkage of the contactor is bad and should be replaced. The other three components should be checked for an open circuit. This can be done easily with an ohmmeter or voltmeter. Do not overlook any problem that could have the high-pressure or low-pressure switches opened (low refrigerant charge, dirty condenser, or a bad condenser fan motor).

Industrial Systems. Commercial and industrial control systems have many more components and are more complex than the simple residential control systems. Therefore, it is more difficult to diagnose and correct the problems in these circuits. First check the power supply to the equipment and the power supply to the control circuit. If there is power, use the schematic wiring diagram and follow a systematic method for troubleshooting the system. The complexity and number of components should not cause you any concern. Each circuit of a diagram can be and should be treated individually.

Service personnel soon learn their own methods of troubleshooting equipment that allow for faster and more efficient service calls. The one item that takes its toll on service technicians is a callback. A callback is a return trip to a piece of equipment that a technician has not fixed properly. This hurts the productivity of the technician as well as the employer because of the added expense, which cannot be charged to the customer. Service mechanics should try to eliminate all possibility of callbacks. If a technician does the right kind of job on the first call, there is no callback.

One procedure for troubleshooting any cooling, heating, or refrigeration system is the following:

1. Check the power supply to the equipment.
2. If no power is available, correct the problem.
3. Make a thorough check of the system to determine which load or loads are not working properly.
4. Locate the loads on the wiring diagram that are not operating properly and begin checking these circuits first.
5. Each device in that circuit is a possible cause of the problem and should be checked until the faulty device is found. (Do not disregard the load in the circuit.)
6. It may be necesary to move to other circuits to check the coils of some of the devices.
7. Never overlook the internal overload devices.

The heating, cooling, and refrigeration industry depends as much on the service aspects as on the selling of new equipment. Service work is a necessity if a company is going to be successful. No customer wants an air-conditioning system if he or she cannot get the proper service and maintenance. Remember, too, that service technicians always represent the employer when they are sent out on the job. They should always try to present themselves in a manner that will sell them as well as their employer. The appearance and attitude of a service technician are just as important as the ability to diagnose and correct the problem that the homeowner is having.

17.7 SERVICE CALLS

Service Call 1

Application: Residential conditioned air system

Type of Equipment: Gas furnace and air-cooled condensing unit

Complaint: No heat, entire system is dead

Service Procedure
1. The technician reviews the work order from the dispatcher for available information. The work order reveals that nothing about the system is operating. The gas furnace is located in the basement.

2. The technician informs the homeowner of his or her presence and obtains any additional information about the problem. The homeowner states that one night the system was working and the next morning it was not.

3. Upon entering the house, the technician makes certain that no dirt or foreign material is carried into the structure. The technician also takes care not to mar interior walls.

4. The technician sets the thermostat to the heat position.

5. The technician measures the line voltage available to the furnace. The technician measures 120 volts, which is the correct voltage.

6. The technician measures the control voltage supplied from the thermostat to the furnace and reads 0 volts, indicating that the heating appliance is not receiving a call for heat.

7. The technician must now check the transformer to determine the available control voltage; checking at terminals C and R of the terminal board on the furnace shows 24 volts. The technician has determined that line voltage is available to the furnace and the transformer is good.

8. The voltage reading from C to W on the terminal board is 0 volts, indicating that the thermostat is not supplying control voltage to the furnace. The technician jumps terminals R and W on the thermostat and then measures the voltage at the furnace terminal board (terminals C and W), which measures 0 volts. It is apparent to the technician that the thermostat cable is bad.

9. The technician now inspects the thermostat cable between the thermostat and the furnace. The technician locates a place in the thermostat cable where the insulation and wire have been cut.

10. The technician splices the thermostat cable together using good wiring practices.

11. The technician operates the system to make sure it is working properly.

12. The homeowner is informed of the problem and the corrective action taken by the technician.

Service Call 2

Application: Commercial refrigeration

Type of Equipment: Walk-in cooler (R-12) with pump-down cycle

Complaint: No refrigeration

Service Procedure:

1. The technician reviews the work order from the dispatcher for available information. The work order reveals that the walk-in cooler is not cooling.
2. The technician informs the manager of the meat market of his or her presence and obtains any additional information about the problem.
3. The technician observes the system to determine which, if any, loads are operating. The evaporator fan motors are operating normally.
4. The technician checks the pressures in the refrigeration system and finds that the discharge pressure is 100 psig and the suction pressure is 5 psig. The technician finds that line voltage is available to the condensing unit, but the contactor is not closed.
5. The technician must determine what switches are in series with the contactor coil. The technician determines that the low-pressure switch is in series with the contactor coil. The low-pressure switch is open, preventing the contactor from closing.
6. The technician knows from past experience that the system is charged correctly and looks further.
7. The refrigeration system is using a pump-down system.
8. The technician must now check the liquid line solenoid circuit. Within that circuit, the thermostat, defrost control contacts, and a manual switch are in series with the liquid line solenoid. The technician checks the circuit and finds all three switches closed.
9. The voltage available to the liquid line solenoid valve is correct, but the solenoid coil is cold to the touch. The technician removes the solenoid from the valve and feels no magnetic pull. To be certain, the technician measures the resistance of the solenoid coil and finds it to be infinite ohms. The solenoid coil is open and must be replaced.
10. The technician replaces the liquid line solenoid coil and checks the operation of the system. The system is operating properly and the temperature is decreasing in the walk-in cooler.
11. The meat market manager is informed of the problem and the corrective action taken by the technician.

Service Call 3

Application: Commercial and industrial conditioned air system.

Type of Equipment: Large fan coil unit with large air-cooled condensing unit

Complaint: Not enough cooling

Service Procedure:

1. The technician reviews the work order from the dispatcher for available information. The work order reveals that the system cools adequately on mild days, but as the temperature increases, the system cools inadequately.

2. The technician informs the maintenance supervisor of his or her presence and obtains any additional information about the system. The maintenance supervisor states the same as in the work order. The maintenance staff has cleaned the evaporator and condenser coils in the last several weeks.

3. The technician checks and finds the system operating properly at about 11:30 A.M. The technician finds that the system is operating normally with an ambient temperature of about 78°F.

4. The technician stops for lunch and reports back at 1:30 P.M. and finds that the compressor is off.

5. The technician installs a set of refrigeration gauges on the system and reads the discharge gauge at 425 psig and the suction gauge at 125 psig.

6. The technician observes the system after it has started and finds that only two out of the three condenser fan motors are operating. The system operates for about 10 minutes and again cuts off. The technician determines that the unit is cutting off due to high head pressure. The technician assumes that the No. 2 condenser fan motor will have to run to maintain the head pressure at a level below the setting of the high-pressure switch.

7. The technician uses the wiring diagram of the unit to find out what cycles the third condenser fan motor. The fan motor is cycled by an outdoor thermostat that is in series with condenser fan contactor No. 3.

8. The technician checks the voltage available to the condenser fan contactor No. 3 and measures 0 volts. The technician next checks the condenser fan motor No. 3 thermostat to determine if it is open or closed. The thermostat is open; the technician checks the setting and determines that the condenser fan motor No. 3 thermostat is good but out of adjustment.

9. The technician adjusts the condenser fan motor No. 3 thermostat so that it comes on and maintains the desired discharge pressure in the system.

10. The technician informs the maintenance supervisor of the setting adjustment made on the thermostat and reminds the supervisor that the condenser fan motor thermostats should not be adjusted by untrained technicians.

Service Call 4

Application: Residential conditioned air system

Type of Equipment: Gas furnace using an intermittent pilot ignition with an air-cooled condensing unit

Complaint: No heat

Service Procedure:

1. The technician reviews the work order from the dispatcher for available information. The furnace is located in a crawl space. There is an access opening at the west end of the house that will be unlocked. No one will be at home; the key to the back door is located over the back door frame.
2. Upon entering the house, the technician makes certain that no dirt or foreign material is carried into the structure. The technician also takes care not to mar interior walls.
3. The thermostat will need to be set to the heat position before the technician goes under the house. The technician turns the fan switch to the "on" position to see if the fan will operate. The fan operates, so the technician has already determined that line voltage is available to the furnace, and the transformer and indoor fan relay are in good condition.
4. The technician proceeds under the house to the furnace and visually inspects the furnace and controls. Nothing out of the ordinary is found.
5. The technician checks the voltage being supplied to the ignition module and discovers that there is no voltage to the module.
6. The technician examines the wiring diagram to determine what is in series with the ignition module. Four limit switches are in series with the module and must be checked; of the four, two are manual reset. The technician locates the manual reset limit switches and finds that one has opened.
7. The technician resets the open limit switch and the furnace immediately starts the ignition process. The main burner lights and in several minutes the indoor fan starts. The technician cycles the furnace several times to make sure it is operating properly and that the limit switch does not open.
8. The technician turns the thermostat to the position requested by the homeowner and makes certain that the house is left in order. The technician leaves a note to the homeowner explaining what the problem was and the action taken.

Service Call 5

Application: Commercial and industrial conditioned air system

Type of Equipment: Large fan coil unit and large air-cooled condensing unit

Complaint: No cooling

Service Procedure:

1. The technician reviews the work order from the dispatcher for available information. The work order reveals that the individual circuit breaker for the compressor is opening each time the compressor attempts to start.

2. The technician reports to the maintenance mechanic of the plant. The maintenance mechanic informs the technician that the compressor tries to start for about two or three seconds and then the breaker trips. The breaker has been reset several times during the last few hours. The indoor fan is operating properly. The three condenser fans are operating properly.

3. From past experience, the technician realizes that the problem could be very serious. The technician examines the condensing unit, looking for broken or burned wires that could be causing the problem, but finds nothing out of the ordinary.

4. The technician resets the breaker and tries to start the compressor but with the same results; the breaker opens.

5. The technician examines the wiring diagram to determine what type of three-phase motor is being used in the condensing unit. The three-phase compressor motor is a part winding motor.

6. The technician decides that the best starting point is to determine if the motor windings are good. The technician measures the resistance of the windings and finds that all six windings read 1.2 ohms, indicating a good compressor motor.

7. The technician now needs to determine if the second contactor in the part winding setup is closing. The technician removes the cover from the contactor and closes the breaker. The compressor again tries to start, but the breaker opens again. The technician notices that the second contactor does not close, which would cause the compressor to overload with high current draw and open the breaker.

8. There are two possible causes of the problem; either the time-delay relay or the second contactor could be at fault. The technician opens the compressor breaker and allows the controls to cycle through the

compressor startup. The technician checks the voltage to the contactor coil and measures no voltage. A resistance check shows that the second contactor coil is good.

9. The technician must determine if the time-delay relay is at fault. The problem is that the time-delay relay coil is receiving 24 volts but is not closing.

SUMMARY

One of the most important tasks that a refrigeration, heating, and air-conditioning technician is called on to perform in the field is to correctly troubleshoot and repair refrigeration, heating, and air-conditioning systems. In order to be an effective and profitable technician for the company, the technician must know how to (1) treat the customer, (2) work in a safe manner, (3) determine problems with the refrigeration cycle, (4) read wiring diagrams, (5) locate individual components that must be checked, (6) troubleshoot individual electrical components, (7) troubleshoot control systems, and (8) leave the work area in an orderly fashion. Technicians represent the company they are working for, and in many cases they are the only company representatives customers will see.

The first element that a technician must consider when assigned to a service call is personal safety. A technician will face many dangers on service calls and must be aware of how to safely perform the necessary tasks. Often it will be difficult to locate a place to interrupt the electrical power source in order to perform some troubleshooting task. The technician should never try to replace or repair electrical components while power is being supplied to the equipment or system. The technician will often be required to troubleshoot components while power is being supplied to the unit because certain troubleshooting operations require live circuits. However, the technician should use extreme care not to become part of an electrical circuit, which would put him or her in danger of electrocution. The technician should make certain that all extension cords and electrical tools are in good condition and all grounding connections are in good condition and properly connected. Technicians will come in contact with many hazards, such as equipment that rotates, and they must avoid getting tangled with the shafts of motors and other devices. When sizing electrical wire, the technician should always follow the *National Electrical Code®*, local codes, and the manufacturer's installation instructions. Probably the two most important words as far as safety is concerned are respect and caution. At least one person on a service vehicle should know how to correctly perform CPR.

The technician will have to troubleshoot at one time or another all the electrical components in a refrigeration, heating, and air-conditioning system. These components include electric motors, transformers, thermostats, pressure switches, capacitors, overloads, and others. The technician must be able to select the correct meter and scale with which to determine the condition of these components.

The technician will have to be able to read wiring diagrams in order to be an effective troubleshooter. The schematic diagram breaks each control system down into a circuit-by-circuit arrangement, making the control system simpler and easier to understand.

Many tools are available to make the technician's job easier, such as troubleshooting charts, manufacturer's service information, wiring diagrams, and electrical test equipment. The technician should use every available aid.

Technicians are expected to correctly diagnose and repair all the problems they are assigned, but seldom does this happen. On occasion there will be callbacks about malfunctions that the service technician thought were repaired but were not. This is a delicate situation, and the technician must also be a customer service agent when this happens. The service technician is often the link between the company and the customer.

 CAUTION When servicing electrical devices the electrical power supply should be shut off at the distribution or main electrical panel. The electrical circuit should be locked out in an approved manner.

REVIEW QUESTIONS

1. What is the most common switch used in air-conditioning control systems?

   a. pressure
   b. thermostat
   c. push button
   d. humidistat

2. The _____ is the most important load in a refrigeration system.

3. Why is it important for a service technician to be able to use volt-ohm meters?

4. A pictorial or a component arrangement diagram shows _____.
   a. the components in a circuit-by-circuit arrangement
   b. the installation connections only
   c. the control panel layout
   d. all of the above

Answer questions 5–7 using the troubleshooting chart in Figure 17.6.

5. Which of the following could be problems if the compressor will not start?
   a. blown fuse
   b. contactor
   c. thermostat
   d. all of the above

6. Which of the following could be the problem if the compressor will not run but the condenser fan motor will operate?
   a. bad run capacitor
   b. bad condenser fan motor capacitor
   c. bad contactor
   d. none of the above

7. What would be the remedy for the question number 6?
   a. replace compressor
   b. replace condenser fan motor
   c. replace condensing unit
   d. replace compressor running capacitor

Answer questions 8 and 9 using the troubleshooting chart in Figure 17.7.

8. Which of the following could be the cause of the compressor cycling on overload on heating?

a. bad contactor coil
b. defective outdoor thermostat
c. damage reversing valve
d. none of the above

9. Which of the following could be a reason for the compressor not to run?
   a. loose electrical connections
   b. compressor internal overload open
   c. locked-down compressor
   d. all of the above

Answer questions 10 and 11 from the troubleshooting chart in Figure 17.8.

10. What could be the problem if the pilot does not stop when the pilot lights?
    a. pilot not covering electrode
    b. thermostat not calling for heat
    c. no spark
    d. none of the above

11. What could be the problem if the main burner does not light and the pilot is operating correctly and the voltage across the MV-MV/PV is 0 volts?
    a. check ignition cable
    b. bad thermostat
    c. change ignition module
    d. change gas control

12. Why is it important for a technician to be able to use a clamp-on ammeter?

Use the diagram in Figure 17.1 to answer questions 13–16.

13. If a technician turns the fan switch to the "on" position and the indoor

fan runs, which of the following conditions can he or she eliminate as problems?

a. transformer
b. indoor fan motor
c. indoor fan relay
d. all of the above

14. If the compressor and condenser fan motor will not operate but the indoor fan motor is operating, what would be the probable cause of the malfunction?

a. bad contactor coil
b. bad thermostat
c. bad condenser fan motor
d. bad compressor

15. If a technician reads 0 volts across the indoor fan relay contacts and the indoor fan motor is not operating but supply voltage is available to the equipment, which of the following would be the problem?

a. bad indoor fan relay contacts
b. bad transformer
c. bad indoor fan relay coil
d. bad indoor fan motor

16. If the thermostat is closed, the compressor and condenser fan motor are not operating, and all other loads are operating properly, which of the following conditions could be the problem?

a. a bad high-pressure switch
b. a bad low-pressure switch
c. both a and b
d. bad thermostat

17. How does the process of elimination play a part in troubleshooting an air-conditioning system?

Answer questions 18–23 from the diagram in Figure 17.2.

18. Which of the following components are not located in the control box of the unit?

a. the defrost board
b. the start thermistor
c. the reversing-valve solenoid
d. the start relay

19. True or False: It is acceptable to use aluminum conductors when installing the unit.

20. Which of the following conditions occurs when the defrost relay is energized if the defrost thermostat is closed?

a. The reversing-valve solenoid is energized.
b. A connection is made between R and W2 in the defrost board.
c. The outdoor fan motor is de-energized.
d. All of the above

21. If the unit in the diagram does nothing when the thermostat is closed, which of the following conditions could exist?

a. FU 2 is bad.
b. The indoor fan relay is bad.
c. The start relay is bad.
d. The high-pressure switch is open.

22. If the unit is operating in cooling mode, the outdoor fan motor is not operating, and all other loads are operating properly, what could be the problem?

    a. bad defrost relay contacts
    b. bad contactor
    c. blown fuse
    d. bad transformer

23. The unit is heating when it should be cooling. Which of the following could be the problem?

    a. the reversing-valve solenoid
    b. the compressor
    c. the defrost board
    d. none of the above

24. Why is hopscotching an important tool when troubleshooting?

25. Why are wiring diagrams an important tool when troubleshooting air-conditioning systems?

PRACTICE SERVICE CALLS

Determine the problem and recommend a solution for the following service calls. (Be specific; do not list components as good or bad.)

Practice Service Call 1

Application: Commercial and industrial conditioned air system

Type of Equipment: Water chiller with air handlers

Complaint: No chilled water

Symptoms:
1. Compressor not operating.
2. Chilled water thermostat, high-pressure switch, low-pressure switch, compressor thermostat, and freeze protector closed.
3. Chilled water pump is operating.

Practice Service Call 2

Application: Commercial and industrial conditioned air system

Type of Equipment: Water chiller with air handlers

Complaint: No chilled water

Symptoms:
1. Compressor not operating.
2. Thermostat closed.

3. Hopscotching through compressor contactor circuit between *L*2 side of contactor coil and load side of freeze protector, there is no voltage; voltage is available at the line side of the freeze protector.

Practice Service Call 3

Application: Commercial and industrial conditioned air system

Type of Equipment: Large fan coil unit controlled by a system switch and time clock with large air-cooled condensing unit

Complaint: No cooling

Symptoms:

1. Line voltage available to the magnetic starter controlling the blower motor and to the air-cooled condensing unit.
2. System switch is closed.
3. Blower motor is not operating.
4. Condenser unit not operating/interlock contacts of magnetic starter are open.

Practice Service Call 4

Application: Commercial and industrial conditioned air system

Type of Equipment: Pneumatic controls (use Fig 17.14)

Complaint: No cooling

Symptoms:

1. Supply air = 20 psig.
2. Receiver controller operating properly.
3. Discharge air temperature showing 85°F.
4. Return air temperature showing 85°F.
5. Air solenoid SV-1 is closed because of high outdoor ambient temperature.
6. Air-cooled condensing unit not receiving signal to start.

Practice Service Call 5

Application: Light commercial conditioned air system

Type of Equipment: Rooftop unit with two compressors (first compressor/first-stage cool and second compressor/second-stage cool)

Complaint: Not enough cooling

Symptoms:

1. Ambient temperature 100°F.
2. Evaporator fan motor operating.
3. First-stage compressor operating.
4. All safety components in second-stage compressor contactor circuit are closed.
5. Second-stage compressor contactor being supplied with 0 volts.

Practice Service Call 6

Application: Residential conditioned air system

Type of Equipment: Electric furnace with four 5-kW electric resistance heaters

Complaint: Not enough heat

Symptoms:

1. Electric furnace supplied with correct line voltage.
2. Control voltage supplied to furnace is 24 volts.
3. Blower motor is operating.
4. Resistance heaters 1, 3, and 4 are operating.
5. Resistance heater 2 not operating.
6. Sequencer is good.
7. All line voltage safety controls in heater 2 line voltage circuit are good.

Practice Service Call 7

Application: Residential conditioned air system

Type of Equipment: Gas furnace with air-cooled condensing unit

Complaint: No heat

Symptoms:

1. Correct line voltage being supplied to furnace and condensing unit.
2. 24 volts being supplied to control system of furnace.
3. Spark being supplied to igniter/sensor.
4. Pilot is igniting.
5. Main burner is not igniting.
6. 24 volts being supplied to gas valve.

Practice Service Call 8

Application: Residential conditioned air system

Type of Equipment: Split system heat pump (Figure 17.20)

Complaint: Heats instead of cools

Symptoms:

1. Correct line voltage being supplied to indoor and outdoor sections of heat pump.
2. 24 volts being supplied to C & O on indoor unit terminal board.

LAB MANUAL REFERENCE

For experiments and activities dealing with material covered in this chapter, refer to Chapter 17 in the Lab Manual.

18

Solid-State Controls and Systems

OBJECTIVES

After completing this chapter, you should be able to

- Identify and describe the operation of basic electronic system components.
- Identify and describe the operation of common one-function electronic controls that are used in the industry.
- Troubleshoot one-function electronic controls.
- Describe the function and operation of an electronic defrost board used in a heat pump.
- Describe the operation of an electronic motor protection module used on motors.
- Troubleshoot electronic defrost modules.
- Troubleshoot electronic motor protection modules.
- Understand and troubleshoot basic electronic control systems used in residential conditioned air systems.
- Identify electronic control systems used in commercial and industrial equipment and structures.

KEY TERMS

Defrost module
Diode
Electronic module
Light-emitting diode
Motor protection module
Rectifier
Semiconductors

Solid state
Thermistor
Transistor
Triac
Varistor
Voltage spike

INTRODUCTION

The refrigeration, heating, and air-conditioning industry has moved rapidly to incorporate solid-state control circuits into many of the control systems currently being used. Vast improvements made in solid-state components have revolutionized the industry, yielding smaller, more accurate, and more diversified control systems. Solid-state control systems allow better control for the entire heating and air-conditioning system. Along with better control, many other advances have been made: in motor speed control, in motor protection, and in defrosting systems for heat pumps and commercial refrigeration.

Major improvements have been made in large rooftop multizone units by using solid-state control systems. Controls for these units in the past were large and bulky and often did not give adequate control to each zone of the building. Electronic controls have almost eliminated erratic control of individual zones in this type of application. Many of these units also incorporate a method for controlling the fan speed, depending on the load of the building, by using solid-state speed control.

Commercial refrigeration control systems have also been affected by the influx of solid-state control systems. Many modern commercial refrigeration systems are controlled by solid-state systems that give better control to the entire system. Defrosting systems for this type of equipment have made the largest step forward by using solid-state controls.

Smaller refrigeration, heating, and air-conditioning systems have also been affected by the wide use of solid-state controls. Most small systems use solid-state components for a specific function, but a few use a total solid-state control system. Many manufacturers use some type of solid-state control for the protection of the compressor motor used in air-conditioning and refrigeration systems. Solid-state defrost controls are included on many heat pumps. The demand for energy efficiency has moved the industry toward better control of furnaces, which is accomplished in the newer systems by solid-state controls. Several new heat pumps that use total solid-state control systems have been introduced to the industry. These systems provide better control and efficiency for the customer.

These new solid-state control systems are not unlike conventional controls in that they require servicing. Although the term "solid state" may scare many service people now in the field, there is no need to be intimidated by the new electronic controls. The new solid-state control systems are nothing more than a group of circuits used to control a system. Furthermore, there is a similarity between these systems and the old conventional systems.

The heating, air-conditioning, and refrigeration technician must become proficient at troubleshooting the solid-state modules that are now appearing in the field. The main thrust of the technician's troubleshooting will be aimed at the module rather than each solid-state component housed in the module. In most cases, the technician will determine that a solid-state module is faulty and will change the entire module rather than trying to troubleshoot the module components. Most heating, air-conditioning, and refrigeration companies are not equipped to repair the module. Therefore, once the determination is made that the module is bad, it is merely replaced with no further testing. With the influx of solid-state modules into the industry, it is imperative that technicians be able to troubleshoot them.

18.1 ELECTRONIC SYSTEM COMPONENTS

Electronic systems have seen vast improvements over the past 40 years. Along with being large and bulky, the old vacuum and gas-filled tubes were made of many delicate parts. The newer solid-state devices are smaller and more durable than the old vacuum tubes. Figure 18.1 shows a size comparison of a vacuum tube and a transistor. The size and durability of solid-state devices have made it possible to incorporate them into many heating, air-conditioning, and refrigeration control circuits.

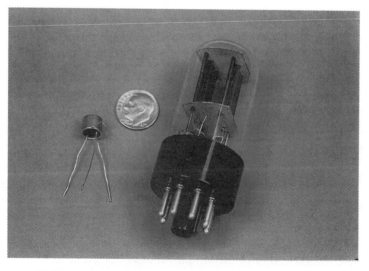

FIGURE 18.1 Size comparison of a vacuum tube and a transistor

Semiconductors

Many terms and definitions are used in electronic control circuits, but it is not necessary for the heating, air-conditioning, and refrigeration technician to become an electronics technician to efficiently service and repair this new breed of control. However, the HVAC technician should have a basic understanding of major electronic components.

Although the terms "solid state" and "semiconductor" may refer to the same device, they indicate different properties of the device. **Solid state** refers to a physical description of the component. **Semiconductor** refers to the electrical conductivity of the materials used in solid-state components.

A detailed description of semiconductor operation is not necessary to understand the function and operation of solid-state modules. We will present only a brief introduction here.

The conductivity of materials plays an important part in the action of semiconductors. For example, good electrical conductors are copper, silver, and aluminum; insulators like plastic, glass, and mica are poor conductors. Semiconductors, like their name implies, are half conductors. Some of the materials used as semiconductors have a resistance almost halfway between that of glass (insulator) and copper (conductor).

Semiconductor devices are versatile. They can provide a wide range of electrical characteristics, making it easier to apply solid-state components to air-conditioning control applications. The one disadvantage of semiconductors is that they cannot stand surges in voltage. In many applications, it is necessary to protect these devices against adverse conditions.

Solid-state devices in their simpler forms date back to the 1930s. In the 1940s, transistors were developed by two scientists at the Bell Telephone Laboratories. The transistor is a combination of two other semiconductor devices: the **thermistor** and the **varistor**. The thermistor is a solid-state device whose resistance decreases with an increase in temperature. The varistor's resistance decreases with an increase in voltage.

The materials most commonly used to make semiconductors are germanium and silicon. At the present time, silicon is used more than germanium. The atomic structure of silicon and germanium is important to the semiconductor materials because it basically controls their operation. So that semiconductor materials can be readily utilized, certain impurities must be added to the basic element to improve the atomic structure, allowing the semiconductor to do its job. The silicon and germanium atoms are extremely pure, and any element added to them is an impurity. The most commonly used impurities are aluminum, arsenic, antimony, boron, and gallium.

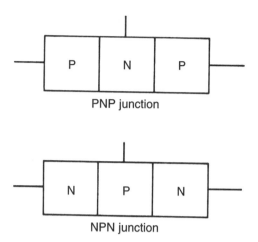

FIGURE 18.2 P- and N- type materials combine to form components

When certain impurities are added to silicon and germanium, the atomic structure of the semiconductor changes. This change provides the semiconductor with free electrons that can be easily moved and thus allow current to flow. Semiconductors with this type of impurity are called *N- (negative) type* materials. If a different impurity is added to the semiconductor material, the atomic structure will be altered to allow electrons into the semiconductor. These vacancies are referred to as *holes.* The holes allow electrons to easily pass through. This type of material is called a *P- (positive) type* semiconductor. When these two types are put together, they become a transistor, rectifier, or diode. Figure 18.2 shows some common examples of P- and N-type materials being combined to form components.

Diodes and Rectifiers

The diode and the rectifier are the simplest of the solid-state components. The only difference between the two is their size; they are basically the same device. A component that is rated at less than one ampere is called a **diode**. A diode is an electronic device that allows current to flow in only one direction. It is often used as an electrical check valve and in some cases as a spark arrestor. A similar component rated above one ampere is called a **rectifier**. Because the common diode and rectifier conduct electricity in one direction much better than they do in the other direction, they can be used to rectify alternating current to direct current.

Diodes and rectifiers are composed of P-type materials bonded to N-type materials. They are made by taking a solid-state material that has free electrons and bonding it to another material that has a shortage of electrons.

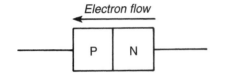

FIGURE 18.3 Drawing of a diode and rectifier

FIGURE 18.4 Symbol for a diode and rectifier

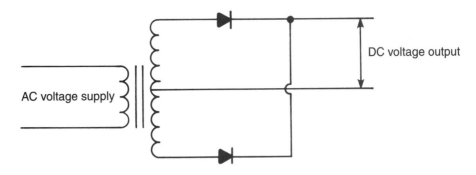

FIGURE 18.5 Rectifier circuit

When the two types of material are bonded together, a solid-state component is produced that will allow electrons to flow in one direction and act as an insulator when voltage is reversed. The electron flow of the device will be from the N-type to the P-type material. Figure 18.3 shows a basic drawing of a diode and rectifier. Figure 18.4 shows the symbol for a diode and rectifier.

Figure 18.5 shows an electrical circuit using two rectifiers to change alternating current (AC) to direct current (DC) voltage. The rectifiers allow current to flow in one direction but will not allow the normal alternating current reversal, thus rectifying the voltage.

Transistors

Transistors are composed of N-type and P-type materials. Their exact composition is dependent on the type of material used in the middle. If N-type material is sandwiched between two P-type materials, the transistor is referred to as a *PNP transistor.* If P-type material is sandwiched between two N-type materials, the transistor is referred to as an *NPN transistor.* Figure 18.6 shows the common symbols for the PNP and NPN transistors. The basic PNP or NPN transistor has three parts: the base, the emitter, and the collector. The transistor can be connected in the circuit in several different ways.

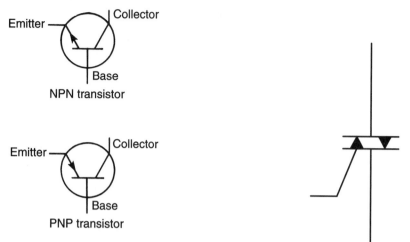

FIGURE 18.6 Symbols for NPN and
PNP transistors

FIGURE 18.7 Symbol for a triac

The transistor needed for a certain job would depend on the circuit, the application, the voltage, the current the device must handle, and the temperature. The transistor is used as a switch or to amplify a signal. In a radio, the transistor will take a weak signal and increase it until there is enough power so that an audible sound can be transmitted to a speaker. In air-conditioning control systems, a transistor will amplify a small signal obtained from a thermistor to provide enough power to do useful work, such as closing a relay or a contactor or driving a small motor.

Triacs

A **triac** is an electronic control that has two diodes in a single device. Triacs are used as switches in AC circuits because they allow current to pass in both directions when a certain current level is reached at the gate. Triacs are commonly used as relays in electrical circuits. The symbol for a triac is shown in Figure 18.7.

LEDs

A **light-emitting diode (LED)** is an electronic device that gives off light when current is passed through it. LEDs are often used in solid-state control boards and in digital readouts. They are used in solid-state control boards to send signals of some type, such as power to board, a blinking light for trouble diagnosis, and other indicators. LEDs can be arranged in a

FIGURE 18.8 Symbol for an LED

FIGURE 18.9 Drawing of an LED

specific way so that they can form numbers or letters. The symbol for an LED is shown in Figure 18.8 and a drawing in Figure 18.9.

Thermistors

A thermistor is a resistor that changes resistance with temperature. There are two types of thermistors: positive temperature and negative temperature coefficient. The resistance of the positive temperature coefficient thermistor increases as the temperature increases. The resistance of the negative temperature coefficient thermistor increases as the temperature decreases. There are many applications of thermistors in the heating and cooling industry, such as sensors for motor overload protection, start assist devices in PSC motors, and sensors to send a signal to an electronic control module. Sensors are nonlinear devices and are only accurate within a specified range.

18.2 BASIC ELECTRONIC CONTROL FUNDAMENTALS

Electronic controls used in the heating, air-conditioning, and refrigeration industry are built as one component. The industry has taken a black box approach to the production and design of electronic controls. These electronic devices are produced as circuit boards, such as the one shown in Figure 18.10, or as modules such as the one shown in Figure 18.11. The electronic circuit boards and modules are replaced as a unit if necessary.

Equipment used in the industry is usually powered by alternating current in the voltage range from 120 to 480 volts. When alternating current is used, there are often **voltage spikes** due to power surges, circuit characteristics, or lightning. A voltage spike is a sudden increase in the voltage supply that is only momentary. Most air-conditioning or refrigeration systems use compressors that require a larger current draw than electronic controls can withstand. Electronic control circuits generally are direct current circuits with their own power supply, usually an AC-to-DC diode rectifier. They operate on a low DC voltage, usually between 5 and 20 volts. The amperage that they draw is in the milliamp or microamp range. If we

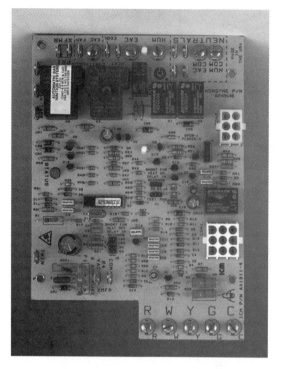

FIGURE 18.10 Electronic circuit board

FIGURE 18.11 Solid-state module

examine the differences between the electronic and normal control circuits, it is not difficult to see why electronic boards and modules must be handled with care. Electronic circuits are sensitive to heat, electronic interference, static electricity, current surges, and voltage spikes.

Electronic controls used in the industry include temperature controls, cycling controls, heat pump defrost controls, gas furnace controls, and motor controls. Temperature controls are in the form of electronic thermostats. Electronic cycling controls are used to prevent a system from short-cycling, to allow fan shutoff delay, and to allow component start delay. Heat pump defrost controls are almost all electronic; they supervise the operation of the heat pump. Gas furnaces incorporate electronic controls in their combustion control systems as well as in the operation of the required fan motors. There are many types of electronic motor controls, such as solid-state relays for fractional horsepower motors, start assist modules for PSC motors, and speed control modules for the operation of variable-speed motors.

18.3 BASIC ELECTRONIC CONTROLS

In the heating, air-conditioning, and refrigeration industry, many solid-state controls are used as one-function devices. A one-function solid-state device is used in a control system primarily to control or protect one function in the operation of the equipment. For example, some of the one-function controls are defrost controls on a heat pump and time-delay devices to prevent a compressor from short-cycling, to provide motor protection, and to act as starting relays on a split-phase hermetic compressor. Although many solid-state devices are used for many varied controls in heating, air-conditioning, and refrigeration control systems, in this section only a few will be covered.

Simple Electronic Temperature Control

A simple control circuit for temperature control would have a thermistor in the input circuit, which varies its resistance with a change in temperature. A change in the resistance of the input circuit would cause a change in current to the transistor base, and the change would be amplified through the transistor. This amplified signal could be sufficient to operate a switch or energize or de-energize a relay or contactor. Figure 18.12 shows a simple circuit for temperature control.

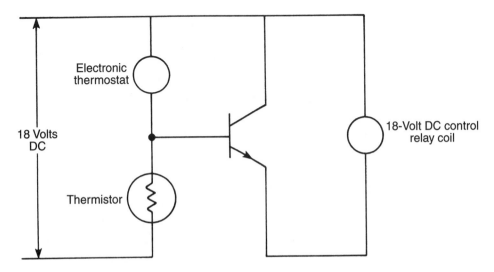

FIGURE 18.12 Temperature control circuit

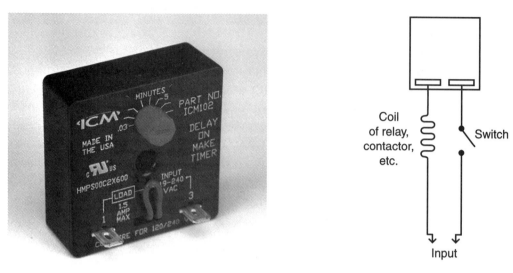

FIGURE 18.13 Adjustable solid-state timer with wiring diagram

Solid-State Timers

One of the most popular solid-state devices is the timer. Timers are available in many different forms and serve various functions in the electrical circuitry of air-conditioning, heating, and refrigeration equipment. Solid-state timers are used in the same manner as the conventional time-delay relay but with much more flexibility. Solid-state timers can delay on the closing or opening of an operating control. They are available with a set delay, used mainly by manufacturers on new equipment, and an adjustable delay, installed by field technicians on equipment for specific functions, such as delaying two heavy loads from starting at the same time or preventing short-cycling of equipment. Most timers are available over a wide range of timing cycles and are suitable for a variety of control circuit voltages. An adjustable solid-state timer is shown in Figure 18.13. These devices are compact and easy to install.

Anti–Short-Cycling Devices

One of the simplest solid-state devices in common use today is a control that prevents a compressor from starting more than once in a certain period of time. These devices can delay the starting time from 30 seconds to 5 minutes between cycles. An anti–short-cycling device is shown in Figure 18.14. This device is like many other simple solid-state devices; it should delay the

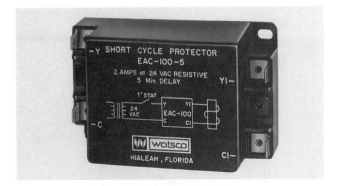

FIGURE 18.14 Anti–short-cycling device

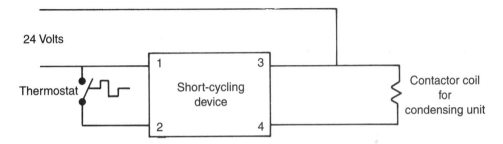

FIGURE 18.15 Control circuit with an anti–short-cycling device

starting of the specified component for the designated period of time or the device is not working properly. Figure 18.15 shows a control circuit with an anti–short-cycling device. These devices use a solid-state material whose cold resistance is high and will not allow voltage to pass. But as the device heats, the resistance decreases, allowing voltage to pass through the device to energize a contactor or relay.

Solid-State Devices for Electric Motors

A solid-state device is now being used to assist the permanent split-capacitor motor in starting. In many cases, this device can be used instead of a potential relay and a starting capacitor. Figure 18.16 shows the start-assist kit. This solid-state device uses a material that when wired in parallel with the running capacitor of a PSC motor (compressor) provides an increase in its starting torque. This material performs much like a small starting capacitor; it momentarily increases the current in the motor starting winding. As the material heats up, its resistance increases quickly to a point where it becomes

FIGURE 18.16 Start—assist component *(Courtesy of Bill Johnson)*

a nonconductor and the motor returns to PSC operation. This device requires a three-minute cool-down period between starts. To check this device, the service technician need only determine if it is dropping out of the circuit or if it is being energized on start-up. If the device is not performing in one of these ways, it should be replaced.

The solid-state starting relay used on small split-phase motor compressors is also simple in its operation. The device is used to drop out the starting winding of a split-phase hermetic compressor motor. One is shown in Figure 18.17; they are simple to install and service. Figure 18.18 shows a device installed on a fractional-horsepower hermetic compressor motor.

FIGURE 18.17 Solid-state starting relay

FIGURE 18.18 Solid-state starting relay installed on a compressor

The solid-state starting relay is actuated on a thermal basis through the use of a semiconductor material. This material is of a given cold resistance. As the compressor starts, the material heats up quickly and becomes a non-conductive material, dropping the starting winding out of the circuit. Some solid-state starting relays are limited to use on split-phase motors, while others can be used on both split-phase and capacitor-start motors. These devices should not be used on rapid-cycling compressors because a cooldown period is required for proper operation of the relay.

For troubleshooting the device, it is only necessary to determine if the device is dropping out the starting winding of the hermetic compressor motor. If the device is not energizing or dropping out the starting winding, it is faulty and should be replaced.

Solid-State Devices for Heat Pumps

Many heat pumps use a solid-state defrost control module that initiates and terminates the defrost cycle of a heat pump. This defrost control keeps the outdoor coil of a heat pump clear of frost, making the unit more efficient. A solid-state defrost module is shown in Figure 18.19. This **defrost module** uses two thermistors to initiate and terminate the defrost cycle and a pressure switch, as a safety, to terminate the cycle only if the module fails to terminate the defrost cycle.

FIGURE 18.19 Solid-state defrost module

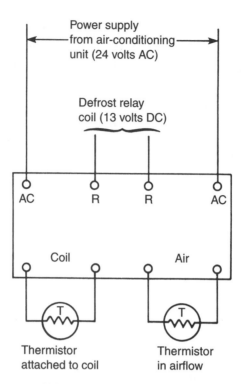

FIGURE 18.20 Wiring diagram of a solid-state defrost module

One of the solid-state defrost modules in present use is very effective in pre-venting a buildup of frost on the outdoor coil of a heat pump. This module controls an 18-volt DC relay that stops the outdoor fan motor and energizes the reversing valve. When the reversing valve is energized, the refrigerant cycle runs in the cooling mode of operation and defrosts the outdoor coil.

This module is shown schematically with the DC relay and the two thermistors in Figure 18.20. The module has eight connections: two to the unit's 24-volt power supply, two to the coil of the 18-volt DC relay, two to the coil thermistor, and two to the thermistor in the airflow across the coil. The thermistors sense the temperature of the outdoor coil and the air temperature across the coil. To initiate the defrost cycle, the coil temperature must be below 30°F. At that time, the module would feed 18 volts DC to the defrost relay, starting the defrost cycle by energizing the defrost relay and putting the unit in the cooling mode. The power is supplied to the module by the AC terminals from the 24-volt AC power supply of the unit.

Once the coil has become defrosted, the air temperature crossing the coil and the coil temperature would be high enough to bring the unit out of the defrost cycle, sending the system back to the heating mode of operation. Figure 18.21 shows the schematic of the heat pump.

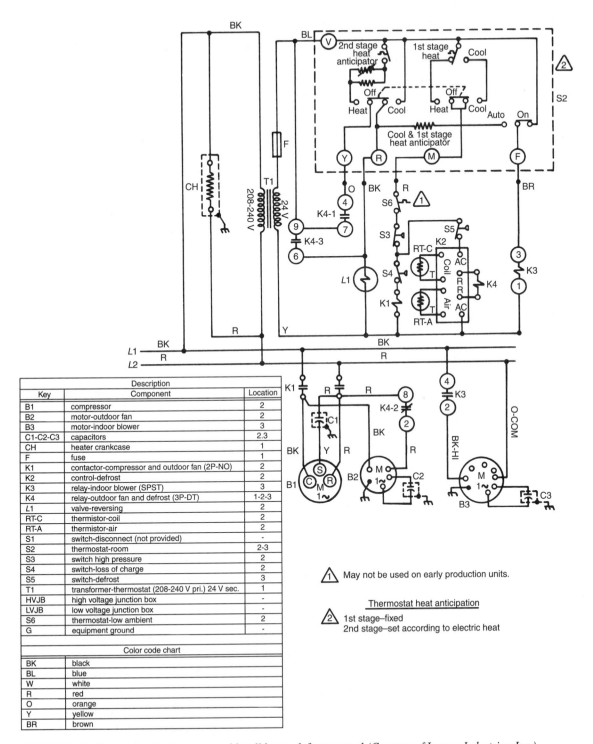

FIGURE 18.21 Schematic of a heat pump with solid-state defrost control *(Courtesy of Lennox Industries, Inc.)*

The troubleshooting procedure for the defrost system is simple and easy to perform. The technician must first determine what the module should be doing and what it is actually doing. If the coil or air temperature is high enough, the system should not go into the defrost cycle. If the system defrosts at a higher temperature than it should, it is a faulty module or has faulty thermistors. To determine which is faulty, the technician should use the manufacturer's specifications. If the resistance of the thermistors is correct for the temperature indicated, then the defrost module is faulty and should be replaced if it is sending 18 volts DC to the defrost relay. If the module is not sending 18 volts DC to the relay and the unit is in the defrost mode, the defrost relay is stuck in the defrost position. If all conditions exist to start defrost and the module is not sending 18 volts DC to the defrost relay, then the module is faulty, assuming that the thermistors have checked out to correspond to the temperature. Troubleshooting solid-state devices is often a process of eliminating the good portions of the particular system and spotting the defective portion.

Solid-State Devices for Motor Protection

Motor protection modules are another area in which solid-state devices are used effectively. These types of safety control systems use thermistors to indicate the temperature of a hermetic compressor motor. The solid-state motor protection is fast acting and sustains periodic and moderate overloads without nuisance trip-outs.

Heat is the main enemy of an electric hermetic motor. The amount of heat that is produced and held in a hermetic motor is the result of load heat and motor efficiency. Some of the input to an electric motor is converted to heat as well as to useful work. Motor life is determined largely by the conditions that are imposed on the motor. Therefore, it is essential that the hermetic motor be effectively protected. The solid-state motor protector can effectively protect the motor within a reasonable period of time.

The most common overload protection used is a single-module, three-sensor system. Figure 18.22 shows the module. The sensors are inserted in the hermetic motor winding, making this type of module more effective than conventional overload systems. The sensor is resistant to oil and refrigerant contained in a refrigeration system.

Figure 18.23 shows the sensor location in a three-phase hermetic motor. The sensors effectively sense the temperature of a hermetic motor.

FIGURE 18.22 Solid-state model, two-sensor

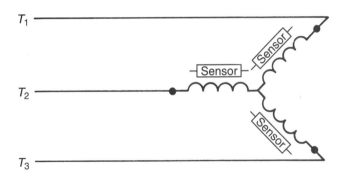

FIGURE 18.23 Sensor location in a three-phase motor

Figure 18.24 is a graph showing a sensor trip-out point and reset point. The resistances of the sensors increase as the temperature rises. This action causes the module to react and break a set of contacts to open the control circuit of the compressor.

The module with its wiring connections is shown in Figure 18.25. The module requires a power supply, the sensor connections, the control circuit contacts, and the manual reset connections. Figure 18.26 shows the connections that are required between the module and other components in the system.

Troubleshooting the overload module and sensors is a basic procedure. It requires only a few simple checks to determine the condition of the module

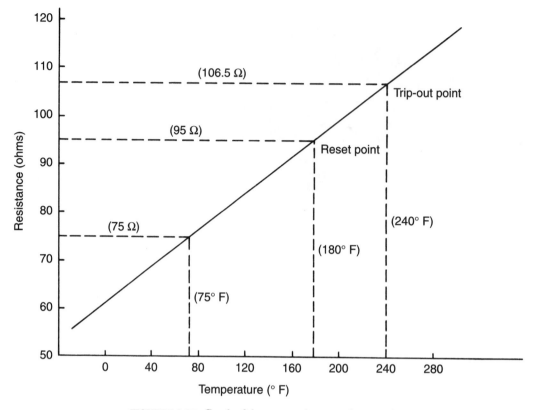

FIGURE 18.24 Graph of the sensor trip-out and reset point

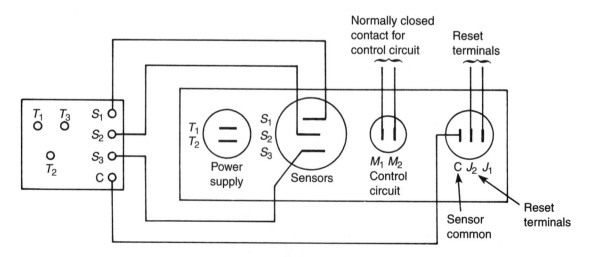

FIGURE 18.25 Wiring connection of motor protector module

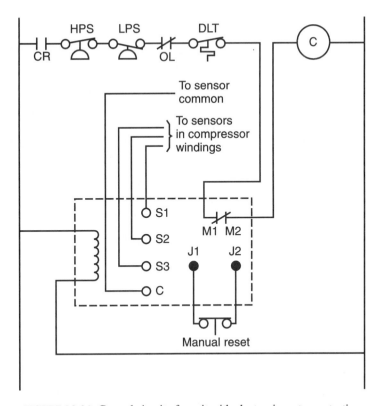

FIGURE 18.26 Control circuit of a unit with electronic motor protection

and the sensors. The first and basic check for the entire system is to determine if the control contacts are open or closed, thus eliminating the total overload protection system. If the control contacts are closed, then the problem is in other components. If the control contacts are open, the problem lies in one of three areas: the module, the sensors, or an actual overload. If it is determined that the motor is in an overload condition, then the motor must be checked to determine its condition. A service technician should always check the motor and its condition before making a decision about the module and the sensors.

To check each of the three sensors mounted in the motor, you use an ohmmeter. When checking the sensors, follow this procedure:

1. Disconnect the power to the circuit.
2. Check the maximum test voltage of 6 volts.
3. Do not short the sensors.
4. Do not bypass the fuses if they are used.

5. Check the resistance readings of the sensors:
   a. If the resistance is above 95 ohms, wait for a reset.
   b. If the sensor is open, install a 75-ohm, 2-watt resistor.
   c. If the sensor is good, replace the module.

The basic checks after determining the problem in the module and sensor circuit are simple and easy. If all the sensors are good, the module should be replaced. If one sensor is bad, replace it with a 75-ohm, 2-watt resistor.

A wiring harness can be used to check the module in a more detailed procedure. Figure 18.27 shows a wiring diagram for the wiring harness.

Most compressor manufacturers have developed an alternative method of protecting the compressor if the sensors become faulty. For example, Figure 18.28 shows a compressor with a discharge line thermostat used for protection.

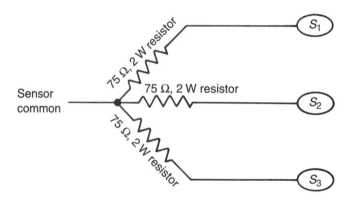

FIGURE 18.27 Wiring diagram for a wiring harness

FIGURE 18.28 Compressor with a discharge line thermostat

Electronic solid-state components can be easily diagnosed for problems. The service technician simply checks the signal being sent back to the control module from the thermistors located in the specific control location.

18.4 BASIC ELECTRONIC CONTROL SYSTEMS USED IN RESIDENTIAL CONDITIONED AIR SYSTEMS

There are many electronic control systems now being used in residential conditioned air systems. Almost all heating systems using fossil fuel as a heating source are using some type of electronic primary control to control the combustion and operation of the burner. Figure 18.29 shows an ignition module used to supervise the operation of a gas burner on a gas-fired furnace. Figure 18.30 shows a cad cell primary control that supervises the operation of an oil burner on an oil-fired furnace. Other warm-air furnaces are utilizing an electronic module that supervises the operation of the gas burner and controls the operation of the blower motor, while other electronic modules control the fan and burner independently. An electronic module used to control the burner and blower motor is shown in Figure 18.31.

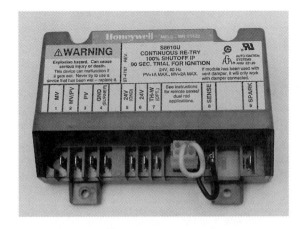

FIGURE 18.29 Ignition module of a gas furnace (*Courtesy of Honeywell, Inc.*)

FIGURE 18.30 Oil burner cad cell primary control (*Courtesy of Honeywell, Inc.*)

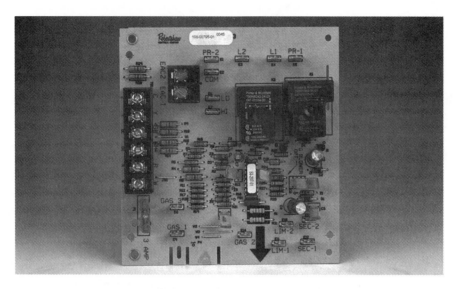

FIGURE 18.31 Furnace electronic module

Heat pumps manufactured today utilize **electronic modules** that control the operation of the electrical loads in a heat pump, including the defrost function. There are many different types and designs used in today's heat pumps. Some manufacturers use an electronic board that controls the entire operation of the heat pump, including defrost, as shown in Figure 18.32, while others use electronic modules specifically for the defrost function, as shown in Figure 18.33. Defrost controls on heat pumps vary from simple controls, having only a time function, to the demand defrost systems that use temperature to determine when defrost occurs. Very few heat pumps can be found without some type of electronic module in their control system.

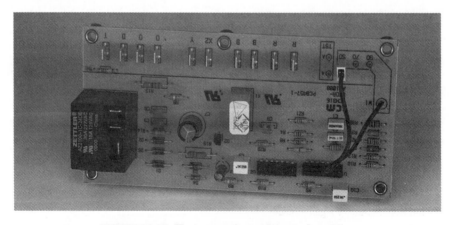

FIGURE 18.32 Heat pump electronic control module

FIGURE 18.33 Heat pump defrost module

Many manufacturers of residential heat pumps use a solid-state control system to control the entire operation of the heat pump, including normal operation, the operation of supplementary heat, and the defrost cycle. These new heat pumps are more efficient with better comfort control.

Figure 18.34 shows the control module of a heat pump. The new heat pumps are usually easy to troubleshoot because most manufacturers have developed a tester or computer program to troubleshoot the entire system.

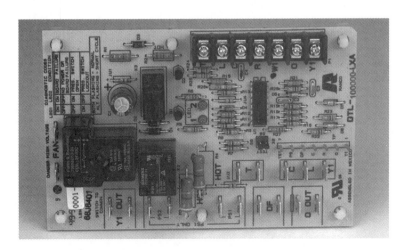

FIGURE 18.34 Control module of a heat pump

18.5 ADVANCED ELECTRONIC CONTROLS

Electronic control systems are increasingly popular on large heating and air-conditioning systems, commercial refrigeration systems combined for more than one interior case, and sophisticated, small, energy-efficient heat pumps. Large heating and air-conditioning systems now use a total solid-state control module for the entire system. Large rooftop units with multizone application especially have gone to the electronic control concept, because electronic controls give higher operating efficiency and better control. Often these systems are incorporated with a variable-air-volume air system, which further complicates control systems because of the varying fan speed. Commercial refrigeration systems that use one condensing unit for several cases, such as those in a large grocery store, are incorporating electronic control for better control of each case and for better defrosting of the entire system.

Large rooftop units are becoming increasingly complicated because of the use of more solid-state control systems. One manufacturer of large rooftop units uses several solid-state control components for the entire system. Large rooftop multizone units usually have many zones controlled from the same basic unit.

Figure 18.35 shows a large multizone rooftop unit installed on a school building. This type of unit has a control thermostat for each zone that controls the zone as well as the operation of the entire system. The thermostat sends a signal to a load analyzer shown in Figure 18.36. This device analyzes the signals and determines the operation of the unit. The thermostat, which is shown in Figure 18.37, also determines the mixture of air that is required by

FIGURE 18.35 Large multizone rooftop unit

each zone. Figure 18.38 shows the damper motors of a multizone rooftop unit. The damper motor controls the air mixture sent to each zone.

Figure 18.39 shows the gas heat operation diagram for this type of unit. A factory service manual is usually required to service this equipment.

FIGURE 18.36 Load analyzer

FIGURE 18.37 Solid-state thermostat for rooftop unit

FIGURE 18.38 Damper motor setup

GAS HEAT OPERATION WITH CONDENSER HEAT ON

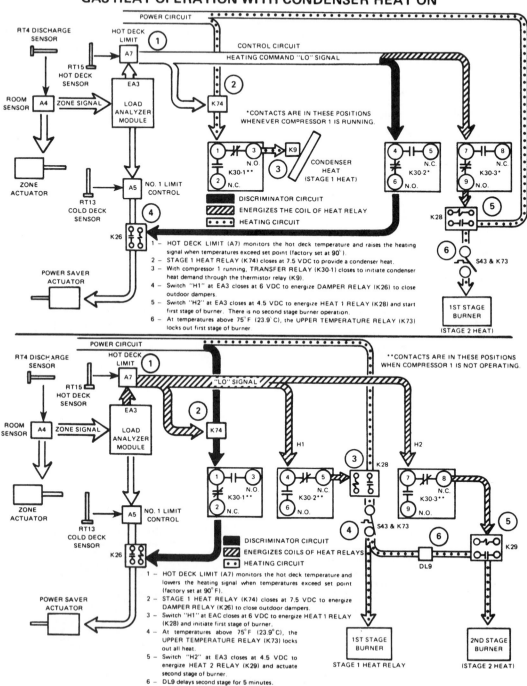

1 – HOT DECK LIMIT (A7) monitors the hot deck temperature and raises the heating signal when temperatures exceed set point (factory set at 90°).

2 – STAGE 1 HEAT RELAY (K74) closes at 7.5 VDC to provide a condenser heat.

3 – With compressor 1 running, TRANSFER RELAY (K30-1) closes to initiate condenser heat demand through the thermistor relay (K9).

4 – Switch "H1" at EA3 closes at 6 VDC to energize DAMPER RELAY (K26) to close outdoor dampers.

5 – Switch "H2" at EA3 closes at 4.5 VDC to energize HEAT 1 RELAY (K28) and start first stage of burner. There is no second stage burner operation.

6 – At temperatures above 75°F (23.9°C), the UPPER TEMPERATURE RELAY (K73) locks out first stage of burner.

1 – HOT DECK LIMIT (A7) monitors the hot deck temperature and lowers the heating signal when temperatures exceed set point (factory set at 90°F).

2 – STAGE 1 HEAT RELAY (K74) closes at 7.5 VDC to energize DAMPER RELAY (K26) to close outdoor dampers.

3 – Switch "H1" at EAC closes at 6 VDC to energize HEAT 1 RELAY (K28) and initiate first stage of burner.

4 – At temperatures above 75°F (23.9°C), the UPPER TEMPERATURE RELAY (K73) locks out all heat.

5 – Switch "H2" at EA3 closes at 4.5 VDC to energize HEAT 2 RELAY (K29) and actuate second stage of burner.

6 – DL9 delays second stage for 5 minutes.

GAS HEAT OPERATION WITH CONDENSER HEAT OFF

FIGURE 18.39 Operation diagram of the heat cycle of a rooftop unit *(Courtesy of Lennox Industries, Inc.)*

18.6 TROUBLESHOOTING ELECTRONIC CONTROLS

Much talk in this chapter has been about troubleshooting some specific electronic controls. When dealing with problems in electronic control circuits, the technician should focus on the solid-state circuit board or the module. At no time will a technician try to troubleshoot individual electronic components on a circuit board or module. The technician should be aware of inputs that are necessary for the correct operation of the module and what output should be expected. If the technician determines that the input is not producing the correct action, then chances are the circuit board or module is faulty and should be replaced.

Many electronic control devices have a fault message system in which a microprocessor will give a number code so that the technician can look up the number in the service material and find the cause listed. Some solid-state control boards have some type of mechanism that signals a problem. This signal is usually a flashing light or something similar. The technician then looks at the fault directory to identify the problem.

Many electronic control systems have no troubleshooting functions, and the technician must determine the problem by using troubleshooting charts and wiring diagrams. The technician should take care not to condemn the electronic controls when other components may be causing the problem. When a technician suspects the electronic control is the problem, he or she should verify that the control is receiving the correct input and producing the right output. In most cases the technician will be looking at specific voltage levels. The product data should give the technician those figures.

Technicians should develop a good procedure for checking electronic controls, using manufacturers' information whenever possible. Electronic controls and control systems have many advantages and are here to stay, so the technician should become familiar with troubleshooting them.

SUMMARY

The heating, cooling, and refrigeration industry is beginning to use many solid-state components and control systems. Solid-state components and control systems allow better control of temperature, better and faster-acting safety controls, flexibility in the speed control of motors, and more efficient defrosting systems for heat pumps and commercial refrigeration equipment. Solid-state controls are used in most phases of the industry, from the large rooftop units to the small split-phase hermetic compressor motors.

Solid-state controls are a vast improvement over the old vacuum tubes. The newer solid-state components are smaller and more durable than the tubes. Thus, manufacturers have been able to incorporate them in the modern heating, air-conditioning, and refrigeration control circuits.

The terms "solid state" and "semiconductor" may refer to the same device. Semiconductor devices are very versatile and can provide a wide range of electrical characteristics, making them easy to use in the control systems in the industry. Solid-state devices have one disadvantage: They cannot stand voltage surges. Therefore, often it is necessary to protect solid-state devices from these surges.

The diode and the rectifier are similar solid-state devices. They allow voltage to flow in one direction more easily than in the other direction. The diode is rated at 1 ampere or less, while the rectifier can handle larger loads. Transistors are used to amplify signals from a thermistor to establish a strong enough signal to cause an action on a relay or contactor. The thermistor is a device that varies its resistance with variations in temperature.

Many solid-state modules are used in the industry that control one function of the equipment. The simplest one-function control is a solid-state device that acts as a starting relay for small split-phase hermetic compressor motors. A solid-state time-delay device has replaced many time-delay relays. Motor protection has become vastly improved because of the use of solid-state safety systems. Many defrosting systems on heat pumps have also begun to use solid-state devices to achieve better and more efficient defrosting of the outdoor coil in winter operations.

Large heating and air-conditioning systems have also gone to solid-state control systems. Such systems are much more sophisticated than the one-function control modules. Larger systems use solid-state control as a total system control for the entire system.

Troubleshooting solid-state devices is relatively simple. The technician must know only what signal must be carried to the module to cause the correct action of the system. The technician must also learn to troubleshoot the module and the thermistors that send the signal. Many of the solid-state control systems have accompanying testers with which to check the systems, simplifying the task for the technician. The larger systems, however, often require a factory service manual.

REVIEW QUESTIONS

1. What is the difference between a vacuum tube and a transistor?

2. Semiconductors are _____.
   a. conductors
   b. insulators
   c. half conductors
   d. none of the above

3. Diodes and rectifiers allow current to _____.
   a. flow in one direction only
   b. flow in both directions
   c. divide paths of flow
   d. none of the above

4. What is the difference between a diode and rectifier?

5. A thermistor is a resistor that changes resistance with _____.
   a. pressure
   b. sound
   c. light
   d. temperature

6. What are the two types of thermistors and how are they different?

7. What materials are used to make semiconductors?

8. What is a transistor?

9. The transistor is used _____.
   a. as a signal
   b. to amplify a signal
   c. both a and b
   d. none of the above

10. Briefly explain the rectification circuit shown in Figure 18.5.

11. Label the following components as one-function (a) or multifunction (b) electronic modules.
    ____ a. defrost module
    ____ b. heat pump control
    ____ c. motor protection
    ____ d. timer
    ____ e. load analyzer
    ____ f. load-sequenced economizer

12. What is the purpose of an anti–short-cycling device?

13. True or False: The conductivity of materials plays an important part in the action of semiconductors.

14. How have the new solid-state devices affected the control circuits in heat pumps?

15. Solid-state timers can _____.
    a. delay an opening operating control
    b. delay a closing operating control
    c. have an adjustable delay
    d. all of the above

16. A semiconductor start assist kit is used on a _____.
    a. PSC motor
    b. shaded-pole motor
    c. three-phase motor
    d. all of the above

17. What is the difference between "demand defrost" and "timed defrost"?

18. True or False: A heat pump with a timed defrost system is more efficient than a heat pump with a demand defrost system.

19. Why is electronic motor protection faster-acting than conventional motor protection?

20. The thermistor used on the solid-state motor protection discussed in this chapter is a _____.

   a. positive coefficient thermistor.
   b. negative coefficient thermistor.

21. Where are the sensors located in respect to electronic motor protection?

22. What is an economizer?

23. How has the emergence of solid-state controls affected large multi-zone rooftop units?

24. True or False: The repair of solid-state modules will be accomplished by repairing the module.

25. What is the proper procedure for troubleshooting a solid-state motor protector?

LAB MANUAL REFERENCE

For experiments and activities dealing with material covered in this chapter, refer to Chapter 18 in the Lab Manual.

Glossary

air-cooled packaged unit: A unit that is made in one complete unit with an air-cooled condenser mounted with the unit or remotely.

alternating current: Electron flow that flows in one direction and then reverses at regular intervals; produced by cutting a magnetic field with a conductor. The most common type of power supply used in the heating, cooling, and refrigeration industry.

alternator: An electrical generator that is used to produce alternating current.

ambient air temperature: The temperature of the air surrounding any device.

American Wire Gauge: Standard unit of measurement for wire.

ammeter: An electric meter used to measure the amperes that are present in a circuit. Ammeters are made as in-line ammeters or clamp-on meters.

ampere: The amount of current required to flow through a resistance of one ohm with a pressure of one volt, measured with an ammeter.

analog meter: A device that uses a meter movement to indicate an electrical characteristic.

anti–short-cycling device: A device that prevents a load from stopping and starting in rapid succession. This device is used mainly to prevent compressors from stopping and starting rapidly.

anticipator: A component of a thermostat that anticipates the temperature of an area and will stop or start the cooling or heating equipment to prevent the thermostat from overshooting the desired temperature.

apparent wattage: The power that is calculated by multiplying the voltage times the amperage of a circuit.

armature: The portion of a contactor that moves; connected to a set of contacts that causes a completed circuit when the armature is pulled into the magnetic field produced by the coil.

atom: The smallest particle of an element that can exist and maintain any identification; can combine with other atoms to form new substances.

back electromotive force: The amount of voltage produced across the starting and common terminals or connections of a single-phase motor.

balance point: The point at which the capacity of a heat pump equals the heating load of the structure. At this point additional heat will be needed to meet the load of the building.

ball bearing: An antifriction device that is used to allow free turning and support of the rotating member of the device. It consists of an outer ring and inner ring with races and has steel balls sandwiched between the rings.

bearing: The part of a rotating machine or motor that allows free turning of its rotating parts with little friction.

bimetal overload: A simple thermal overload that breaks the power supplying a small motor directly. When it reaches a high enough temperature, the bimetal opens by warping; when the overload cools, the bimetal will warp back and close the circuit.

bimetal relay: An overload device that opens a set of contacts by a temperature that corresponds to the current draw of a load.

breaker: A device usually used in a breaker panel that is capable of being used as a disconnect switch and an overload protection for the circuit it is supplying power to.

breaker panel: An electric panel that houses breakers used to distribute power to circuits in the structure.

cad cell: A light-sensitive cell that changes its resistance in the presence of light. In most cases, the resistance decreases as the cell views the light. The cad cell is used to detect the ignition of an oil burner.

callback: A return trip to a piece of equipment that was not properly serviced the first trip.

capacitive reactance: A resistance caused by using capacitors with motors; when a capacitor is put in a circuit, it resists the voltage change, causing the amperage to lead the voltage.

capacitor: A device that consists of two aluminum plates with an insulator between the plates; used to boost the starting torque of single-phase motors. The two types of capacitors are electrolytic or starting, and running or oil-filled.

capacitor-start–capacitor-run motor: A high starting torque and good running efficiency motor; it uses a starting capacitor to increase starting torque and a running capacitor to increase running efficiency.

capacitor-start motor: A motor that uses a starting capacitor to increase the starting torque.

capacity control: The action of limiting the capacity of a piece of equipment to meet the needs of the load. Capacity control is used in light commercial and larger commercial and industrial applications.

cardiopulmonary resuscitation (CPR): The action taken by others after a cardiac arrest has occurred in which breathing and massage of the heart are attempted for the victim.

charge: A condition in which an imbalance of protons and electrons exists in an atom.

circuit breaker: See *breaker.*

circuit lockout: A method used to lock out a circuit that a technician is working on to prevent someone from restoring electrical energy to the circuit.

clamp-on ammeter: A type of ammeter with jaws that clamp around one of the conductors supplying power to the load.

clock thermostat: A thermostat that is used to control the temperature of a structure with a time function that can be set or programmed by the occupant of the structure. Clock thermostats can have multiple programs adding to their flexibility.

closed circuit: A complete path for electrons to follow.

coil: An electrical device that is used to convert electrical energy into a magnetic field.

combustion chamber: The part of a fossil fuel appliance where the combustion occurs.

compound: A substance that is a combination of atoms from at least two different elements.

compressor: A device that is used in a mechanical refrigeration system to compress the refrigerant. Most compressors are rotated by an electric motor. The motor may be external or be an integral part of the compressor.

condensing unit: A portion of a split air-conditioning or refrigeration system that is mounted outside and contains the compressor, condenser, condenser fan motor, and controls for these compo-nents; most used today are air-cooled. It takes the cool suction gas from the evaporator, compresses it, condenses it to a liquid, and forces it back to the evaporator.

conductor: A wire that is used for the path of electric flow. Most electric conductors are copper or aluminum.

constant discharge pressure: Maintaining the discharge pressure at a constant pressure; this is accomplished by controlling the amount of heat rejected by the condenser by controlling the condenser fan motors.

contactor: A device that opens and closes a set of electric contacts by the action of a solenoid coil; composed of a solenoid and the contacts.

contacts: The part of a relay that opens and closes to allow for the flow of electrical energy.

continuity: A complete path for electrons to follow in a circuit or component.

control circuit: A circuit that controls some load in the entire control system, whether it be a relay or contactor coil or a major load.

control relay: A relay that is used to control a circuit or circuits in an air-conditioning system.

current: Electrons flowing in an electric circuit, measured in amperes.

current electricity: Electricity that results from the electron being displaced and moving back to the atom.

current overload: An overload that opens a set of contacts on high current draw and allows them to close when the current draw has decreased. It usually is a pilot duty device.

current relay: A relay that is opened or closed by the starting current of an electric motor. The relay allows a starting capacitor and starting winding to drop out or drop in the starting circuit.

current-sensing lockout relay: A relay that will lock out a control system in the event of high current.

cut-in pressure: The pressure at which a pressure control will close, starting the device it is controlling.

cut-out pressure: The pressure at which a pressure control will open, stopping the device it is controlling.

cycle: One complete cycle of alternating current is the production of a positive and negative peak.

de-energize: To stop the electron flow to an electric device.

defrost control: A control that is used to initiate the defrost cycle of a heat pump or initiate the defrost cycle in a commercial refrigeration freezer.

defrost cycle: A cycle of operation of a heat pump that is designed to defrost the outside coil during the heating operation.

defrost module: A solid-state module that is responsible for controlling the defrost of a heat pump. Defrost modules come in many different forms, such as the time/temperature module, which only considers one condition, or a demand module, which considers several conditions.

dehumidifier: An appliance that is used to remove humidity from the air.

delta system: A three-phase electrical supply system that is determined by transformer hookup. The delta system due to the transformer hookup gives two usable low-voltage legs plus a high leg. The delta system is used when a large number of three-phase loads are used.

delta transformer: A three-phase transformer that has the ends of each of its windings connected together to form a triangle. The delta transformer produces a high leg on one of its power legs.

delta winding: A winding layout of some three-phase motors, where the beginning of the windings is connected to the ending of the windings.

dielectric: The substance that is between the plates and fills the case of a capacitor. The substance is a nonconductor of electricity.

differential: The difference between the cut-in and cut-out point of a control. This can be applied to thermostats, pressure switches, and most controls.

digital meter: A meter that uses a digital display to indicate an electrical characteristic.

diode: A semiconductor that allows voltage to pass in only one direction.

direct current: An electron flow in only one direction; used in the industry only for special applications such as solid-state modules and electronic air filters.

direct drive: A method of transferring the rotating motion of a motor to a device that must be turned. This type of hookup connects the motor directly to the device that must be turned and rotates at the same revolutions per minute as the motor.

disconnect switch: A switch that is used to disconnect the power supply to a piece of equipment; it is sometimes referred to as a *safety switch.*

distribution center: An electric panel used to distribute electric supply to several places in a large structure; can be of fusible or circuit breaker design.

double insulated: A method used by appliance manufacturers that places a double insulating shield between the user and the appliance.

effective voltage: Alternating current, with its many reversals and peaks, never peaks at a constant value. The effective voltage is the working voltage of alternating current. The effective voltage is 0.707 times the peak voltage.

electric circuit: A path for electrons to follow; the circuit may be open or closed, depending on the position of its switches.

electric energy: Energy that is produced by a movement of electrons. The energy can be produced by chemical, light, thermal, or mechanical means.

electric control system: A control system that is operated by an electrical source. An example of an electric control system is the control of a window unit.

electrical resistance heater: A designed piece of wire that produces heat when supplied with electrical energy and usually manufactured to be inserted into an air flow. Many times resistance heaters are used as supplementary heaters in a heat pump system.

electric meter: A device used to measure some electrical characteristics of a circuit such as the voltage, amperage, resistance, or wattage.

electric power: The rate at which electricity is being used, measured in watts.

electric pressure: Another term used to refer to electromotive force, potential difference, and voltage.

electric switch: A device that opens or closes to control some load in an electric circuit. It can be opened or closed by temperature, pressure, humidity, flow, and manual means.

electrical shock: A condition in which a person becomes part of an electrical circuit. Many times this condition causes serious injury or death.

electricity: Energy that is capable of producing an electron flow. An unbalanced condition that results when an electron can be easily displaced from an atom.

electrodes: The dissimilar metal conductors in a battery that produce a small difference of potential.

electrolyte: The chemical paste between the electrodes of a battery and some capacitors.

electromagnet: A magnet produced by coiling wire around a metal core.

electromechanical clock thermostat: A thermostat that can be used to set back the temperature at specific times.

electromechanical control system: A control system that uses electromechanical controls to maintain the temperature of the conditioned space.

electromotive force (emf): The difference of potential that forces electrons through a resistance.

electronic control system: A control system that uses electronics to control the temperature of a structure.

electronic module: A solid-state module that controls the operation of a system, equipment, or single function of a conditioned air system.

electronic self-diagnostic feature: A feature in a control system that automatically diagnoses problems in the heating and air-conditioning system.

electrons: Particles that orbit around the nucleus of an atom and have a negative charge.

element: A substance that has weight, takes up space, and cannot be broken down by chemical means.

energize: To apply voltage to an electric device.

energy efficiency ratio (EER): The means of measuring an air conditioner for its efficiency by stating how many Btus of cooling are available from one watt.

factory-installed wiring: The wiring in-stalled in a piece of equipment at the factory; usually the connections between the components in the control panel and the system components in the unit itself.

factual diagram: A wiring diagram that is a combination of the pictorial and schematic diagrams.

fan switch: A temperature-controlled switch that starts and stops a fan motor on a gas furnace depending on the temperature.

fault isolation diagram: A type of troubleshooting chart that isolates problems in a control system.

field of force: The area around a magnet that is affected by the strength of the magnet.

field wiring: The wiring that must be installed in the field by the installation mechanic.

flow diagram: A block diagram that outlines the operation of a heating and air-conditioning unit.

flux: The magnetic lines of force of a magnet that connect the north and south poles of the magnet.

free electron: Electrons that are easily removed from the outer orbits of atoms.

frequency: The number of complete cycles per second of alternating current.

full-load amperage: The amp draw of a load when operating at full load conditions.

fuse: A device that breaks a circuit when its ampere rating is exceeded: constructed of two ends or conductors with a piece of wire that will melt and break the circuit on an overload.

fusible disconnect switch: A disconnect switch used to interrupt the power supply to a load and also to provide fuse protection.

fusible load center: An electric panel that supplies circuits with power and protects them with fuses.

gas furnace: A fossil fuel appliance that is designed to heat air and supply it to the structure.

gas pack: A unit that heats in the winter by using gas as its fuel and cools in the summer by using electric power.

gas valve: A valve that opens and closes upon a call for heat from the thermostat. Some gas valves have more than one valve built into the body. These additional valves are used for safety.

ground: A conducting connection between an electrical circuit or equipment and the earth.

ground fault circuit interrupter (GFCI): An electrical device that will open a circuit, preventing current flow to the circuit when a small electrical leak to ground is detected.

grounded: The electrical condition that exists when a current-carrying conductor comes in contact with a ground.

grounding adapter: An adapter used between a grounded appliance and a nongrounded receptacle. This practice is not recommended.

head pressure: The discharge pressure of a refrigeration system, sometimes called *high-side pressure.*

heat pump: A refrigeration system that reverses the flow of refrigerant in the normal refrigeration cycle, which allows the unit to cool in the summer and heat in the winter.

heater: An electric load that converts electric energy to heat.

hermetic compressor motor: A motor that is designed for single- and three-phase operation and is totally enclosed in a shell with refrigerant and oil.

hertz: The number of complete cycles per second of alternating current; more widely accepted than the term "frequency."

hopscotching: A troubleshooting procedure for electric circuits that is accomplished by jumping from one component to another.

hot surface ignition: A method of lighting a main gas burner with a hot surface igniter.

hot-wire relay: A relay that is opened or closed by a thermal element that senses the starting current of the motor. The relay allows a starting capacitor or starting winding to drop out or drop in the starting circuit. This type of relay also has a built-in means of overload protection.

humidistat: A device that is used to control humidity; it uses a moisture-sensitive element to control a mechanical linkage that opens and closes an electric switch.

high-pressure switch: A pressure operated switch that opens or closes on a rise or fall in pressure on the high side of a refrigeration system. This type of switch can be used as a safety control that would open on when the pressure reached an unsafe condition or operate some electrical component by closing on an increase of pressure.

ignition module: An electronic module that is designed to supervise the lighting of a pilot or main burner.

impedance: The sum of the resistance and reactance in an alternating current circuit.

indoor fan relay: An electric relay that starts and stops an indoor fan on cooling, electric-heating, and heat pump systems.

indoor fan relay package: A package that incorporates a control transformer, indoor fan relay, and low-voltage control terminal board.

induced magnetism: The magnetism induced around a current-carrying conductor.

inductance: A property of an alternating current circuit by which an electromotive force is produced in it by a variance in current.

inductive load: A load that starts with a larger ampere draw and reduces it as the load starts normal operation. The increase in the ampere draw initially is due to inductance.

inductive reactance: The opposition to the change in alternating current flow that produces an out-of-phase condition between voltage and amperage.

installation and service instructions: A written set of instructions that explains the proper installation procedures and services procedures for a specific model of equipment. This set of instructions are usually packed with the equipment when shipped from the factory.

installation diagram: A diagram that shows little internal wiring but gives specific information as to terminals, wire sizes, color coding, and breaker or fuse sizes.

insulator: A material that retards the flow of electrons or electricity.

interlock: The action of stopping and starting a component only when another component has started. A good example of an interlock control is not allowing a water chiller to operate until the chilled water pump has begun operation.

internal compressor overload: An overload that is embedded in the windings of a motor. Some internal overloads break the power to the motor directly, while others merely open a set of contacts that is wired into an electric control circuit.

internal pressure relief valve: A valve placed in the discharge side of a hermetic compressor that would open and relieve the pressure if it exceeded a certain point.

kilowatthour: The rate at which electric energy is being used at a specific time. Most electric utilities bill their customers in this method.

law of electric charges: Like charges repel and unlike charges attract.

light commercial air-conditioning system: Air-conditioning equipment that is used in the light commercial phase of the industry, usually 25 tons or less.

light-emitting diode: A diode that will produce light with electrical energy flows through it.

limit switch: A safety that is designed to open in the event of excessive temperature in an appliance.

line break overload: An overload that breaks the power going to the motor and is most commonly used on small motors.

line voltage: The voltage being supplied to the equipment at the power supply.

line voltage control system: A control system that utilizes line voltage to control an air-conditioning system or equipment.

line voltage thermostat: A thermostat that is used primarily to break line voltage to a load to control the temperature.

"live" electrical circuit: A circuit that is being supplied with electrical energy.

load: Electric devices that consume electricity to do useful work, such as motors, solenoids, heaters, and lights.

load analyzer: A device that analyzes the load of a structure and determines the correct action of a heating and cooling unit.

locked rotor amperes: The current a motor uses the instant it starts while the rotor is in a stationary position.

lockout relay: A high-impedance relay that has a normally closed set of contacts and is used to lock a control system out when a safety control opens. A lockout relay must be reset in order for the equipment to operate.

low-pressure switch: A pressure operated switch that opens or closes on a rise or fall of pressure on the low side of the refrigeration system. This type of switch can be used as a safety control, to stop loads in the event of low pressure or operating control, to operate loads to maintain a certain temperature.

low-voltage control system: A control system that uses low voltage, usually 24 volts, to operate the controls of an air-conditioning or control system.

low-voltage thermostat: A thermostat that is designed to interrupt a 24-volt power supply to electrical loads depending on the temperature.

magnetic field: The area around a magnet in which the effect of the magnet can be felt.

magnetic overload: An overload device that senses the current draw of a load by the magnetic field produced, which is proportional to the current draw. The device will open a set of contacts on high

current draws and allow them to close when the ampere draw returns to normal.

magnetic starter: A device that opens and closes its contacts when a solenoid is energized. A means of overload protection is provided. It is the same as a contactor except for the overload protection.

magnetism: The ability of two pieces of iron to be attracted to each other by physical means or electrical means.

matter: The substance of which all physical objects consist.

measurable resistance: The actual resistance of a circuit or component measured with an ohmmeter.

mechanical linkage: The linkage that connects the contacts to the armature, enabling the contacts to close or open when the coil is energized.

microfarad: The unit of measurement used to measure the strength of a capacitor.

module: An electrical device that is used to control one or more functions of a control system.

molecule: The smallest particle into which a substance can be divided and still maintain the properties of that substance.

molten-alloy relay: An overload device that opens a set of contacts by thermal energy. This type of device allows the temperature produced by the starting current of a load to be transferred to a molten-alloy device. When it reaches a certain temperature, it will melt the solder around the device, causing it to slip and open the contacts; when it cools it will harden again and the relay must be manually reset.

motor: A device used to create a rotating motion and drive components that require rotating motion. Electric energy is changed to mechanical energy by magnetism, which causes the motor to turn.

motor protection module: A solid-state module that is responsible for the protection of a large electric motor. This type of control usually has sensors mounted in the motor windings.

multistage control system: A control system that is used to control heating or cooling elements of an air-conditioning or heating system at different temperatures.

multistage thermostat: A thermostat that is used to control different stages of a heating and cooling system.

multizone: A heating and cooling unit that is equipped to condition more than one zone in a structure.

National Electrical Code®: A set of standards published by the National Fire Protection Association that specifies the minimum standards that must be met for the safe installation of electrical systems.

negative charge: The result of electrons joining atoms.

neutron: The neutral particle in the nucleus of an atom.

nonfusible disconnect switch: A disconnect switch used only to interrupt the power supply to a load.

noninductive load: A load that has only resistive qualities with no inductive qualities. An electric heater and incandescent lighting are two common types of noninductive loads.

normally: A term that refers to the position of a set of contacts when the device is de-energized.

normally closed: The position of a set of contacts or other electric devices that are closed when the device is de-energized.

normally open: The position of a set of contacts or other electric devices that are open when the device is de-energized.

nucleus: The central part of an atom composed of protons and neutrons.

ohm: The amount of resistance that will allow one ampere to flow with a pressure of one volt.

ohmmeter: An electrical meter used to measure the resistance of a circuit or electric component.

Ohm's law: The relationship between current, electromotive force, and resistance in an electric circuit: $I = E/R$.

oil safety switch: A switch that is used to open the circuit when the oil pressure in a compressor is below an acceptable level.

one-function solid-state device: An electronic device that performs only one function such as an electronic time-delay module, defrost module, and anti–short-cycling module.

open: The condition that exists in an electrical circuit when there is no complete path.

open circuit: A circuit without a complete path for electrons to follow.

out of phase: A condition in which the voltage and current are not working together.

overload: A device that is used to detect a high ampere draw of some electric load and break the controlling circuit, stopping the load.

overshoot: The additional heating or cooling that has been delivered to the conditioned space after the thermostat contacts have opened.

packaged air-conditioning unit: A system built with all components in one unit except for the field wiring.

parallel circuit: An electric circuit that has more than one path for current flow.

peak voltage: When the voltage reaches its peak in an alternating current circuit.

permanent magnet: A piece of material that is magnetic by physical means. Iron, nickel, cobalt, and chromium are materials that can easily be magnetized and will maintain their magnetism for a period of time.

permanent split-capacitor motor: An electric motor, widely used in the industry, that has a moderate starting torque and good running efficiency.

phase: The number of currents alternating at different times in an alternating current circuit.

pictorial diagram: A wiring diagram that shows the actual internal wiring of a unit, much like a picture taken of a control panel. It is also called a line or label diagram.

pilot: A flame that is standing or established to light the main burner of a gas valve.

pilot assembly: An assembly that holds the pilot burner and the method of ignition.

pilot duty: A term used to refer to an electric device that indirectly controls a major load because of its large ampere draw but controls it directly through a device that is capable of carrying the load.

pilot duty overload: An overload that senses the load of the circuit or power-consuming device and breaks a set of contacts that is isolated from the sensing element.

pneumatic control system: A control system that uses air to control the temperature of a structure.

pole: One set of electric contacts either in an automatic device or a manual switch. Electric devices such as relays, contactors, switches, and breakers can be purchased with one or many poles.

positive charge: The result of electrons leaving an atom.

potential coil: A coil energized by a voltage being applied to it. It can be designed to operate on 24, 110, 208/240, or 480 volts. These coils are used on relays, contactors, and magnetic starters.

potential difference: Two points that have a difference in electric charge; the electric difference between two points in an electric circuit.

potential relay: A relay that uses the back electromotive force of a motor to drop out the starting apparatus when the motor reaches 75% of full speed.

power circuit: An electrical circuit that supplies electrical energy to a load or equipment.

power factor: The ratio of true power to apparent power, usually expressed as a percentage.

pressure switch: A device that opens or closes a set of contacts when a certain pressure is applied to the diaphragm of the switch.

primary control: An electrical control used to supervise the operation of an oil burner.

programmable thermostat: A thermostat that can be programmed to set up and set back the temperature of the structure for certain periods of the day or week.

proton: A positively charged particle in the nucleus of an atom.

pump-down control system: A control system that closes a solenoid to allow the compressor to pump all the refrigerant from the low side of the system into the high side. This system is used on large air-conditioning systems and some commercial refrigeration systems.

push-button switch: A switch that can be opened or closed by pressing buttons on the switch. Push-button switches come with a wide variety of purposes and labeling.

range: The operating ranges or limits of a control.

reactance: The resistance that alternating current encounters when it changes flow.

rectifier: A device that will allow electrical current to pass in one direction but stops electrical current from flowing in the opposite direction. This device is commonly used to rectify AC voltage to DC voltage.

relay: A device that opens and closes a set of contacts when its coil is energized. The relay is much like the contactor except for its smaller size.

relief valve: A device that will open on a rise in pressure and release pressure to return a closed system to a safe operating condition and close when the pressure has decreased.

reset point: The point at which an electrical control will close its contacts after an unsafe condition has corrected itself.

resistance: The opposition to the flow of electrons.

resistive load: See noninductive load.

reversing valve: A valve used to reverse the refrigerant flow in a heat pump.

rooftop unit: A heating and cooling unit that conditions a structure; it is mounted on the roof after adequate reinforcement has been built into the roof.

rotor: The rotating part of an electric motor.

running (oil-filled) capacitor: An electric device that is used to momentarily store electrons and create a second phase in the starting winding circuits of single-phase motors. This type of capacitor is designed to stay in the circuit whenever the motor is running as a means of heat dissipation.

safety device: Any device that is in a control system for the purpose of making the operation of a major load safer.

schematic diagram: A diagram that lays out the control system circuit by circuit and is composed of symbols representing components and lines representing their interconnecting wiring.

seasonal energy efficiency ratio (SEER): An equipment efficiency rating that takes into account the start-up and shut-down for each cycle.

semiconductor: A conductor whose electrical conductivity is between that of an insulator and a metal.

sensor: A device that produces a signal that changes with a temperature change.

sequencer: An electrical device that is used to control electric resistance heaters. Sequencers can have up to five sets of contacts that open and close at different time intervals. In most cases, sequencers are 24 volts.

series circuit: An electric circuit that has only one path for electron flow.

series-parallel circuit: A combination of series and parallel circuits.

set point: The point at which a control will open and close.

shaded-pole motor: An induction type of motor that does not incorporate an ordinary type of starting winding. It uses a band on one side to obtain a short-circuit effect that produces a rotating magnetic field. This motor has a low starting torque.

short: The condition that exists in an electrical circuit when there is no resistance.

short circuit: An electric circuit that has no resistance.

short-cycling: A term used to refer to a condition that occurs when a load is stopping and starting too frequently.

signal light: A light that is used to show when some electric component or circuit is energized by illuminating the light.

sine wave: A graphical representation of alternating current; a graph showing the sine function of all angles from 0 to 360 degrees.

single phase: An electrical power supply that supplies two hot legs of electrical energy to a circuit.

sleeve bearing: An antifriction device that allows free turning and support of the rotating member of a device. It consists of a solid piece of bronze or babbit that is round and drilled to the diameter of the shaft. The bearing is sometimes called a *plain bearing* or *bushing*.

sliding armature: An armature that mounts between two slots in a contactor frame and moves up and down the slots when the contactor is energized.

snap action of a thermostat: The closing of a set of contacts of a thermostat with a snapping motion rather than with a light contact.

solenoid: A device that, when energized, will create a magnetic field and cause some action to an electric component. It opens and closes to control some element of a heating, cooling, and refrigeration control system.

solenoid valve: A valve that opens or closes by a solenoid coil being energized to pull a steel core into the magnetic field of the solenoid.

solid-state relay: An electronic device constructed from semiconductor material, used to control electrical loads.

spark ignition: A method of creating a spark, igniting a gas pilot or main gas burner.

split-phase motor: An electric motor that has a running and starting winding. This is an induction type of motor.

squirrel cage rotor: The rotating part of an electric motor; its name is derived from the similarity of its appearance to a squirrel cage. This type of rotor is used in split-phase, capacitor-start, shaded-pole, and three-phase motors.

stack switch: A primary control that supervises the operation of an oil burner by sensing heat in the stack after the oil burner has ignited.

staged system: A system that has more than one mode of heating or cooling operation.

staging thermostat: A thermostat that is designed to open and close more than one set of contacts to control several modes of heating or cooling operation.

star transformer: A three-phase transformer that has the ends of each winding connected to a common point. The star transformer produces a balance of all hot legs to ground.

star winding: A winding layout of some three-phase motors in which the ends of the windings are connected together.

starting capacitor: An electric device that is used to momentarily store electrons, creating a second phase in the starting windings of single-phase motors. This type of capacitor is designed to stay in the circuit only a short period of time.

starting relay: A relay that is used to energize or de-energize the starting components of a single-phase motor.

static electricity: Electricity that results from the electron being displaced and not returning to the original atom; usually results from friction.

stator: The stationary part of an electric motor.

swinging armature: An armature used in a contactor that is mounted on a line and moves up and down in a swinging motion.

switch: A device for making, breaking, or changing the connection in an electric circuit.

system lag: The difference in temperature between the point at which the thermostat closes and the point at which the thermostat starts to rise or fall.

thermal overload: An overload device that senses the current draw of a load by the heat produced, which is proportional to the current draw.

thermal relay: An overload device that determines current flow and opens a set of pilot duty contacts when an overload is indicated.

thermistor: A semiconductor that has a temperature coefficient of resistance that corresponds with a designated temperature. This device is widely used as a signal device in a control circuit.

thermocouple: A device that is made of two dissimilar metals that produce a small voltage when heated.

thermostat: A device that responds to a temperature change by opening or closing a set of electric contacts.

thermostat controlling element: The portion of a thermostat that reacts to temperature change by opening or closing the contacts through a mechanical linkage. The two types of elements are bimetal and bulb.

three phase: An electrical power supply that supplies three hot legs of electrical energy to a circuit.

three-phase motor: An induction type of motor that has a very high starting torque and requires no special starting apparatus. The motor must be operated on a three-phase current.

three-prong plug: A electrical plug on an appliance that has a ground. The round or oval prong on the plug is the ground.

throw: This refers to the number of positions of the movable contacts that will complete a circuit.

time clock: A clock that opens and closes contacts at specific time(s). Time clocks can be 24-hour or 7-day clocks.

time-delay relay: A relay that delays its closing for a certain period of time.

torque: The starting power of an electric motor.

transformer: A device that decreases or increases the incoming voltage to the desired voltage.

transistor: A semiconductor device used for the control and amplification of a signal from one circuit to another circuit.

triac: A bidirectional electronic component that has an on or off and is used to control AC voltage.

trip-out point: The point at which an electrical control will open its contacts in the event of an unsafe condition.

troubleshooting chart: A chart furnished by equipment manufacturers to guide the technician in troubleshooting a specific type of equipment.

troubleshooting tree: A type of troubleshooting chart that follows a sequence to troubleshoot a control system.

varistor: A semiconductor with a voltage-sensitive resistance.

V-belt: The belt that connects the pulleys of a motor and the device that must be rotated and transfers the rotating motion from the motor to the device. V-belts can be purchased in several widths and almost any length.

volt: The amount of electric pressure required to force one ampere through a resistance of one ohm.

voltage: The difference in electric potential between two points.

voltage drop: The amount of voltage lost through any type of switching device or conductor.

voltage spike: A sudden and temporary increase of voltage that can damage electronic circuits.

voltmeter: An electric meter used to measure voltage.

water chiller: A refrigeration system that cools water that is pumped into other parts of the system to maintain the desired condition in a specific area.

water-cooled packaged unit: A unit that is made in one complete unit with a water-cooled condenser as an integral part.

watt: One ampere flowing with a pressure of one volt. The unit measurement of power.

wiring diagram: A systematic method of laying out the wiring that is interconnecting the control components within the control system; three types are schematic, pictorial or line, and installation.

wye system: A three-phase electrical supply system that is determined by transformer hookup. The wye system due to the transformer hookup gives three usable low-voltage legs. The wye system is used when a large number of low-voltage, single-phase circuits are needed.

wye transformer: A three-phase transformer that has a common junction point and forms a Y. This transformer hookup allows for a completely balanced load when using all hot legs and ground.

zone: A section of a structure that has a heating and cooling load.

Index